The Growing-Block View

Mind, Meaning and Metaphysics

The Mind, Meaning and Metaphysics series publishes cutting-edge research in philosophy of mind, philosophy of language, metaphysics and epistemology. The basic questions in this area are wide-ranging and complex: What is thinking and how does it manage to represent the world? How does language facilitate interpersonal cooperation and shape our thinking? What are the fundamental building blocks of reality, and how do we come to know what reality is?

These are long-standing philosophical questions but new and exciting answers continue to be invented, in part due to the input of the empirical sciences. Volumes in the series address such questions, with a view to both contemporary debates and the history of philosophy. Each volume reflects the state of the art in theoretical philosophy, but also makes a significant original contribution to it.

Titles in the series include:

Knowledge and the Philosophy of Number, by Keith Hossack
Names and Context, by Dolf Rami
The Metaphysics of Contingency, by Ferenc Huoranszki
Fragmenting Reality, by Samuele Iaquinto and Giuliano Torrengo
Unreal Beliefs, by Krzysztof Poslajko
Language, Cognition and the Way We Think, by Nikola A. Kompa
Frege's Pragmatics, by Thorsten Sander

The Growing-Block View

Philosophy of Time, Change and the Open Future

Graeme A. Forbes

BLOOMSBURY ACADEMIC
LONDON • NEW YORK • OXFORD • NEW DELHI • SYDNEY

BLOOMSBURY ACADEMIC
Bloomsbury Publishing Plc
50 Bedford Square, London, WC1B 3DP, UK
1385 Broadway, New York, NY 10018, USA,
29 Earlsfort Terrace, Dublin 2, Ireland

BLOOMSBURY, BLOOMSBURY ACADEMIC and the Diana logo are trademarks of Bloomsbury Publishing Plc

First published in Great Britain 2025

Series design by Louise Dugdale

A catalogue record for this book is available from the British Library.

A catalogue record for this book is available from the Library of Congress.

ISBN: HB: 978-1-3505-0428-8
ePDF: 978-1-3505-0429-5
eBook: 978-1-3505-0430-1

Typeset by Deanta Global Publishing Services, Chennai, India

Contents

List of Figures viii
List of Tables ix
Preface x
Acknowledgements xiii

Part I Change

1 McTchange and temporal variation 3
2 Experience of McTchange 23
3 Enduring McTchange 53
4 The science of McTchange 79
5 Relativity and McTchange 103

Part II: The Arrow of Time

6 Time, entropy, memory and causation 129
7 Agency, fate and the open future 159
8 Thank goodness that's over! 187

Part III: The Growing Block

9 Temporal ontology 223
10 The Growing-Block view 253

References 275
Index 295

Figures

1.1	Bob in the flip-book	6
5.1	Transformations which preserve different features	106
5.2	Reichenbachian simultaneity	111
5.3	Past and future light cones	112
5.4	Two cannonballs getting closer as they fall	115
5.5	A line represented from two perspectives	116
5.6	A causal set	120
9.1	A range of temporal ontologies	236

Tables

3.1	Accounts of Persistence	57
8.1	A Simple Decision Problem	190
8.2	A Decision Problem with Utilities	191
8.3	Potentiality and Actuality for the Tenses	208

Preface

> *What we should do in this case, as with the others, is to begin from the appearances, and first work through the puzzles about them. Thereby we should be able to prove (as far as possible) perhaps all of the established views about these experiences; but if not all, then most of them, and the most authoritative. For if the difficulties are resolved, and the established views remain standing, then our position would have all the proof it needs.*
>
> —Aristotle, *Nichomachean Ethics,* 1145b1

This book is about time; what makes time interesting and what time is. It attempts to articulate a version of what Kristie Miller (2008) calls 'the intuitive view', that is, the view on which time passes and there is a metaphysical asymmetry between past and future.[1]

It has three parts. Part I asks why we should think that time does something that space does not, namely, *pass*. Having argued that we should think that time passes, in the appropriate sense, Part II will then ask why we should think that the past differs in some metaphysically interesting way from the future. Given the various widely noted asymmetries inherent in our treatment of earlier and later times, why should we treat these asymmetries as being explained by an asymmetry in our metaphysics? Having argued that our metaphysics should reflect various asymmetries between past and future, Part III will ask why we should accept the Growing-Block view – the view on which the past exists, the future doesn't and the passage of time is causation bringing about events in accordance with the laws of nature. In particular, why should we accept the Growing-Block view as a theory of time *given* that we want a view of time on which (i) time passes and (ii) there is a metaphysical asymmetry between past and future? After all, rival theories of time accept that time passes and that there are metaphysical asymmetries between past and future; why *this* theory?

I have structured the book so that it builds in a linear fashion. The order takes us from the armchair to the philosophy of physics, and then from the philosophy of physics out to the very human world in which we plan, fall in love and die (amongst other things). Only after we've traversed those very different topics are we ready to consider the ontological questions that have been the focus of most of the literature on the metaphysics of time. Because it's in trying to make sense of how things (in the broadest possible sense of the term) hang together (in the broadest possible sense of

[1] Indeed, in some sense, this book, Part I in particular, is an extension of a debate I had with Miller at the third conference of the International Association for the Philosophy of Time in Winston-Salem, North Carolina, in 2014.

the term)[2] that the philosophy of time has value, and temporal ontology only matters to the extent that it is part of a project that attempts to address time as it occurs in our best science and in our lives.

This book is a work in metaphysics of time. It is concerned with what A. W. Moore (2012) characterised as 'our most general attempt to make sense of things'. It is not a work in philosophy of physics, though I will be discussing some philosophy of physics. Neither is it a work in psychology, though I will be discussing some psychology. In some sense, it is a work of what is known as 'armchair philosophy', in that it doesn't represent an empirical enquiry that I have carried out, besides living in the world, remembering, anticipating, hoping and wishing, ageing and procrastinating. But, of course, even when ensconced in my armchair, I think philosophy should engage with the best empirical evidence we have available. I just don't think that gathering such evidence settles many of the questions in which we might reasonably be interested.

Equally, this book isn't a work of conceptual analysis, though it will look carefully at some important concepts, including change, persistence, causation and simultaneity. I don't think that we can simply describe how we think about the world and conclude that's how the world is. 'Sense-making' inevitably involves some reflective equilibrium. We may not get to save all of our pre-theoretical commitments, but we should strive to keep the best and the most important. One pre-theoretical commitment I will confess now is that the world you and I share is the real world. There is no *more real* reality hidden behind it. So, insofar as physics describes reality, physics gives a description – a description of a particularly abstract sort within a particular explanatory project – of the world you and I share. Insofar as our experience is subjective and idiosyncratic, it is for the most part a subjective and idiosyncratic experience of an objective, shared world. Reality is not exhausted by our experience of it, or by a physics formula. Reality is the thing that both are partial attempts to capture. Given our comparative success in getting by in the world, we have reason to think our concepts are *some* guide to what reality is like, but given the tenacity of the metaphysical debates – arising independently in different cultures over the centuries – we have reason to think that some revision of our thinking in light of recalcitrant experience might be required too.

The view I defend is based on common sense; my aim is to make sense of the world we share, rather than to *disrupt* it. But the view I will defend will go beyond outlining the commitments of common sense. Common sense doesn't provide a *theory* of time; it just provides some broad-brush commitments, each of which is compatible with a number of metaphysical interpretations.[3] Though issues about the methodology of metaphysics will be revisited during the course of the book, I know that I am unlikely to persuade everybody. If you think that all there is to giving a theory of time is to be

[2] Cf. Sellars (1963:1).

[3] Baron, Miller and Tallant (2022:22) use a series of experimental studies – The Sydney Time Studies – to argue that 'appeals to common sense should be banned in the philosophy of time'. This is because participants in experiments give a range of answers about what common sense requires and don't seem to produce consensus. While I agree with them that folk concepts are often 'messy indistinct and incomplete' (p.9), I think common sense can provide, and indeed *must* provide, the starting points of inquiry. The methodological ban should be on taking common sense to provide a precise set of metaphysical commitments, rather than on beginning inquiry with a shared lived experience of how to get by as creatures like us in a world like this.

done by theoretical physicists, or to be based in deference to philosophers of physics, then you should think of this book as an articulation of a sophisticated version of a folk theory of time. I think that the view I defend here is more than a mere folk theory, but the importance to science of well-articulated but false views should not be underestimated either. If I can only persuade you that my view is interesting but false, I won't have wasted my time.

The book may surprise some readers. I have, quite deliberately, adopted an approach that may strike people expecting a metaphysics monograph as unusual. One reason is that much of the discussion of temporal ontology, including the specific account that I defend, is consigned to the final, and shortest, part of the book. The most interesting debates mostly happen before we get at explicitly *ontological* issues. Secondly, threaded throughout the book are discussions of methodological considerations about why we should take some hypotheses seriously rather than others, and what we should look for in a good theory. The aim is in part to get to grips with some puzzles about time but also to get to grips with what we are doing when we attempt to make progress on such puzzles. I'm deliberately trying to be fairly synoptic in my approach to time and to bring in ideas from a range of areas of philosophy – areas that might otherwise get siloed – potentially at the cost of depth in any specific debate. That's a trade-off I'm prepared to make.

As Wilfred Sellars (1962:527) observed, 'the "problem of time" is rivalled only by the "mind-body problem" in the extent to which it inexorably brings into play all the major concerns of philosophy.' Indeed, the comparison with the philosophy of mind recalls a quote from Thomas Nagel (1986:12):

> There is a persistent temptation to turn philosophy into something less difficult and more shallow than it is. It is an extremely difficult subject, and no exception to the general rule that creative efforts are rarely successful. I do not feel equal to the problems treated in this book. They seem to me to require an order of intelligence wholly different from mine. Others who have tried to address the central questions of philosophy will recognise the feeling.

I most certainly recognise it. I have perhaps tried to tackle too much. The value of this book is intended to be in the package we end up with across a range of topics in philosophy, rather than as a series of rigorously defended journal articles collated for the reader's convenience. I suspect one lifetime is too short to plumb the depths of the philosophy of time, so I shouldn't expect one monograph to do so. I hope that this book can at least draw some disparate debates into discussion with one another, even if I cannot be exhaustive in each of them. Like F. Scott Fitzgerald's Nick Carraway, I am that most limited of all specialists, the 'well-rounded man'.

Acknowledgements

I would like to acknowledge those who have helped me put this book together, through what turned out to be a very challenging time; buying a flat, being head of department during a failed management restructure, non-functional computer system and a global pandemic, as well as heartbreak, rejection, a career change and bathroom renovation saga that left me without a functioning shower for 102 days.

I must acknowledge the University of Kent for giving me a job from 2013 to 2022 and permitting me study leave in the springs of 2019 and 2020. I must also remark on the colossal act of vandalism that is the closure of the Philosophy Department at Kent; may the great whirligig of time bring its revenges on those responsible. I thank the Mind Association for generously awarding me the Mind Six-Month Research Fellowship, and my colleague R. A. Briggs for inviting me to spend my spring 2019 study leave at Stanford, and for exceptional generosity and friendship since we met in 2009. Without these, this book simply would not have been written.

This book is dedicated to the four individuals who have, between them, most shaped what I think the methods and aims of philosophy are: James Tartaglia, Stevie Makin, Chris Hookway and Eric T. Olson. Whatever skill or insight I have acquired is to their credit, though my flaws are, of course, mine alone.

I should also like to thank the various friends, colleagues and organisations who have hosted me, talked to me, made helpful suggestions, and/or read draft work related to this book. The friends and colleagues: Taymaz Azimi, Harriet Baber, Marius Backmann, Ray Briggs, Rosa Cao, Abigail Connor-Bourne, David Corfield, Thomas Crowther, Zoe Drayson, Nikk Effingham, Fiona Jones, Alise Kirtley, Jason Konek, Hannah H. Kim, Veli-Pekka K. Parkkinen, Simon Prosser, Hans Maes, Ann C. Thresher, Giuliano Torrengo, Alex Trofimov, Michael Wilde, Nathan Wildman and Jon Williamson. A number of friends and colleagues had no involvement with this book but helped with other concurrent challenges immeasurably. I would like to thank them here too.

Thanks to Jess Leech for suggesting I submit to Bloomsbury, and to Mark Textor, Colleen Coalter and Aimee Brown at Bloomsbury for accepting and publishing the manuscript (though I still prefer the original title 'The Times They Are A-Changin''). Thanks also to two anonymous referees for their comments and suggestions, but most of all for seeing the strengths of the book that I hoped it would have. Thanks, additionally, to Ben Smart and the University of Johannesburg for giving me an affiliation when I became freelance (gforbes@uj.ac.za).

The organisations that hosted me: the Australian National University, University of California, Davis; University of California, San Diego; Durham University; University of Glasgow; University of Iowa; University of Johannesburg; Keele University; the Centre for Reasoning at the University of Kent; London School of Economics; the

Centre for the Philosophy of Time at the University of Milan; Macquarie University; Jesus College, Oxford University; the Varieties of Agency workshop, the Philosophy Department, the Stanford Humanities Centre and The Tanner Philosophy Library at Stanford University; the Centre for Time at the University of Sydney; the University of West of England; the University of Western Australia and the University of Witwatersrand.

The conferences at which I developed this work and the audiences who contributed to it: the 2011 PERSP Space and Time workshop on the Open Future in Barcelona, the Joint Session in Stirling in 2012, the PERSP A's and B's in the Philosophy of Time conference in Barcelona 2013; the third IAPT in Winston-Salem, 2014; the 2014 Joint Session in Cambridge; the 2014 Warwick Time, Tense, and the Psychology of Relief Conference; the 2015 Philosophy of Time Society meeting at the Pacific APA in Vancouver; the 2015 Joint Session in Warwick; the 2016 AAP at Monash University; the fourth IAPT in Gargnano 2015; the 2018 Joint Session in Oxford, eighth Wild Coast Philosophy Symposium, 2019, Chintsa, South Africa; the Phenomenon of Temporality Workshop in Dundee 2021, the Properties and Times workshop in Lisbon 2022.

Those who gave permission to reprint work in epigraphs: Sophie-Grace Chappell (translation of epigraphs by Aristotle to the preface and Chapter 7), the Editor of the Proceedings of the Aristotelian Society (epigraph by Broad to Chapter 1), Imprint Academic (epigraph by G. E. M. Anscombe, Chapter 2), Gallimard and Routledge (epigraph to Chapter 4 by Maurice Merleau-Ponty and the translation by Colin Smith) and Jimmy Lenman (translation of the epigraph by Kongzi to Chapter 8). The rest of the epigraphs are in the public domain, but I thank the respective authors for writing them.

Finally, I'd like to thank Graeme R. Forbes of Boulder, Colorado, for the good humour with which he has taken being confused with me, and for the opportunities I have benefitted from when I have been confused with him.

Part I

Change

1

McTchange and temporal variation

I feel it in my bones that Mr Braithwaite's theory of past, present, and future misses something that is essential. And yet I must confess I cannot state to my own satisfaction exactly what it lacks. But at the risk of talking nonsense, I will say this.
—Broad (1928:188)

If one asks the question 'does time pass?', one is likely to be met with confusion or, better, a flippant response – 'ask me again in ten minutes'. It tells you something about the philosophy of time that one of its central questions seems so obviously answered in the affirmative. But of course, once one has affirmed that time does pass, the follow-up question: 'what is it for time to pass?' looks less easy to resolve. It is easy to have sympathy for Augustine's line: 'If nobody is asking me, I know; if somebody asks and I try to explain, I don't know.'[1]

I am going to argue that time's passage involves a commitment to a difference between space and time. Time does something that space does not – it *passes*. The central concerns of this chapter are to outline what I mean by temporal passage and then to defend this understanding of temporal passage from claims that it is incoherent or confused. This will set us up for a defence of the passage of time throughout Part I of this book.

There are roughly three positions one can take on the question 'does time pass?': one can deny that time passes, adopting an error theory about time, although this position is comparatively rare.[2] The bulk of the debate occurs between those who claim that they believe time passes. One can, for example, claim that time passes, but the 'passage' here is metaphorical, 'anaemic' or reducible to variation across a temporal dimension.[3] Alternatively, one can claim that time passes in a 'robust' sense. This robust sense is one where time does something that space does not – it *passes* – but this passage is an

[1] Augustine of Hippo, trans. Buckner (2006), Book XI section xiv. The original Latin is '*Quid est ergo tempus? Si nemo ex me quaerat, scio; si quaerenti explicare velim, nescio.*'

[2] Error theorists include McTaggart (1908), and might also include Barbour (1999). See Baron and Miller (2015a) for a discussion, and Baron, Miller and Tallant (2022) for an argument that our best theories in physics might, for all we know, be error-theoretic about time.

[3] Defenders of this view include Braithwaite (1928); Callender (2017); Dyke (2003); Mellor (1998); Oaklander (1990); Price (1996); Smart (1980)

irreducible feature that is not to be solely explained in terms of variation of one thing relative to another.[4] It is this robust sense of passage that is the focus of this chapter, and Part I of this book. This robust sense of passage comes from a particular understanding of change – associated in particular with J.McT.E. McTaggart – that is not reducible to mere temporal variation. §1 will introduce change in McTaggart's sense. §2 will be concerned with worries about the coherence of this McTaggartian understanding of change, initially advanced by McTaggart himself, but now a perennial feature of debates around time. I will argue that, despite arguments to the contrary, there is nothing incoherent or contradictory about the conception of change we need, it's just that we can't explain it in terms that don't presuppose it.

§3 will consider another objection to this understanding of change: that, as a species of change, it must happen at some rate. This argument can be used to suggest that there is a conceptual confusion at the heart of this view of change, but also to suggest that a commitment to additional time-dimensions is required, relative to which change takes place. But this objection is to misunderstand the distinction which C. D. Broad (1923) made, between changes *in* time and changes *of* time. While changes in time might require rates, it does not follow that changes of time require rates. Indeed, given that we are understanding change as a peculiarly temporal phenomenon with no spatial analogue, there is simply nothing that could count as a rate in a way that speaks to the temporality of change. The sense of change we are after is not to be analysed in terms of variation of one thing relative to another. In §1.4 I will briefly clarify how my terminology compares to other usages in the literature – this may be of interest to specialists, but can be skipped by the more casual reader – before moving on to a worry about the debate about change itself. In §1.5 we will consider whether the debate between the robust and anaemic views of change is actually substantive, or is in the end a merely verbal dispute. As Craig Callender (2012) puts it, the debate makes no difference; the views are metaphysically, empirically and explanatorily equivalent.

I think that there is a genuine dispute here, but I concede that Callender lays down a gauntlet, which the rest of Part I will be concerned with picking up the methodological challenge. In order to defend the view, we must pay careful attention to what hangs on the debate. Throughout the book, we will come back again and again to considerations of why, even if we agree about how things appear, we can have a fruitful metaphysical debate about how to understand time. To anticipate those later chapters, the debate does make a difference to our ability to make sense of the world in various ways; even if it is not a dispute that can be settled empirically, it is one that matters.

[4] Defenders of this view include Bourne (2006); Broad (1928); Cameron (2015); McCall (1994): Markosian (2004); Zimmerman (2007)

1.1 McTaggartian change and mere temporal variation

There is certainly something difficult about the passage of time; its recalcitrance as a philosophical topic since Heraclitus,[5] and independent occurrence in Greek, Indian and Chinese philosophy,[6] suggests that the topic is no mere fad.

The central claim is this:

Time/space asymmetry: Time does something that space does not: time 'flows' or 'passes'.[7]

Our litmus test between views is whether they give us a difference between space and time that accepts time as doing something/having some feature that has no spatial analogy. The question of how we combine space and time to get the thing physicists talk about – space-time – will be discussed in Chapter 5. Let's put those complications aside for now.

Naturally, talk of 'flow' or 'passage' is metaphorical. These are, by now, largely dead metaphors, but we should not lose sight of the fact that we tend to describe time by appealing to things in time. This makes our discussion tricky because when we dispute the nature of time itself, we are automatically also disputing the things in time we had appealed to metaphorically. Metaphors though they are, they have good pedigree. There was certainly widespread use of metaphors of flow in the ancient world: Richard Sorabji (1983:42, fn40) lists Plato, Aristotle, Simplicius, Iamblichus, Seneca, Boethus of Sidon, Gregory of Nyssa and Marcus Aurelius as employing such language.

I'm going to call those views that accept this time/space asymmetry 'dynamic views' and those that accept that there is something deserving the name 'time', but deny this time/space asymmetry 'static views'.[8] Those views that deny that anything is worthy of the name 'time' are error theories of time, and are neither static nor dynamic.[9]

Central to our interest in whether or not there is something time does that has no spatial analogy is change. What makes the flow of time *dynamic* is there is a change of a certain kind, a change constitutive of time's flowing or passing. What makes the A-property of presentness an A-property is the evanescence of presentness; there is change in what is present or how the present is. C. D. Broad (1959:766) described this change as time's 'rock bottom peculiarity', and it is a dispute about whether there is change of this kind that is at the heart of disputes between dynamic/static views.

5 Cf. Plato *Cratylus* 402a. Strictly speaking, Heraclitus is talking about things in time here, not time itself.

6 See e.g. Kalupahana (1974) for a discussion of the role of time (*kaala*) in Theravada Buddhism. Also see Confucius *Analects* 9.17. Both of these traditions are plausibly independent of one another and of the Ancient Greek one.

7 It is an artefact of this formulation that it suggests time is a thing in its own right. One might think that time is something that supervenes on relations between events. My formulation here shouldn't be taken to commit me either way on this issue. It's just snappier like this. I should add that there may be other disanalogies between space and time, such as time having a direction. This directedness of time will be the concern of Part II.

8 Huw Price (1996:13) has objected to this naming convention as being misleading because it suggests a time frame in which the universe doesn't change. I note his disapproval, but intend it to indicate that time does not differ from space in the sense outlined.

9 See Baron and Miller (2015) for a discussion of Error theory, and Baron, Miller and Tallant (2022) for a (qualified) defence.

Figure 1.1 Bob in the flip-book. Taken from Leininger (2018: fig. 4).

We go from the question 'what is it for time to pass?' to the question 'how do we characterise the change that is the passage of time?' then. But even if we manage to characterise the relevant change, there will be more to do to explain what it is for time to pass. That concern will be the task of the rest of the book. For the rest of this chapter, however, we will focus on change.

Is change just the temporal equivalent of variation over space? Or is there something about change over time that means it has no spatial analogy? Let's call the first of these options 'mere temporal variation'.[10] I will shortly flesh out our second option a little, but let's call it 'McTaggartian change' – or better, to avoid cumbersome word-wrangling – 'McTchange' (the 'T' is pronounced), after McTaggart's (1927:15) rejection of mere temporal variation.[11]

Crucial for McTchange is that change, in this sense, isn't relative to anything else. It's not that a proposition is true at time *t*, and false at time *t*' (that would just be a mere temporal variation), but that the proposition changes truth-value *simpliciter*. This commitment to propositions that can change truth-value is what Mark Richard (1981) describes as 'temporalism', so if there is McTchange, temporalism is true.

Temporal variation and McTchange are discussed by Bradford Skow (2015) and Lisa Leininger (2018) under the names 'anaemic change' and 'robust change', respectively. Skow is amongst the defenders of the view that (higher-order) temporal variation is enough for time to pass (i.e. he defends a static view). Leininger explains why this isn't enough (see Figure 1.1).

> Consider the following situation: there is a flip book starring Bob the stick figure. The flip book shows a kind of change in Bob's location from the left side of the book (on the first page) to the right side of the book (on the last page). Some of the pages of the book look like the following:
>
> There is even animation when one flips rapidly through the pages! One can see Bob change location from one side of the page to the other when the pages are flipped. Using Skow's definition of movement, when the book is closed, Bob's variation in location still satisfies the definition of movement: there is a thing

10 Cf. Sider (2001), Skow (2015). Though the view is commonly attributed to Russell (1915), e.g. by Blake (1925; 1926); Braithwaite (1928); Geach (1969). It is also known as 'at-at' change or Russellian change.

11 Though, as discussed above, McTaggart did not think there was any McTchange. Broad (1923); William James (1911); A. N. Whitehead (1929) were prominent defenders of McTchange who were McTaggart's contemporaries.

> (Bob), who moves on the pages, with respect to the book as a whole. But Bob does not genuinely move through the book; that is, Bob does not undergo any motion. Another way to understand this is that there is no robust change in Bob's location in the book. Bob is always on the left side of the page on page one and he is always on the right side of the page on page thirty. The sum totality of facts about Bob stays the same when the book is considered as a whole. (p.116)

There is no McTchange in Leininger's flip-book, because Bob's position never changes.[12] And it seems like we don't want Bob's position to merely change from our perspective; we think our perspective to be tracking something more general. We think 'now' doesn't work like 'here'.[13] Bob's 'movement' is always relative to some pages in the flip-book, but that means that it is just a question of occupying different regions of space at different times. Movement, the thought goes, requires *flow*, and Bob lacks flow, or, in Leininger's words, 'A-ness'.

Most attempts to explain McTchange involve showing that something lacks McTchange. The test of whether you can get rid of the change by making it relative to something else seems to be an effective litmus test for McTchange, and *a fortiori* a dynamic view of time. But these are still murky waters.

1.2 Is McTchange coherent?

There's a question about why we would believe in McTchange. So far, I've only appealed to a commitment to some difference between space and time; between our treatment of 'here' and 'now'. Subsequent chapters will focus on building the metaphysical case for McTchange. In this section, I want to head off an objection: McTaggart's view that McTchange was/is/will be self-undermining.

McTaggart (1908; 1927) argues that the A-series, a sequence of times defined in terms of their relationship to the present, generates a contradiction, vicious circle or an infinite regress. This is probably the most famous argument in the philosophy of time. Plenty of ink has been spilt on this subject, and I shall confess now, I think David H. Sanford's (1968) eight-page response is basically right.[14] I won't, therefore, dwell on McTaggart's formulations of his argument. Instead, I will focus on a more a contemporary statement of the idea that there is an incoherence to McTchange, from Kit Fine.

Fine (2005:297) provides a challenge for defenders of McTchange – 'standard realists' – which he claims to be a version of McTaggart's argument:

12 Cf. McCall (1994)'s critique of Smart (1980:7).

13 I haven't argued in favour of McTchange yet; this is by way of motivation. I'll return to these issues in subsequent chapters.

14 As Sanford (1968:373) says, 'McTaggart's argument is a *reductio ad absurdum:* a contradiction follows from the assumption there is an A series [i.e. a series of events defined in terms of their being past, present or future]. The same assumption from which a contradiction is said to follow surely cannot be disallowed in the attempt to show that it really does not follow' (italics original). See Ingthorsson (2016) for a fuller discussion of McTaggart's argument.

> For suppose we ask: given a complete description of tenseless reality, then what does he [the standard realist] need to add to the description to render it complete by his own lights? The answer is that he needs add nothing beyond the fact that a given time *t* is present, since everything else of tense-theoretic interest will follow from this fact and the tenseless facts. But then how could this solitary 'dynamic' fact, in addition to the static facts an anti-realist is willing to accept, be sufficient to account for the passage of time? We naturally read more into the realist's tense-logical pronouncements than they actually convey. But his conception of reality, once it is seen for what it is, is as static or block-like as the anti-realists, the only difference lying in the fact that his block has a privileged centre. Even if presentness is allowed to shed its light upon the world, there is nothing in his metaphysics to prevent that light being 'frozen' on a particular moment of time.

If there were a complete description of the world, as Fine imagines, it would be possible for a being with the right powers to know this description, and thus to know everything once and for all. And if they knew everything once and for all, then the world, as comprehended by this being, would involve no McTchange. Yet no such once-and-for-all omniscience would be possible if time's passage were real. The reason is McTchange, in order to *count* as McTchange, would need to be irreducible such that even an omniscient being would have to change what they knew. Suppose, to use Helen Beebee's (2000) phrase, we're interested in the content of 'God's Big Book of Facts'. What we learn from McTaggart and Fine's arguments is that God's Big Book of Facts will constantly require a new edition because the facts it sets out to describe have changed. But the fact that there are outdated editions of a book doesn't mean that the current edition isn't the correct one. Because each edition is inconsistent with every other, they are not combinable into any single, all-encompassing omnibus edition. It is not, as Fine suggests, that the defender of standard realism is committed to a complete description of reality that fails to be dynamic, but that the defender of passage is committed to there being no description that is correct and complete *once and for all*.[15]

Natalja Deng (2012:20) takes a different lesson from Fine's argument. Rather than merely showing that we can't look for dynamism in the property of presentness, she argues that 'we should conclude that theories of passage cannot deliver the dynamicity of what I [Deng] call "our intuitive picture" of passage'. Deng's argument is that *nothing* can give us the kind of intuitive picture of passage, where there is something time does that space does not do, and so we should stop grasping for it. I think this is the most compelling version of the McTaggartian worry: that the thing we're after is unreachable. To borrow Ramsey's (1929:238) phrase: 'what we can't say we can't say, and we can't whistle it either.'[16] As Deng puts it:

> When we imagine [a change in which time is present, or in which tensed facts obtain], we imagine more than the tensed facts that obtain at present, to the effect

15 This is reminiscent of Dummett's (1960) view that what McTaggart's argument shows is that there cannot be a complete description of reality *sub specie aeternitatis*.

16 D. H. Mellor used the phrase in conversation in this context, but its original context is quite different. See Diamond (2011).

> that certain other times were present and others will be. The act of imagination itself unfolds over time; first we imagine a certain time being present and certain others being past and future, *but then we also imagine the next time being present and certain others being past or future.* (p.26)

Deng is right that thinking a description of the world in terms of tensed properties isn't going to give us a description of McTchange, except insofar as McTchange is presupposed by the very idea of a tensed property. For any single description of the world, we can imagine that description being frozen. And our intuitive picture of the passage of time – the thing that time does, but space does not – involves an act of imagination that itself takes some time. So such an act of imagination cannot give an account of, or explain the passage of time. Alternative views, such as Fine's 'non-standard realism', Deng argues, offer no more than a 'conceptual gesture' in the direction of such dynamicity.

1.2.1 Response-dependent properties

We can't explain McTchange without an appeal to McTchange. This is no doubt unsatisfactory for those who deny there is any such thing, though not a problem; we have a good test of whether one subscribes to McTchange or not. And it is a feature of a number of other properties that we can only explain what they are by pointing to exemplars of them. McDowell (1985:111) describes secondary qualities as properties:

> the ascription of which to an object is not adequately understood except as true, if it is true, in virtue of the object's disposition to present a certain sort of perceptual appearance: specifically, an appearance characterizable by using the word for the property itself to say how the object perceptually appears.

So, colour properties are secondary qualities, on McDowell's view, if you can't *understand* what it is to ascribe redness to something unless whether something is red depends on whether it perceptually appears (looks) red to certain observers under certain conditions. A number of philosophers have argued that we should exploit an analogy between colour properties and temporal passage and have argued McTchange is a secondary quality (e.g. Price (2011)). But for some, the analogy between colour properties and McTchange is used as evidence that McTchange is not objective but *mind-dependent* (e.g. Grunbaum 1968). We needn't accept this. Colour properties may well be objective despite the fact that we can't understand what it is for something to be red without having a grasp of what it is for something to look red. The claim is not that colour properties depend on the mind, but that our ability to use colour properties in our classification practices, or pick out colour properties with our concepts, depends on our minds working the way they do. But this is going to apply to almost everything that is not defined in mathematical terms.

Rather than 'mind-dependence', which suggests a lack of reality (cf. Smith 2011), we can use Mark Johnston's (1989) phrase. He says colour properties are 'response dependent'; our grasp of the relevant concepts depends on us being (or standing in the right relationship to) a community who uses those concepts. 'Colour terms', Johnston says:

> like terms for other secondary qualities, can be introduced by ostension. For example, we can rely upon the neophyte's quality space and a cannily chosen collection of foils and paradigms to make salient a way things look, going on to introduce the neophyte to a colour concept by saying 'Something is red just in case it is such as to look the way these paradigms look from here now and not the way those foils look from here now, *but* look that way to standard perceivers under standard conditions'. (p.143)

Colour properties and causal properties, for McDowell, are both response-dependent in this sense. James Woodward (2003) treats causation as similarly response-dependent. McTchange might seem like a poor candidate to be introduced by ostension since any change, for example, of a poker cooling down, is going to be an instance of McTchange, if there is such a thing, and will not be an instance of McTchange, if there is not such a thing. We cannot distinguish by ostension between the changes in things that involve McTchange and those that do not; what is in question is precisely how to characterise such changes. But we do have a paradigm and a foil. There is McTchange in what time it is, but there is no McTchange in what place it is (unless it is also a McTchange in time). Armed with McTchange as something that time does, but space does not, gives the neophyte something to go on.[17]

The fact that we cannot grasp what McTchange is without appealing to McTchange is no more worrying for claims of the objectivity of McTchange than the analogous claims for causation or colour are for their objectivity.[18] Indeed, nothing so far pushes us either way on those claims, nor demands we treat all three cases alike. You may disagree with these analyses of colour or causation, but they do not leave colour or causation beyond our grasp or entail that it is dependent on the mind.

1.3 At what rate does time pass?

Perhaps there is another threat to the coherence of time's passage. Eric T. Olson (2009:3) puts an objection to the passage of time with characteristic clarity:

> A traditional objection to the dynamic view says that if time passes in this way, it must pass at some rate, yet we cannot say what the rate would be. I claim that we can see easily enough what the rate of time's passage would have to be, but there can be no such rate. This objection appears to refute decisively not only the dynamic view, but any tensed theory of time.

17 This foreshadows my argument in Chapter 2.

18 I take up this comparison again in §6.5, and discuss some of the respects in which time and causation are plausibly more objective than qualities like colour and taste.

This argument, like Fine's argument, is meant to be an attack on the coherence of McTchange. But I will argue that, like Fine's argument, the argument proceeds by misunderstanding what the disanalogy between space and time is meant to be.

We can set out this argument as a *modus tollens*, which we can call 'the passage argument':

1. If time passes then there must be some rate at which it passes.
2. There can be no rate at which time passes.
3. Therefore, time does not pass.

The passage argument is the substance of an objection to the claim that time passes, presented by a number of philosophers who defend a static view of time and deny that there is any McTchange. Their number includes J. J. C. Smart (1949), Huw Price (1996) and Eric Olson (2009).

There are two responses that have been made to the passage argument by those who think that time passes. One response is to accept premise 1, but reject premise 2, and thus deny the conclusion. I shall call this group the 'pro-rate' group. Their number includes A. N. Prior (1958), Tim Maudlin (2002), Ian Phillips (2009) and Michael J. Raven (2011). They argue that there is a rate at which time passes, and that rate is one second per second. The second response is to deny premise 1 and accept premise 2, and thereby deny the conclusion. Their number includes C. D. Broad (1959) and me. I shall call them (us!) the 'anti-rate' group. Contra the pro-rate group, they argue that the notion that there is a rate at which time passes is incoherent, for reasons that I shall address and defend in what follows. Ned Markosian (1993) argues disjunctively for both responses. Obviously, what the pro-rate group and the anti-rate group have in common is that they defend a dynamic view of time, on which there is McTchange.

It is striking that, though all members of the pro-passage group agree that time passes, the pro-rate group think the request for a rate is trivial, and the anti-rate group think it nonsensical. The former group find the answer so obvious as to be surprised that the topic has arisen, and the latter find it absurd that someone would think that was a coherent question. The anti-passage group tend to be baffled by either response, finding the request for a rate at which time passes neither trivial nor incoherent. I will argue that the anti-rate group are correct in rejecting the argument by virtue of rejecting premise 1, while retaining their commitment to the claim in premise 2, and thereby have an attractive response to the 'at what rate does time pass?' objection raised by the anti-passage group. I will argue that the pro-rate group fails to successfully defend the claim that there is a rate at which time passes. This is because whatever we come up with a rate *of*, it won't be whatever passing or flowing it is time does that space does not do.

1.3.1 The pro-rate group

The pro-rate group deny premise 2. There is, they claim, a rate at which time passes. That rate is one second per second. Much of the critical response from the anti-passage group has been focused on arguing that one second per second is not a rate. Eric Olson, defending a static view against the pro-rate group, puts the argument as follows:

> One second per second is one second divided by one second. And when you divide one second by one second, you get one. Not one of anything, just one . . . . But if we ask at what rate something is changing – whether how fast the temperature is rising or at what rate things grow older – 'one' is not even a potential answer, right or wrong.[19]

If you divide something by itself, Olson argues, this just gives you the number one – a dimensionless unit. A dimensionless unit looks to be in the wrong category to provide us with a rate, because rates are measures of change of one thing in terms of another. To draw out the oddity of one as a proposed rate, he claims that the answer 'one' cannot be given in response to the question 'How fast . . . ?', but rather only to 'How many . . . ?' or 'What number?' This is problematic because 'How fast . . . ?' is precisely the question that is raised in asking about the rate of time's passage.

In response, the pro-rate group has produced examples intended to demonstrate that one second per second doesn't give you the dimensionless unit – one – or if it does, that one can still be a rate. Tim Maudlin (2002) provides the example of an exchange of liquorice, measured in feet, for tile, also measured in feet. Maudlin asks us to imagine a barter system in which I wish to trade liquorice for floor tile. If both are measured in feet, there will be a rate of so many feet of liquorice per foot of tile. Maudlin argues that if a rate expressed in feet per foot makes sense, then a rate of seconds per second makes sense, too. Similarly, Michael J. Raven (2011) shows that we routinely use rates in which the units aren't stated because they are the same. Mortality rates are expressed in such a manner: a plague may have a mortality rate of 10 per cent (or 1 in 10, or 0.1). If you worry that the rate of time's passage *necessarily* being one is the problem, consider the mortality rate for, well, mortals. It is a necessary and sufficient condition for x being mortal that x will, eventually, die. This gives us a rate of mortality that is necessarily one. If we do not find it problematic that this mortality rate is necessarily one,[20] then we cannot claim to find the notion of a rate being necessarily one problematic in itself.

1.3.2 The anti-rate group

By contrast to the pro-rate group, the anti-rate group will think that such a response misses the point. As C. D. Broad puts it:

> In the case of a qualitative variation it is sensible to ask: At what speed is it taking place? We know that the speed of some such changes is a matter of empirical investigation. But there is no sense in asking: At what speed does a certain phase

[19] Olson (2009:5). This is itself a paraphrase of van Inwagen's (2002) formulation.

[20] Problematic, that is, beyond the issue of dealing with our own mortality. We will revisit attitudes towards our own mortality in §8.1.4.

[of existence], which was present, retreat into the past? And there is no sense in the suggestion that some might do this faster than others. (1959:766)

Broad thinks that the pro-rate response – that time passes at a rate of one second per second – is too conciliatory to the anti-passage group. The problem, as Broad points out, is that the anti-passage group are treating passage as just another species of variation, like the variation in temperature at different points of a mountain. In such a case, a rate might be appropriate, for example, 10°C/km. It might also be appropriate to ask for the rate of change in the temperature over the course of the day. The day might warm at 1°C an hour until noon, for example. Broad argues that the change that occurs as time passes is of a different category, however. The change that occurs as time passes is not another case of variation of one thing relative to another, but a change *of* time, rather than a change *in* time. When a light changes colour from red to green, this change takes some period of time. That is, the light changes *in* time. The change from one event being present to the next event being present is in a different category because it is time, rather than some variation relative to time, that has changed. Such a change requires no second dimension relative to which it changes. Broad (1923) calls this change that occurs as time passes 'Absolute Becoming', which is a form of what I have been calling McTchange.

To see why the change that occurs as time passes cannot be a change in one thing relative to another thing, and so must be a non-relative form of change, let us consider the view that introducing a second time dimension would allow us to make sense of the rate of time's passage.

1.3.3 The second-dimension 'solution'

If one tried to present a rate for the flow of time that did have a change of one quantity relative to another quantity, one might try to do so by introducing a second time dimension, relative to which time could pass in the first time dimension. On such a view, time passes at a rate of one second on the first time dimension for each second that passes on the second time dimension. This strategy, defended by both Graham Nerlich (1998) and Bradford Skow (2011a; 2015), will not work. Either time passes (i.e. there is McTchange) in the second time dimension, or it does not. If time passes in the second temporal dimension, the question 'At what rate?' looms menacingly. If time does not pass in the second temporal dimension, the second temporal dimension is not genuinely temporal since it lacks McTchange and, according to the pro-rate group's own commitments, it can hardly be a *temporal* dimension if time cannot pass in it. The pro-rate group is defending a view on which there is a distinctive feature of time that space lacks – namely, that time passes – so it is inconsistent with that view to introduce a temporal dimension in which time does not pass.

If passage of time is meant to be distinctive from variation across space, and so qualitative variation (e.g. temperature changing with altitude) is contrasted with the change that takes place as time passes (McTchange), then understanding passage as change of time in one dimension relative to a second time dimension is no help. Either the distinctiveness of time disappears, or we are still to explain what makes the second time dimension a temporal dimension, rather than a dimension only in the sense that colour or temperature can be considered dimensions. It should be clear that a third time dimension will help no more than the second did. An infinite regress of such dimensions will not go any way to make them *temporal* dimensions. Although an infinite regress could offer a *technical* solution to providing a rate of some kind, it won't offer an explanation of what makes McTchange a change in time. As Broad (1938:280) commented, in response to McTaggart's argument that such a regress was required: 'When one finds oneself launched on an endless series of this kind it is generally a sign that one has made a false move at the beginning.' But it is hard to see what, other than a second time dimension, the change that takes place as time passes might be a change relative to.

A primitive sort of change that doesn't take place relative to anything else might well provoke confusion amongst the anti-passage group when the rate at which it takes place is requested. As Broad (1959) says in the passage quoted earlier, there is no sense in asking such a question. I think that, by asking for a change of one thing relative to another, the anti-passage group are missing the point. The pro-rate group are trying to provide a rate that might satisfy the defenders of a static view of time, but a rate that is not one thing changing relative to another thing will never be satisfying for them, and it is not clear that a rate of passage relative to something else could work. Rather than continuing to try to provide a rate, the pro-rate group should join the anti-rate group in claiming the request for a rate at which time passes misses the point. Failing that, the pro-rate group should tell us what changes as time passes, and how this kind of change is different from variation in space, or variation in other non-temporal dimensions. It will not do to answer that the difference from variation is space is that it is variation along the temporal dimension because a defender of passage needs to explain what makes the dimension a temporal one. If there is to be a contrast between change and temporal variation such that time passes but space does not, this contrast needs to do some work.

One possibility the pro-rate group might consider is that time changes relative to the events within time. Maudlin (1993), for example, thinks that time is a substance and so something in addition to events. Someone holding such a view may think sense can be made of a rate at which events happen relative to time passing. She may think that rate to be one second$_{\text{of events}}$ per second$_{\text{of time}}$, for example. I wish to remain neutral about whether time is a substance. But even if someone believes time and events to be separate, and believes that one, e.g. clock-ticking event happens for each second that passes, this will not satisfy the anti-passage group. The anti-passage group would still desire to know at what rate time passes and at what rate events happen. To see this, consider the view on which time is something in addition to events, but passage, as we have understood it, does not take place. We can, on such a view, still have a rate of how many clock tickings there are per second duration, without believing in passage at all. So, even if time is something in addition to events, the anti-passage group would still demand an explanation of the *passage* of time, rather than a rate of events happening to time's passing. To offer such a rate would be no more than an attempt to change the subject.

1.3.4 Time's passage needs no rate

So far, I have argued that trying to find a rate – that is, comparing change in one quantity relative to change in another quantity – at which time passes is doomed. The pro-rate group talk as if time does pass relative to something else, but I have argued that, if they are motivated by treating time as metaphysically different to space, they cannot really believe that there is something relative to which time passes. That is to say, I have argued that a request for a rate at which time passes makes no sense, since it relies on a misunderstanding about what the passage of time must be like.

Olson considers the response that it makes no sense to ask at what rate time passes. He says, 'any semantic theory austere enough to prevent us from stating the rate of time's passage is likely to make the dynamic view itself unstatable'.[21] The thought at work here seems to be as follows. The anti-rate group claims that the change that takes place as time passes (McTchange) is unlike other sorts of change in that it is not change relative to anything else. McTchange is a central feature of an account of passage; indeed, it seems to be constitutive of passage. Claiming that such change is *sui generis*, not requiring a rate, seems to prevent one from saying what passage is. It certainly rules out giving a reductive analysis of the passage of time in terms of Russellian change in one thing relative to another.

If stating the view means providing this kind of reductive analysis, then I agree that the view is unstatable. We discussed Deng's (2012) argument to this same conclusion above. The passage of time, to those who defend it, is a feature of the universe that cannot be explained in terms of other, non-temporal things. Analysis must stop somewhere, and McTchange seems a most plausible candidate to be a feature of the universe for which a reductive analysis may not be given.

I know of no way to defend a dynamic view that doesn't involve a primitive appeal to McTchange. That is, there is no way I know of to defend passage in a way that will provide a rate of change of one thing in terms of another that will satisfy the anti-passage group. This, I claim, is what all of the pro-passage group should say. The change that takes place as time passes – McTchange – is not of the same category as other changes, and so, unlike them, does not have to happen at a rate.

1.3.5 The difference between the dynamic and static views

I can imagine an exasperated defender of a static view having another attempt at the question. They might argue as follows:

> I agree with you that rivers flow. I agree with you that locomotives move. But I don't believe in (robust) passage. I'm trying to understand this extra thing – 'McTchange' – that you have but that I don't. You say that there is perspective-independent change, but can't say at what rate it occurs. Either you agree with me (and this passage talk is just a metaphor) or you just say opaque and baffling things. I just don't understand what we disagree about!

[21] (2009:8).

Here's a final way of trying to get at the nub of the issue. My interlocutor, a defender of a static view, and I disagree about what it is for things to change. She says we agree about rivers flowing and locomotives moving, but we do not. By my lights, she does not believe in rivers flowing *at all*. For a river to flow, a locomotive to move or for you to understand this paper, defenders of a dynamic view claim, time would need to pass. For these ordinary sorts of change – changes in things relative to time – are, according to the pro-passage group, changes relative to the change that takes place as time passes (McTchange). It is not that my view is the same as that of my interlocutor, except my view has this mysterious feature – 'passage' – tacked on, but instead we have a profound disagreement about change in general. This was the response to Fine's frozen present argument, above; passage isn't a matter of having a static world and a property of presentness. The defender of passage will echo McTaggart (1927:27–8) by claiming that just as spatial variation (e.g. a poker's being hot at one end and cold at the other) isn't sufficient for change, temporal variation (e.g. a poker's being hot on Monday but cold on Tuesday) isn't sufficient for change, unless there is also a change from Monday being present to Tuesday being present. A poker that is red at one end and blue at the other, the pro-passage group claim, does not change because it just has the colours it has on the sections of the surface it has them. For the poker to change colour, time must pass, and, the pro-passage group believes, the passage of time is constituted by a particular sort of change: a change not relative to anything else.

Ordinary cases of change, like a river flowing, only count as cases of change because they are change relative to the passage of time. As Broad puts the point, 'the notions both of qualitative change and of qualitative invariance plainly *presuppose* that of Absolute Becoming, in the sense of that phrase which I have indicated.'[22] For something to change its qualities over time, or to stay the same (i.e. to be qualitatively invariant), it must change, or remain the same, relative to the change that takes place as time passes. Maudlin agrees. He says, 'Of course, rivers only flow and locomotives only move because time passes.'[23] Both figures are not only committed to the passage of time but think that this brute feature of reality is presupposed by the sorts of change from which we draw the need for a rate. Whatever we are coming up with a rate of, it is not going to get at the irreducibly temporal phenomenon but at variation of one thing with respect to another.

1.4 Terminological comparisons and alternative frameworks

Some readers of this book will be surprised by the absence of some familiar terminology. The literature in the philosophy of time is replete with similar but not quite identical distinctions between A-theory and B-theory, tensed and tenseless views, and standard and non-standard realism. For those who care, here is how my terminology relates to other distinctions in the literature. Those who aren't bothered by these distinctions can skip this section.

[22] (1959:766–7).
[23] (2002:260).

The dynamic/static distinction is related to, but not quite the same as, the A-theory/B-theory distinction and the tensed/tenseless distinction, which are also closely related to each other. The relation is close enough that for most purposes they can be taken as interchangeable. The A-theory/B-theory distinction goes back to the work of McTaggart (1908; 1927). He introduced a distinction between the A-series of time (events ordered in virtue of their relationship to the present) and the B-series of time (events ordered in virtue of their relationship to each other). Also worthy of note is the C-series (events ordered in virtue of a non-temporal relation), which will re-enter the picture in Chapter 6. R. M. Gale (1968) then introduces the distinction between A-*theory* and B-*theory*, built on this McTaggartian terminology.

The A-theory is so called because it is committed to the irreducible existence of properties ('A-properties') that depend on a relationship to the present (e.g. properties of being now, of future, last week or next year). The A-theory claims that (i) time passes and (ii) time's passage irreducibly involves A-properties. The B-theory accepts (i) but denies (ii). Any time it looks like we're talking about A-properties, the B-theorist will claim, can be understood as shorthand or disguised talk of 'B-properties' (e.g. properties of earlier than or later than or contemporaneous with). McTaggart was himself an error theorist about time. He accepted (ii), but argued that (ii) leads to a contradiction, so he argued that we must reject (i) (more on this in §1.2).

The tensed/tenseless distinction replaces talk of A-properties with talk of 'tenses' and B-properties with talk of 'dates'. But these are either just different labels for the same things, or tenses are descriptions of A-properties and dates are descriptions of B-properties. The tensed/tenseless distinction is one about whether tenses are ineliminable (as on the A-theory) or eliminable (as on the B-theory). Historically, the distinction was one of whether tenses could be eliminated from *language* (Gale 1968), but then it became widely accepted that tenses were really useful in language but could be eliminated from truthmakers (Mellor 1998), or truth-conditions (Torre 2009). So, if there is a distinction between A-theory/B-theory and a tensed/tenseless view of time, the former is concerned with whether there are irreducible A-properties in our ontology, and the latter is concerned with whether there are ineliminable appeals to tense in our theory.

How do these relate to the Dynamic/Static distinction? Well, the thought is that tenses are things that have no spatial counterpart; space isn't tensed. 'Now' (tense *par excellence*) is unlike 'here', because 'here' is just an indexical; it only picks out somewhere relative to a speaker. 'Now' (if the tensed theory is correct) picks out an objective evanescent present, but 'here' doesn't pick out an objective but perpetually changing spatial location.[24] This present, being evanescent, changes, in a way. And this change is the flow of time.

Kit Fine (2005) distinguishes between standard and non-standard realism about tense. So far, I have presented only standard realism about tense, and non-standard

[24] One can, of course, have the reference of 'here' be continually changing. Satellite navigation systems are helpful precisely when we do. But we tend not to think that people in other places lack an objective property that we have. Or, at least, if we do, it tends to get criticised as parochialism in a way that thinking one's contemporaries are different from past or future generations simply in virtue of being around now isn't so parochial. This is a theme we'll come back to in §10.2.

realism is presented as an alternative framework. We had a story about how standard tensed views relate to dynamic views; since the tensed facts McTchange, a view committed to tenses must be dynamic.

Fine proposes two non-standard realisms about tense: Relativism and Fragmentalism. Relativism is to be sharply distinguished from the view that things have the property of being present relative to each time they are at (i.e. this position isn't meant to make 'now' a straightforward analogue of 'here'). Rather:

> reality is constituted, at each time t, by the fact that t is present. This is quite different. The facts that belong to reality are genuinely tensed though their belonging to reality is a relative matter whereas, for the anti-realist, the facts that belong to reality are tenseless while their belonging is an absolute matter. (p. 280)

Reality, on this view, is relative to the time at which you assess it. We give up any idea of describing reality as a whole, so the idea that change in tense is merely relative to that whole goes. The most general descriptions of the world it is possible to offer are absolutely tensed descriptions; it is simply that there is no unique most general description of the world to offer.[25]

Fragmentalism is a form of non-standard realism in a similar way; rather than claiming there are multiple equally compatible most general descriptions of the world, the fragmentalist denies that the world is consistent. It comes in partially overlapping tensed fragments, each of which are absolutely true, but are mutually inconsistent. So, there is a fragment where this year is present, and many, many other fragments where it isn't. And each of these fragments are true descriptions of reality; they just don't add up to a coherent whole. Fragmentalism captures the thought that if time passes, there cannot be a complete description of reality, but it doesn't automatically capture the thought that the truth about the world is changing. It is merely fragmented.

These two non-standard forms of realism present a challenge for the relation between tense and McTchange. They give us a sense in which what time it is varies in a way that has no spatial analogy. This is why they have some claim to being realist views about tense. But it's hard to see the place of metaphors of flow here. If reality is always relative to some time of assessment or fragmentary, we do not yet have an account that gives us a change *of* time or commits us to temporalism. Fine's non-standard realism has subsequently been developed, for example by Lipman (2018) and Iaquinto and Torrengo (2022). Even if we order the times, or the fragments, into a B-series, where some are earlier to others, that doesn't seem to give us a sense of flow. Non-standard realism won't accept an account of change of the type Russell or Skow gives, but equally, it doesn't seem to give change of the kind McTaggart, Broad or James had in mind. But of course, that is just to observe that Finean non-standard realism is not standard realism. So I might dismiss Finean non-standard realism as giving an example of tensed views that are not dynamic.

[25] I'm going to be using 'reality' and 'the world' more or less interchangeably; nothing is real that isn't worldly, and nothing in the world isn't real.

Iaquinto and Torrengo insist that their 'Flow Fragmentalism' *is* dynamic, however, and anticipate the objection that the sequence of mutually incompatible fragments they are committed to fails 'the update test'. That is, when they put the view in their 'official idiom' the view doesn't require updating what is true of reality; there is no coherent account of reality as a whole to update on the fragmentalist picture, and the individual fragments don't change.

The central disagreement between the view I present and Iaquinto and Torrengo's view is whether the update test is the appropriate test for the passage of time. Just as it is possible to have precise theories of vagueness, one can have theories of changing reality that don't themselves change, they argue. To insist on the update test, they urge, is to fail to take seriously that there can be non-standard versions of realism about tense.

It is noteworthy that Iaquinto and Torrengo, as Fine did, start with a view of tense and try to account for passage on such a view. Many of their criticisms of standard realism are ones I agree with, despite my view likely counting as standard realism by their lights, simply because I start with McTchange rather than tense. In fact, what I start with is a distinction based on the update test; a static view is one that fails the update test, and a dynamic view – a view with McTchange – is one that passes the update test. But is this *fair*? After all, they do accept that there is a difference between space (non-fragmented) and time (fragmented), and explicitly reject the view of change as mere temporal variation (at least in the sense that there is a unitary description of reality with change being a difference between two time indices). This question of fairness deserves a fuller treatment than it will get here. I think the update test gives a very clean way of dividing the views, and beginning with tense is problematic as a starting point, for exactly the reasons that motivate Fine to develop non-standard realism. Whether fair or not, the reason for insisting on the update test is that it speaks to the phenomena I want to describe: that there is something inescapable about the need to describe the world while time ineluctably evicts you from one situation into another.

1.5 Is there a substantive disagreement here anyway?

During the course of this first chapter, we have introduced an intuitive distinction between space and time: that time does something space does not, namely, it passes. Passage, in this sense, involves a perspective-independent change that is not relative to anything else: McTchange. We have considered two objections to the very idea of McTchange: McTaggart's argument that the A-series involves a contradiction and the argument that if time were to pass, there would be some rate at which it passes, but since there is no rate, time does not pass. Having responded to those arguments, I would like to finish the chapter by considering another objection to dynamic views of time. This is the objection that there is not a substantive dispute here anyway.

We might, following Jenkins (2014:11) talk of a 'merely verbal dispute' about some subject matter as 'occurring when the two parties do not disagree about the subject matter(s) of their (putative) dispute, but merely present the appearance of doing so owing to their divergent uses of language'. The worry is that the philosophy of time might be a dispute about how to use language, rather than a dispute about the world.

This kind of objection arises perennially; for example, in Clifford Williams (1996); Laurence Lombard (1999); Thomas Crisp (2004); Yuval Dolev (2007); and Craig Callender (2000; 2012).

Sometimes the question arises because of objections to particular ways of specifying the debate. If one thinks the contrast is between time as viewed internally, in a tensed way, against time viewed externally, in a tenseless way, it may seem that there is no disagreement here; it is just a difference analogous to that between a first and third personal perspective to describe the same physical phenomenon. There's a danger that metaphysical-sounding claims like 'only the present exists' become trivial, since they might amount to the claim that only the present *presently* exists, which is only denied by error theorists about time. Many participants to these debates are very sensitive to such worries, however. Craig Bourne (2006:10) and more recently David Ingram (2018:16), for example, are careful to frame the debate such that the language in which the debate is framed is shared across the different positions they consider.[26] And many of the objections are due to a partial consideration of the positions available; Oaklander (2001) criticises Williams (1996) for missing out a number of contributors to the debate whose positions aren't so easily susceptible to being recast as notational variants of rival views. Tallant (2009) similarly criticises Dolev (2007) for missing out positions that don't fit the analysis that there is a merely verbal disagreement.

There are a few quick lessons here for those arguing that a disagreement is merely verbal: try not to overgeneralise from a small number of cases that there is no disagreement; and try to explore a range of different ways of setting out the central disagreement. For there are certainly some ways of setting up the debate where there is nothing of interest going on. D. C. Williams (1951) talks scathingly, and at length, about metaphors that various philosophers appeal to for the something extra that time has that space lacks; 'the surge of process', 'the formality of taking place', 'the present flash and urge', 'genuine creative flux' and 'the mode of potency and generation'. Such notions were criticised by Williams for not adding anything to our understanding of time. If we are to frame the debate between dynamic and static views as one of whether there is McTchange, the role of McTchange in the theory we end up with had better be reasonably clear. McTchange can't amount to a magical pixie dust that we sprinkle on a view to make it a dynamic view. McTchange must be doing some theoretical work that integrates with the rest of the theory.

Craig Callender (2012) has provided one of the best arguments that the philosophy of time provides a merely verbal disagreement, arguing for this conclusion in three steps. These steps take the form of questions that should be put to the various models of time we consider:

MD: Are they metaphysically distinct?
OD: Are they observationally distinct?
ED: Are they explanatorily distinct?

[26] This topic gets revisited in §9.1.

Callender's steps are to answer each of these questions in the negative and conclude that there is not a genuine disagreement. In Chapter 9, I will be considering various models of time in more detail, and I will look carefully at what makes those models distinct from each other. For now, I want to discuss the three questions Callender raises; they can help us understand what we are trying to achieve in offering a model or theory of time. And since this book aims to do just that, making clear what it sets out to achieve is no bad thing.

First, I want to assert that two metaphysical views can be observationally equivalent, but nonetheless distinct. This is no more than a rejection of verificationism. One reason for rejecting such an inference is suspicion that the relationship between 'observations' and 'theories' is straightforward enough to support the inference. W. V. Quine (1953:41) argued that observations didn't offer confirmation and disconfirmation of statements, but statements 'face the tribunal of sense experience not only individually but only as a corporate body'. What any particular set of observations shows is going to depend on other hypotheses, particularly about whether we think the observations were competently acquired from a trusted source.

One might push the point to ask whether metaphysical equivalence implies observational equivalence if we consider all possible observations, rather than merely actual observations. Nonetheless, we might still allow that there could be some metaphysical disputes that would remain unsettled without being merely verbal. This might be the case if they served to explain different things. C. S. Peirce (1998:183) thought that induction only made sense if we believed that there were 'general principles really operative in nature'. But we couldn't determine through observation alone that there were real generals, rather than just particulars. His thought was that belief in real generals made a difference to how we conducted 'inquiry' – the catch-all term for any activity of finding out, be it rigorous and scientific or more quotidian. Thinking that there are real generals allows us to make certain inferences (especially inductive ones) that allow us to have a better – more useful! – theory of the world. But were it not for the fact that without being able to carry out such inductive inferences we simply could not get by in the world, Peirce would dismiss such an appeal to usefulness as mere wishful thinking. The indispensability of real generals is what Peirce thinks justifies our appeal to them on pragmatic grounds.[27] Whether or not one accepts his views on the relation between induction and generality, such a theoretical role for real generals would justify their inclusion in our metaphysics. Peirce's example shows that one can (take oneself to) have an explanatory difference without an observational one.

The relation between metaphysical equivalence and explanatory equivalence is hard to pin down. Explanations are often thought to be context-specific communicative acts, and metaphysics is thought to be mind-independent. But I'm not sure if metaphysical differences are distinct from those differences that would occur in *good* explanations. If metaphysical differences are something else, I'm not sure what 'metaphysical' means, except, perhaps, as a placeholder for whatever differences metaphysicians are interested

[27] See Misak (2011) for a discussion of indispensability arguments in the pragmatist tradition. See Haack (1992) for a contemporary defence of Peirce's view on universals as required by science.

in.[28] I can imagine cases where the same explanation, such as Hooke's law, might apply to any elastic object equally, even though there are different ways of making springs. Leaf springs are 'metaphysically' different from coil springs; they are simply different sorts of things. Nonetheless, those 'metaphysical' differences are likely to show up in explanations elsewhere (like how much room they take up and their shape). For now, I will remain neutral about whether there are any metaphysical distinctions that make no explanatory difference, but I will return to this in Chapter 9.

Can there be cases where metaphysical equivalence is compatible with explanatory differences? It's possible that different explanations can be given of the same world. Someone who looks at a car crash and blames the road layout and someone else who blames the braking system of the car may not be disagreeing in their account of reality, merely in their explanatory projects. A metaphysical difference, I take it, will be a difference not merely in human purposes, but in the character of the world that the human explanatory context engages with.

Callender and I agree that observational differences are not the place to settle this debate. And we agree that there had better be metaphysical differences for there to be a genuine disagreement. When we come, in Part III, to deal with different views of time in detail, they had better amount to *different* views of time. Callender's most challenging contention is that the different metaphysical theses don't offer competing explanations:

> It may be that one model of time best explains the temporal data, and the lack of experimental vindication is nothing to apologise about. This comforting picture of the debate would be fine if any of the three models [of temporal ontology that Callender has been discussing] actually helped explain the temporal data. I submit, however, that we are not in that situation. (2012:86)

Callender, it should be noted, is arguing that different models of *existence* are explanatorily impotent (his phrase). I haven't talked about existence (yet). But I am unquestionably sympathetic to his point. Indeed, I argued that giving a story about what times exist wasn't enough to explain certain features of time's passage in my (2015).

One of the reasons for setting up this chapter in terms of a debate between McTchange and temporal variation is that this seems the kind of debate that doesn't presuppose too much philosophical baggage. We have a question about whether time is different to space in that time undergoes McTchange, and space does not. Before we even get to questions of existence, in Part III, we shall spend the remainder of Part I looking at the case for accepting that there is McTchange, and then in Part II considering whether there is a metaphysical asymmetry between past and future. Once we have a clear picture of the temporal data we are trying to explain, we can then see whether the different views of temporal ontology make any difference.

[28] Maybe we shouldn't think of metaphysicians as offering explanations.

2

Experience of McTchange

> *He [Wittgenstein] once greeted me with the question: 'Why do people say that it was natural to think that the sun went round the earth rather than that the earth turned on its axis?' I replied: 'I suppose, because it looked as if the sun went round the earth.' 'Well,' he asked, 'what would it have looked like if it had* looked *as if the earth turned on its axis?'*
>
> —Anscombe (1959:151)

Part I is a defence of McTchange. The previous chapter was a defence of the conceptual viability of the change that constitutes the passage of time: McTchange. But why might we be inclined to accept that there is McTchange? Here's a candidate answer: it really *seems* like there is. At the end of the previous chapter, we discussed whether we should expect different metaphysical theories to be observationally distinct. I argued that we should not. So this leaves a challenge: why would it seem like there is McTchange, if it's not something we can observe? This chapter will be about whether experience of McTchange provides evidence that there is McTchange, and so evidence that a dynamic view is true. The answer will be that experience doesn't straightforwardly provide evidence that there is McTchange, though the hypothesis that there is McTchange wouldn't be nearly so plausible if we didn't have the kinds of experience we have.

We're going to start, in §2.1, with the thought that it seems like time passes and look at some examples of the different things we might mean by that. Then we're going to consider a literary example: Captain Dunbar from Joseph Heller's *Catch-22*. Captain Dunbar will provide us a naïve view that can't possibly be right. This will push us into a more sophisticated view in §2.2, which will encounter an objection from Simon Prosser (2007; 2013a; 2013b; 2016): we can't discriminate the passage of time from other phenomena. I will respond to this objection by arguing that we should think of McTchange as an enabling condition of experience, rather than a phenomenon that we detect the presence or absence of. This will push us onto another objection from Prosser: that the passage of time is unintelligible because we can't fix reference to the correct enabling condition. I will argue that beginning with the asymmetry between space and time, such that there is something time does that space does not – passing or flowing – does allow us to make McTchange intelligible. In §2.3 we will consider how we are thinking of experience and the extent to which we should think of McTchange as a feature of perceptual experience or of some broader and richer conception of experience. In §2.4, we will consider the range of experiences that seem to be involved in temporal experience. Finally, in §2.5, we will consider some puzzles about the very possibility of temporal experience. Nothing in this chapter will provide experiential evidence in favour of McTchange, but it will lay the groundwork for a defence of McTchange to come. For we will have seen that our experience is in need of explanation, and McTchange may potentially be a good explanation of our experience being the way it is.

2.1 Naïve starting points and Captain Dunbar

It is commonly supposed that experience *does* give us reason to believe in a metaphysical view of time that involves McTchange. As we'll see, things aren't so straightforward. Consider Schuster (1986:695): 'the flow of time, or passage, as it is known, is given in experience'. Those who deny that there is passage in the robust sense that requires McTchange often cite experience as a reason to think there is passage in a robust sense, for example Le Poidevin (2007:76): 'We just see time passing in front of us.' But the proposed relationship between experience and time is not always made clear. MacMurray (1928:145) prevaricates between evidence in experience and requirement in explanation:

> We must ask, 'In what connection does Time enter into our experience of the world?' or, perhaps better, 'Why is the idea of Time indispensable in any attempt which we make to describe the nature of our world?' To this question the answer can hardly be in doubt. It is the fact of change which demands the idea of Time for its elucidation or description.

Those questions don't seem to be equivalent. McTchange may well be required in our explanations even if we are not in a position to observe McTchange *as such*. Dean Zimmerman doesn't appeal to *experience* at all:

> My reason for believing the A-theory is utterly banal (some philosophers reading this essay will want to say 'insipid'): it is simply *part of commonsense* that the past and future are less real than the present; that the difference between events and things that exist at present, and ones that do not, goes much deeper than the difference between events and things near where I am and ones that are spatially far away – in Australia, for example. (Zimmerman (2007:221 emphasis original))

While I concede that we do commonsensically think there is some difference between space and time, we ought to be able to do better than an appeal to common sense to settle metaphysical questions – common sense varies in reliability across domains.[1] It

[1] See, e.g., Kahneman, Tversky and Slovic (1982) for a discussion of well-known failures of common sense. Also consider Kant's admonition against uncritical trust in common sense pronouncements: 'To appeal to ordinary common sense . . . is one of the subtle discoveries of recent times, whereby

is common to find evocative descriptions of experience, as of the passage of time, such as this example from L. A. Paul (2010:333):

> I step out of the house and feel the cool breeze on my face. I feel the freshness of the cool breeze now, and, as the breeze dies down, I notice that time is passing – I need to start walking or I will be late for class. . . . Reflection on the qualitative character of such experiences suggests that events occurring now have a characteristic property of *nowness*, responsible for a certain special 'feel'.

But equally, it is by no means universal that we are taken to arrive at a belief in an important *metaphysical* difference between space and time through observation. As Peter K. McInerney says:

> Defenders of B-series (earlier-later) time admit that ordinary experience includes the 'past, present, future' portion of A-series (past-present-future) features; it is these for which they wish to provide token-reflexive definitions. However, since they deny that passage is real, they usually also deny that the passage of entities is actually experienced. Temporal passage is considered to be at best a feature of a particular theory of time. (McInerney 1991:182)

There are lots of temporal data to be explained, but it is not obvious that the metaphysically significant distinction between McTchange and temporal variation can be identified in or settled by experience. That there is McTchange can't be part of the luminous content of experience.[2]

Would empirical data on people's experience help? In a word: no. Recent empirical investigations have been conducted into whether people at large take themselves to experience the passage of time, to implicitly or explicitly endorse something like McTchange, and whether they subscribe to a more specific metaphysical picture.[3] The results show that around 70 per cent of people claim to 'feel time passing', or something similar. They also show that there is no precise temporal ontology that those who claim to feel the passage of time stably subscribe to.[4]

As Tina Röck (2021:146) observes, using a phenomenological rather than experimental approach, we have different modes of experiencing time, using metaphors of 'the timeline', 'the flow of the timestream', and 'time as unfurling and growth'. The empirical evidence shows that in the population tested, we have a variety of such experiences present, and a clear majority with experiences *as of* passage (though a sizeable minority not reporting such experiences). At best, what the empirical evidence shows is that people think about time in different ways, and that people's experiences of time differ amongst each other, but broadly reflect the descriptions that have generated

the dullest windbag can confidently take on the most profound thinker and hold his own with him' (2008:9).

2 I will consider issues of content internalism/externalism in §2.3. For now, I will talk about luminous content as that part of perceptual content whose nature is apparent to us.

3 For example, Latham, Norton and Miller (2020); (2021); Shardlow et al. (2021).

4 This will be a key point in Chapter 9, when we discuss various temporal ontologies.

discussion amongst philosophers for the last two and a half thousand years. The empirical data, of course, only include data about those who were able to complete the studies. Those with radically different experiences of time would have been unable to take the surveys, with their sequential structure, filling-in one screen and then the next.

I will argue that, if we mean *observational* evidence, we have no experiential evidence of McTchange. As Marc Wittman says, 'If we think about it at any length, we have to admit that we cannot really perceive time. We do not have a sense of time in the same way we have a sense of color, sound and temperature' (Wittmann 2014/2016:123). Our observations are consistent with there being some timeless perspective relative to which it merely seems like there is McTchange, or with the change we experience really being irreducible from our most general attempt to make sense of things. There is, as Callender (2012:82) puts it, 'a one-to-one mapping between experiences in the world of becoming [i.e. McTchange] and in a world without'. But I shall argue that this is no surprise. And a good understanding of how we experience time will help us see what a view of time is supposed to explain.

To help us in thinking about this, we will enlist a fictional character: Captain Dunbar.

2.1.1 Dunbar's challenge

Dunbar, the character from Joseph Heller's novel *Catch-22*, is desperate to prolong his life. He hits upon a strategy: he will live as boring a life as he can. Time passes more slowly when you are bored, he reasons, so a day during which he is constantly bored will take longer than one during which he is excited. If he only has *n* days left to live, those *n* days will take longer if he is bored.

Dunbar is deluded. This strategy will not mean he lives longer. Days merely *seem* to pass more slowly when you are bored. Dunbar is confusing how time seems to him with how things *really are*. If it even makes sense to think of there being a rate at which time passes (I argued in the previous chapter it does not), such a rate would be unaffected by an individual's mood. This, I take it, is what most metaphysicians would say in response to Dunbar's case.

Dunbar's case is clearly ridiculous. He takes a subjective feeling of boredom to indicate that he is living longer, when people normally don't take that to follow. Normally, we don't think that the passage of time is something we infallibly detect. But perhaps that's because Dunbar is failing to calibrate his experience properly. Perhaps the passage of time is more like noticing something feels wet or looks red. In such a case, the thought goes, we can't be wrong that they *look* red or *feel* wet, but under some circumstances, these perceptual experiences can be non-veridical.

The view that certain experiences are a more-or-less reliable guide to the world being a certain way looks more sophisticated than Dunbar's view. It seems plausible that feelings of boredom distort our sense that time passes. But the relationship between our phenomenology and the very general claims of a dynamic metaphysics is, I argue, not the sort of relationship that holds between our phenomenology and the claim 'I am wet' or 'that is red'. We should not expect that sort of experience to favour the view that time passes.

2.2 Prosser's arguments that we don't experience time pass

Many philosophers have offered arguments that we should reject claims that the passage of time is part of the representational content, or the phenomenal character, of experience.[5] I will look, in particular, at two arguments by Simon Prosser, one from indiscriminability (the logically weaker argument) and one from the uniqueness of perceptual relations (the logically stronger argument). Prosser's arguments, combined, are supposed to lead to the conclusion that the passage of time is not only illusory but also unintelligible. I'm going to concede the argument from indiscriminability – we shouldn't expect McTchange to be the kind of thing we discriminate in experience, any more than we expect Dunbar to be discriminating between fast and slow-flowing time – but contest the argument from unintelligibility. We can make sense of McTchange without thinking we can discriminate instances of time passing from instances of time not passing.

2.2.1 The argument from indiscriminability

Let's begin with Prosser's argument from indiscriminability. Prosser argues against views which take experience to provide perceptual evidence for time's passage, after the manner that we perceive that a ripe tomato in front of us is red.[6] He argues that experience doesn't favour such views over the view that time does not pass.

In brief, Prosser's argument from indiscriminability goes like this:

1. For phenomenology to have x as part of its phenomenal character, the brain must be able to act as a *detector* for x.[7]
2. To act as a detector for x is to have a capacity to discriminate between cases where x is present (or obtains) and cases where x is not present (or does not obtain).
3. Because time passes whenever a brain is functioning, the brain cannot discriminate between cases where time passes and cases where time does not pass.
4. The brain is not a detector for time's passage (from 2 and 3).
5. Therefore, the passage of time cannot be part of the phenomenal character of experience.

The thought here is that if the feature of the world we are purported to be having an experience of changes, the phenomenal character we experience should not remain unchanged if the feature of the world in question is part of the content of our

[5] Paul (2010); Callendar (2012); Skow (2011b); Dainton (2011; 2012); Prosser (2007; 2013a; 2013b; 2016); Hoerl (2014); Frischhut (2015). Skow considers a version of the link between passage and experience – the Argument from Presented Experience – which is along the lines of the argument I will endorse later.

[6] (2013b:162).

[7] The word 'detector' is from Prosser (2016:33). Though he does not use it in his (2013a), or (2013b), the idea of the brain acting as a detector, if not by that name, does appear in both places.

experience. If you are looking at a ripe tomato, you have a certain experience, and, typically, if I remove the ripe tomato then the phenomenal character of your experience changes. In order to experience a ripe tomato, according to Prosser, it is required that your experience changes based on whether your brain is affected by the presence or absence of the tomato. Likewise, if you are to experience time's passage, your brain should be affected by the presence or absence of time's passage.

In some sense, I think Prosser's argument works. It is successful in defeating views that are committed to the idea of a brain as a detector of McTchange. But I shall argue that the defender of McTchange should never have thought of the brain as being a McTchange detector in the first place. Prosser's argument invites us to think, as we began to in the previous chapter, about the relationship between observation and metaphysics.

Dunbar thinks he is in a position to perceive (in Prosser's sense) differences in the rate of time's passing, but the other defenders of a dynamic view will reject that. But if what the dynamic metaphysician were trying to discriminate between were cases where time passed and cases where it didn't, then, it is clear, the defender of a dynamic metaphysics would be unable to perceive time's passage at all since they could not isolate a feature of their experience as being uniquely temporal. According to the defender of a dynamic metaphysics, *all* awareness is awareness of time's passing. So, if we thought that the positive reason for thinking time passes was that time's passage was part of the phenomenal character of experience (such that our brains *detected* time's passage), Dunbar would be in a *better* position than the defender of dynamic metaphysics! For Dunbar, at least, can detect variability in the passage of time. If Prosser is right, Dunbar offers a better account of time's passage than other dynamic metaphysicians. But that can't be right! Dunbar is an absurd and risible character, precisely because he is so obviously wrong about the passage of time. We need to explain where the difference lies between the relationship phenomenology has to metaphysics (for reputable metaphysicians),[8] and Dunbar's naïve reading of metaphysics from his experience.

The defender of a dynamic metaphysics, then, should think that attempts to perceive time's passage (in Prosser's sense) are anathema to their project. This is not just true of Prosser's formulation of the argument, but true of the general project of trying to show that time's passage is something we can 'read off experience' because of some special feature it has (even if it is merely a reliable, and not infallible, detector of this feature). If McTchange were the sort of thing we could perceive, because we could have some (coherent) experiences with it, and some without, we would have two options. Either dismiss McTchange, along with Dunbar's deluded view of the world, as confusing how things seem with how they really are, or, if working brains could be differentially influenced by McTchange under similar circumstances, we should treat it as something to be detected by physicists, like gravity. It is clear, however, that most defenders of dynamic metaphysics don't think that McTchange should be dismissed like Dunbar's view and don't think that McTchange is something to be detected, like gravity, by physicists with expensive equipment.[9]

8 Or at least the more rigorous disreputable ones.

9 In Chapters 4 and 5, we'll explore the relationship between McTchange and Physics in more detail.

Let's explore that dilemma a little. On the first horn, we have a distinction between how things seem and how they really are. Both defenders of static and dynamic views are committed to an appearance/reality distinction, such that appearance is sometimes, but not always, a good guide to reality. The dispute between Prosser and defenders of dynamic views is one about whether experiencing McTchange is like experiencing a red tomato. In the previous chapter, I suggested an analogy with secondary qualities. Seeing a red-looking tomato is good evidence that there are red things, conditional on colours being response-dependent properties that are features of the world. But those who think redness is projected onto the world by human observers aren't going to be able to settle the metaphysics of redness by appeal to the brain as a detector. Both parties agree that things sometimes *look* red. Similarly, appeals to experience aren't going to settle the metaphysics of change; both parties agree that sometimes things appear to change – but what's at stake is whether this is McTchange or temporal variation. Physicists will be no help because they measure change using clocks. Clocks do measure time's passage (if it passes), but they do not count as *detecting* it either, since they don't discriminate between McTchange and the absence of McTchange, for the same reasons our brains do not.

Prosser does concede that there are other options for the defender of time's passage. He says, 'At best it might be argued that passage is an essential element of all physical causation; but this would apply to all phenomenal characters equally.'[10] Elsewhere he says, 'At most, the passage of time would be a kind of 'background' or 'enabling' condition, necessary for causation to be possible at all.'[11] I think this is the approach the dynamic metaphysician should accept. McTchange, if there is any, will be a precondition of experience, simply because change is a precondition of a functioning brain. McTchange, if there is any, is an enabling condition, much like causation, but not something that we detect as such.[12]

2.2.2 The argument from unintelligibility

Simon Prosser moves from the argument from indiscriminability to the argument from unintelligibility:

> We cannot experience time as passing, and this undermines our ability to speak or even think about time passing. We do not have any understanding whatsoever of what the world would be like if time passed, for the nature of experience is no guide to this at all. (Prosser 2016:55–6)

This picks up a theme from the previous chapter, that there is a problem in even so much as explaining what McTchange is. If McTchange isn't a feature of experience, then it is unclear how we could so much as refer to it. If we can't refer to it, we could

10 (2013b:162).
11 (2013a:76).
12 Meditation is sometimes purported to provide *atemporal* experience (Metzinger 2020). Even such experiences last a certain duration and rely on brain processes involving constant change. That some states don't involve detecting McTchange does not show that they detect an absence of McTchange.

not even entertain the thought that there is McTchange, and then asking the question 'Is there McTchange?' is asking no question at all.

A crucial point to note about the argument from unintelligibility is that it is dependent on the argument from indiscriminability. If the passage of time is not, as I have conceded, something brains can detect, then how is it that the owners of those brains can refer to it?

> The passage of time is just the wrong kind of phenomenon to have a selective influence on, or connection to, a specific mental state, and therefore is not the kind of phenomenon that could be an object of experience. (Prosser 2016:46)

If we can't refer to or think about McTchange, we can't argue in favour of it. If we can think about McTchange, then we need some account of how it is intelligible. And here (quite reasonably) Prosser is adamant that something satisfying by way of explanation is needed. The defender of McTchange 'must provide an adequate explanation of the *mechanism* that would allow passage to be perceived, with enough detail to make it at least plausible that some such story could be true' (2016:47).

The threat that McTchange is unintelligible is a serious one. Hopefully, some of the work in the previous chapter has partially defused the worry. We should not be worried that McTchange is an irreducible concept, or that it is one that depends on our being temporal creatures in order to grasp, since those are not barriers to its coherence. But the challenge Prosser lays down is one of reference; how can we so much talk about McTchange?

In order to answer this question, we need to consider the relation between experience and the things experienced more generally, and then we need to consider what gets things to count as being provided by 'common sense'. These will be the concerns of the next two subsections.

2.2.3 What we experience

If we talk about the *objects* of perceptual experience, we're thinking of experience as putting us in a relation to something experienced. The objects of experience are whatever we are related to when we experience something. We might contrast this with the *contents* of experience; the contents of experience are the ways experience represents the world as being. We can then ask whether McTchange is an object of experience and/or whether McTchange is represented in experience (i.e. it is part of the contents of experience).[13] These two questions matter for different reasons.

If we are interested in what our evidence is that there is McTchange, we might wish to appeal to the contents of experience; for that would give us a really straightforward story: 'What is your evidence for *p*?' 'My evidence for *p* is that experience represents the world such that *p*.' This would be defeasible evidence, but it would explain why people thought there was a strong *prima facie* case that *p*. If, alternatively, we are interested in what we refer to by 'McTchange' (assuming we succeed in referring at all), we might be

[13] This distinction is taken from Speaks (2009).

interested in the objects of experience; this would give us a story about how we stand in a relation to McTchange: 'It's what I see every time I see something change.' There's more to be said about both of these stories.

The argument from undetectability seems to be about the contents of experience; everyone has been assuming (Prosser supposes) that experience *represents* the passage of time, and Prosser has argued that such a representation isn't to be found. Experience represents qualitative change, for example, but doesn't thereby represent such change *as being McTchange*. Or at least, if experience does represent change as having the metaphysical baggage of McTchange, this is theory-laden observation and doesn't constitute evidence of McTchange; the representation simply couldn't be connected to McTchange in the right way to provide such evidence.

The claim we would need to get McTchange in the *content* of experience, in a way that helps, is that humans act as detectors for McTchange; that we reliably represent McTchange when there is McTchange and not otherwise. But Prosser has shown that we are not detectors in that sense. I will say more about what ways time is featured in the content of experience in §2.4, but we can happily deny that McTchange is part of the *contents* of experience without undermining the claim that there is McTchange, that we can think and speak about it, and that we have reason to believe this.

To get a reason to believe that there is McTchange, we merely need to claim that the evidence for McTchange is something like inference to the best explanation – as McInerney (1991) claims – rather than perceptual evidence. We are still left with the claim that we can't so much as *think* about McTchange in order to ask if we have evidence to believe in it. The content of experience might play a role here; it is useful for allowing us to point at things, and thereby refer to them, rather than merely providing a description and hoping that something satisfies it. Jeff Speaks, writing about the nature of perceptual content in general, says:

> Whatever one's view of content, there is some distinction to be captured between thinking about an object de re and merely thinking about it under a description. Given this distinction, it seems clear that there are some external particulars about which I am in a position to entertain and express de re thoughts, and that there are some external particulars with respect to which I am not in such a position. It also seems clear that an object can become available for de re thought by my having a perceptual experience of that object. (Speaks 2009:560)

One of the reasons, Speaks writes, that we should be interested in the content of experience is that plausible content of our experiences enables such *de re* reference to the object of experience. Perception represents things as being a certain way and enables us to entertain and express *de re* thoughts about those *res*.[14] Now, if as a result of theory-laden observation, McTchange is part of the content of experience (for about 70 per cent of people), provided that there is the *right kind* of connection between the representation and McTchange, we can refer *de re* to McTchange. So, I see some qualitative change and it is represented in my experience as being a qualitative

[14] (lit. 'things'). '*De re*' means 'of the thing'.

McTchange; if nature has been kind and all change is McTchange, then I can refer *de re* to McTchange. This story can hold, even if perceptual experience doesn't give us *evidence* of McTchange rather than mere temporal variation.

Prosser's unintelligibility argument is meant to block this move, because we can only have theory-laden observation if we can refer to the theoretical posit. And we have, he thinks, no way of picking out the relevant theoretical posit. But perceptual experience isn't necessary for *de re* reference, and we can successfully refer to things under a description – *de dicto*. Prosser (2016:55) admits this:

> We can refer to many imperceptible entities, such as abstract objects, future events, distant objects that lie outside our light cones, and so on. But in every such case we have some non-perceptual means by which we can identify the reference (i.e. distinguish it from other things). . . . There is, for [such cases], an answer to the questions: 'which feature are you talking about?'. The answer takes the form of a definite description stating what distinguishes the feature from everything else.

When I introduced McTchange, in the previous chapter, I appealed to a difference between space and time. McTchange is the change, not relative to anything else, that is constitutive of time passing. Passing is the thing time does that space does not, that is aptly characterised through metaphors of flow or passage. That's a definite description.

Prosser's unintelligibility argument entirely depends on the claim that no such definite description can be given:

> But do we have any such way of identifying the feature of the world supposedly denoted by 'the passage of time'? . . . It is hard to see what other identifying description [than some perceptual phenomenon] could be given that would fare any better. Consequently, the linguistic expression 'the passage of time' fails to denote; and thus, in one important sense, we cannot even talk about the passage of time. (2016:55)

We can talk about the passage of time by contrasting it with space. So, just as Prosser's argument from undetectability wasn't a problem for McTchange if defended as it should be, his argument from unintelligibility does no damage to McTchange understood in terms of a difference between space and time. It's not that we are detectors for time because we detect time when it is there and its absence when it is not, but that we detect a difference between time and space. We can use the metaphors Röck discusses, of flow and of growth, to make intelligible the thought that qualitative change differs from mere temporal or spatial variation.

When I introduced McTchange, in the previous chapter, I appealed to a difference between space and time. McTchange is the change, not relative to anything else, that is constitutive of time passing. Passing is what time does that space does not. One may deny that there is anything that fits that description, but Prosser aims to deny that such a description is unintelligible. It *is* intelligible, however.

2.2.4 Common sense and sense experience

The difference between McTchange and mere temporal variation is concerns an absence of something, not the presence of something. Looking for something that is present in some, all or roughly 70 per cent of experiences will not help move the debate forward. For McTchange is change that is *not* relative to anything else. It is equally compatible with any experience of change that the change experienced is not relative to anything else, or relative to something else not experienced. The debate is about whether there is, to use an expression from Putnam (1990), a 'God's eye view' from which all change has been made relative to something unchanging.[15] Modifying Putnam's (1990:28) remarks:

> What I am saying, then, is that elements of what we call *'change'* and *'temporal perspective'* penetrate so deeply into what we call 'reality' that the very project of representing ourselves as being 'mappers' of something *'atemporal'* is fatally compromised from the very start.

The debate is about whether our embedded perspective on the world is a mere appearance that obscures a McTchangeless objective alternative, or if McTchange is so central to our grip on the world that the very attempt to offer a view of the world abstracted from a McTchanging perspective in time is doomed to stop being a view of *the world.*

Given that we should expect the manifest content of experience to be explanatorily impotent with respect to whether there is some unexperienced, 'objective' characterisation of the world, where all change is temporal variation, how do we make sense of appeals to experience or (even more opaquely) 'common sense' that there is McTchange?

Let's put aside our concern with McTchange for a moment and consider an example of competing hypotheses where we clearly favour common sense, from Chris Hookway (2002): whether diamonds are hard when nothing touches them. In this case too, we see the distinct hypotheses we are considering go beyond what might be detected in experience.

> Given our experiences of scratching diamonds so far, consider the following possibilities:
>
> 1. • All diamonds are hard.
> 2. • All diamonds are soft until tested, at which point their hardness increases in proportion to the pressure with which they are scratched.
> 3. • All diamonds are hard except for the one that was crystallised in soft cotton and destroyed without being tested for hardness.
> 4. • All diamonds tested before the year 2010 are hard, and all diamonds not tested before that time are soft.

[15] This way of framing the problem will return again in §§4.5.3, 4.5.1 and §9.5.

> At any time in the future our evidence will be compatible with the first three of these if it is compatible with the first – and it will also be compatible with a suitably altered version of the fourth. . . . Once we accept that it is laws, above all, that require an explanation, we can note that no explanation is readily available of why or how diamonds steadily become harder when scratched; nor of how the untested and now destroyed diamond can possess the inner structure required to be a diamond yet still not be hard. (2002:56)

All four hypotheses seem to be compatible with anything our brains might detect. But 1 seems like it's part of common sense. Hookway argues that it is because 1 is better as a candidate law of nature, but what's important for us is just that there's a reason to choose this hypothesis over the alternatives.

We might ask whether we have any *evidence* for 1. Here are two answers: no, but we are entitled to proceed on the hypothesis that 1 is true insofar as it has been a useful auxiliary hypothesis in our investigations; or yes, 1 has been confirmed by the success of our theories. Let's look at these answers in turn. You might think that 1 is an inference based on pragmatic factors – it has more theoretical virtues like elegance or simplicity – but that these aren't reasons to think that 1 is more likely to be true, merely that 1 is part of a more elegant theory. We have no evidence, after all, that the world is designed to fit with our preferences for theory selection. Nonetheless, we might think, when choosing between empirically equivalent hypotheses, such pragmatic factors might give us practical reasons – though not *epistemic* reasons – to prefer some hypotheses over others. That is, we might think that, in the absence of available evidence, we can use (without believing) hypotheses that would give our theory more theoretical virtues, even if we don't take those theoretical virtues to be truth conducive.[16] We might call this method of hypothesis generation 'abduction', after Peirce's introduction of the term to contrast with induction and deduction.[17]

Let's take the idea that abduction generates hypotheses for testing. Now, it will always be the case that there are many (infinitely many?) hypotheses compatible with our evidence that we simply do not test or investigate. Choosing to be indifferent between these hypotheses would be paralysing; we could not gain any evidence for them by testing one because we would need to select one to test. We would make (Buridian) asses of ourselves.[18] It can be no objection, then, to selecting one hypothesis over others that we have no positive evidence of some hypothesis compared to some rivals; selecting some defined set of hypotheses to contrast at least allows inquiry to continue. The objection comes when we select a hypothesis that looks to be not

16 Nolan (2014) has argued that we *should* take theoretical virtues to be epistemic if we are to value them at all. I wish to remain neutral on this point, but *even if Nolan is wrong*, we shouldn't think that appealing to theoretical virtues is unmotivated.

17 Peirce (1903); See Mohammadian (2019) for a discussion.

18 Buridan's ass is an example of the impossibility of choosing between equally rational alternatives; an ass equidistant between two bales of hay starves to death. The example has been associated with Jean Buridan since at least 1597, but Zupko (2018) points out that this example cannot be found in Jean Buridan's extant work, and a similar example featuring a human can be found in Aristotle's *De Caelo* 295b32. Ullmann-Margalit and Morgenbesser (1977) argue that the solution in such cases is to pick, rather than choose. My 'select' is intended to be neutral between picking and choosing.

worth investigating. We can investigate hypotheses that are *probably* false but would be interesting if true. However, we should not investigate hypotheses that are manifestly *ad hoc*, explanatorily impotent or just ridiculous.

Given Quine's observations about the holism of confirmation, mentioned in §1.5, we might think dividing our hypotheses into those for which we have evidence and those for which we have merely pragmatic support is a dichotomy that cannot be supported. We might then take those hypotheses generated through abduction and that have stood the test of time, as having been confirmed. As Hookway (2002) puts it, they change from 'hopes' to 'commonsense certainties'. Or we might, as Nolan (2014) argues, claim that theoretical virtues confer epistemic justification in any case.[19]

This gives a sketch of how we could arrive at a belief in *some* hypothesis that we don't have perceptual evidence for (contrasted with empirically equivalent hypotheses). Although we lack observational evidence, we can nonetheless quite reasonably accept, and possibly even gain confirmation for such a hypothesis. But this does not yet explain why roughly 70 per cent of people appeal to experience in the case of time's passing. Our arrival at good hypotheses is, in the case of 1 and many other cases, something that tends not to be deliberate. I submit that we would want to appeal to experience here *in some way*. Perhaps, as suggested above, we experience the world as operating according to laws, and our observation of diamonds is thereby theory-laden. If so, we do have experiences, in a broad sense of experience, *as of* the operation of laws (or, in our original case, *as of* McTchange).

There is much to say about our experience of time; there are many relevant temporal phenomena that can work somewhat independently of one another. Before we survey those in §2.5, I want to consider a case where many of the same considerations apply: we seem to be dealing with an enabling condition of experience, which we don't seem to have perceptual evidence of, but which seems to be explanatorily useful in our investigations: causation.

2.3 Comparison with experiencing causal connexions

We saw above various appeals to seemings or experience to explain the passage of time. I have argued that we do not detect passage in the sense involved in Prosser's argument.[20] But it should be clear from the sources I quoted at the start of the chapter that there is some relationship between our experience and our belief in McTchange. If we did not have the sorts of experience we have, people wouldn't find McTchange so plausible.

It looks like there is a general problem about the relation between experience and the kind of disputes metaphysicians are often interested in. Compare the dispute about

[19] Forbes 2023b discusses these methodological points at greater length.

[20] I am denying what Christoph Hoerl (2023) calls an 'empiricist' approach to the flow of time. I'm cautious, however, about being described as a 'rationalist' about the flow of time. The role of abduction, outlined in the previous section, is meant to maintain a connection to experience that might deserve the term 'pragmatist' rather than rationalist.

whether we experience McTchanges with a claim about causal realism; if we did not have the sorts of experiences we had, causal realism would not have the plausibility it has. This is true even though, as Hume observed, 'any two objects or actions, however related, can never give us any idea of power, or of a connexion betwixt them' (*Treatise*, III, 14 1888:166). It is widely accepted that we can't tell, just by looking, whether two distinct events are connected by a causal relation. We make errors where we think there are connexions that do not in fact exist, and we fail to see connexions which do. Magicians and illusionists trade heavily on our ability to be convinced of a causal relationship where there isn't one, and to fail to notice causal relationships that actually explain how a trick is done. Our inability to detect metaphysically interesting causal properties in magic tricks carries over to tensed facts, as D. H. Mellor (1998:16) pointed out: 'We cannot for example refute someone who claims to see the future in a crystal ball by pointing out the visible pastness of the image: there is no such thing.' Nonetheless, causal realists and defenders of a dynamic view alike can enjoy an evening at a magic show with little risk of falsifying their theories.[21]

The similarities between causal connexions and McTchange are quite striking. G. E. M. Anscombe (1971:69) responds forcefully to the Humean charge that we cannot observe causal connexions:

> [Hume] confidently challenges to 'produce some instance, wherein the efficacy is plainly discoverable to the mind, and its operations obvious to our consciousness or sensation'. Nothing is easier: is cutting, is drinking, is purring not 'efficacy'?

One could equally respond to Prosser by giving the following examples of McTchange: movement, qualitative change and substantial change. This might seem like it is missing the point, but I don't think Anscombe is. She continues:

> But it is true that the apparent perception of such things may only be apparent: we may be deceived by false appearances. Hume presumably wants us to produce an instance in which *efficacy* is related to sensation as *red* is. It is true that we can't do that; it is not *so* related to sensation.

Anscombe is making a point about what we should expect perception, defined in roughly Humean terms, to deliver.

> Someone who says [we can never observe causality in the individual case] is just not going to count anything as 'observation of causality'. This often happens in philosophy; it is argued that 'all we find' is such-and-such, and it turns out the arguer has excluded from his idea of 'finding' the sort of thing he says we don't 'find'. (Anscombe 1971:68)

Prosser's treatment of perception as being analogous to a detector seems very Humean in this regard. We can concede to him that, if we have *this* approach to perception,

[21] See Fraps (2014) for a discussion of how magicians manipulate subjective awareness of time.

we won't find evidence of casual connexions, nor will we find McTchange. But, as Anscombe points out, we won't even perceive billiard balls travelling towards each other (How are we to detect that they are not virtual billiard balls? How are we to detect that they are the same billiard balls throughout?). Just as we find it obvious that diamonds are hard when no one's touching them, we find it obvious that billiard balls are not holographic-until-touched, and persist through time, though I can't see how we would detect such properties.

Causal connexions are not merely another case where we appeal to common sense to defend them. They seem, like McTchange, to be enabling conditions for experience. All experience that is perception *of* anything will – the causal realist will claim – presuppose some causal connexion. At the bare minimum, perception relies on a causal connexion between something and someone perceiving it. Most interestingly, for our purposes, there is some basis in our experience for thinking that there are causal connexions, even if we don't 'detect' or 'find' such connexions in experience.

We have experiences of efficacy or intervention on the world. We can capture what causation is, as distinct from other enabling conditions for experience, by appealing to a definite description: causal connexions are what separate interventions from the mere accidental succession of one thing by another.[22] The causal realist, such as Woodward (2005), can draw upon our experience of making things happen as a reason to think that there is a difference between genuine causal relationships and mere regularities. Prosser (2011) himself has defended 'affordances' – possibilities for causal interaction – as being part of perceptual experience. Likewise, if we did not have the sorts of experiences of movement, change, anticipation, recollection and endurance that we have, then a dynamic metaphysics would not be so plausible.[23] But the fact that certain sorts of experiences make a view plausible does not mean that we can *tell* in virtue of having them that some controversial metaphysical claim is true. Paul (2010) and Le Poidevin (2003) both accept that our experiences lend *plausibility* to a dynamic view, even though they both accept that we can't tell on the basis of our experience that a dynamic metaphysics obtains, since they both reject dynamic metaphysics.

Consider Wittgenstein's question: 'What would it have looked like if it had *looked* as if the earth turned on its axis?' We might offer causal and temporal analogues. 'What would causation have looked like if it had looked as if we didn't intervene on the world?' and 'What would change have looked like if it had looked as if there were no McTchange?' respectively. These questions are meant to challenge us to say what difference our experience makes; to take up Callender's concerns from the last chapter about whether metaphysical theories are distinct. The thought is that, given that the earth does turn on its axis, it must look as if the earth turns on its axis.[24] So the denier of necessary connexions and the denier of McTchange will say that the world looks as if

22 Some authors restrict 'interventions' to things performed by agents, and others allow that rocks can intervene. This definite description should be read neutrally on this point. *At least* in the case of agents, there is some difference between police investigating, pursuing and catching a criminal and the hapless Inspector Clouseau coincidentally discovering the culprit by serendipity.

23 The case of endurance will be taken up in the next chapter.

24 Anscombe (1959:151) reports 'My reply was to hold out my hands with the palms upward, and raise them from my knees in a circular sweep, at the same time leaning backwards and assuming a dizzy expression. "Exactly!" he [Wittgenstein] said.' Perhaps a reply, 'seeing all time as you might see

there are no necessary connexions, or that it looks as if there is only temporal variation. In short, the world *looks just how we would expect it to, if our theory is true*. It is difficult to overstate how unproductive a dialectical move this is.[25]

Perhaps the problem is one of whether we have reasons to believe in causal relations, or in McTchange. If there are causal connexions in the world, experience which (reliably) correctly represents things as causally connected plausibly gives us justification for believing in causal connexions. If there is McTchange, experience which (reliably) correctly represents things as, for example qualitatively changing gives us justification for believing in McTchange. But, experience can provide such 'reliabilist' justification without *luminously* doing so. Such justification does not require our being able to settle, *internal* to the experience itself, what the correct metaphysical account is.

One might think that there is an important disanalogy between time and causation; we deal, in the causal case, with causal relationships between different individual billiard balls, but the passage of time isn't a relationship between distinct individual things. This disanalogy can be overstated, however. We certainly see some causal relations as more salient than others, but the background enabling conditions of those causal relations are still causally relevant.[26] Causation is still an enabling condition to experience, and all perception involves our standing in some form of causal relation to the world. Similarly, we have, as we will discuss in §2.5, a variety of distinctly temporal experiences where we notice, for example, the change or persistence of individual things. Some aspects of the passage of time are more salient than others.

Before moving on, let's consider what's driving us down a *cul-de-sac*. Let's assume that none of us are error theorists about time or causal explanation.[27] We look at clocks and think, 'it's nearly 2.30 pm, I must get to my dentist appointment', and anticipate not being able to talk properly afterwards because the local anaesthetic will cause our speech to be slurred. These sorts of inferences aren't the ones in dispute in the relevant metaphysical debate. Quite the contrary; these are the data for which we seek metaphysical explanations. So, it should be no surprise that both parties who attempt to explain the data think that their theories offer some explanation of that data. It would be bizarre if things were otherwise.

Nonetheless, given some purported explanation of the data, we might think that some counterfactual claim holds true, such that if the situation were different, our experience would be different. We must not confuse the indicative ('if there are no necessary connexions then my experience would be just like this') with the counterfactual ('were there to be no necessary connections, my experience would be just like this'). The former is trivially true; the latter almost certainly false. If we think that necessary connexions *are* required for causal explanations, then we would, of course, take our experience to rely on them. Things are no different for McTchange;

a stretch of the Rocky Mountains' would be an equally appropriate reply in the temporal case (cf. Vonnegut 1991:70).

25 As is made clear in Anscombe's report of the exchange (cf. fn. 19), this is not quite the dialectical move Wittgenstein was making.

26 In §§7.2.3, 7.4.3 and §10.3.

27 Baron, Miller and Tallant (2022) argue that, for all we know given our best physics, the error theory is true. I deal with the worries about physics in Chapter 5.

if we accept that there is McTchange, the counterfactual 'if there were no McTchange then I would have no experience at all' will almost certainly be true. Our experience depends on McTchange, then. But the test is one of whether we need McTchange to explain the relevant data. Given that there are some very clever people who have come up with ingenious explanations that assume there is no McTchange, we are going to find that the choice is between competing explanations rather than one theory that clearly can explain the data and another that simply cannot. For the remainder of this chapter, I shall outline some of the temporal data and finish with a puzzle. Nothing in this chapter is going to be decisive either in favour of or against, McTchange being needed to explain our experience. Chapters 3 and 4, though not decisive, are where arguments in favour of McTchange will be advanced.

2.4 On the content of experience

We've considered general reasons not to expect there to be some feature of experience which is *the passage of time*, or *McTchange*, such that we (or our brains) can 'detect' it. But the word 'experience' could mean any number of things. In the next section, I'll survey some of the things it might cover in more detail, but for now, I will focus on making some important distinctions about what 'experience' might mean. I'll proceed by running through a few questions: What is it for an experience to be temporal? What is it for an experience to have content? What is it for an experience to be perceptual? Finally, what is it for an experience to have temporal perceptual content? We'll see that debates about many of these terms only serve to muddy the waters, and that we can bypass most of the points at issue. There are views of perception and experience that are amenable to me, but nothing in this section will constitute an argument for a particular view of experience.

At the start of any such exercise is a problem: if McTchange is an enabling condition for experience, the idea of 'temporal' experience (as opposed to some other kind) seems a little odd. For experience of frantic activity is just as much temporal experience as experience of a sensory deprivation tank. So, the idea of focussing on specific experiences and comparing them to other, *atemporal*, experiences isn't likely to get very far – indeed, this is what we learnt from considering Prosser's arguments. We might, however, pick out various features of experience that we take to be of more interest; our experiences of qualitative change are going to be of more interest than our experiences of depth, for instance.[28]

Experiences have content if they represent things as being a certain way. As Prosser (2016:51) claims, '[w]e can understand this as equivalent to the claim that experiences have correctness conditions.' As we saw above, Prosser's arguments depended on the content of experience, both because having something in the content of experience allows for *de re* reference or attitudes towards it, and because it provides something that can be veridical or falsidical. Only experience that represents things as being a certain

[28] We will discuss a variety of distinctly temporal experiences in §2.5.

way can misrepresent things or be accurate in its representation. But this leaves stories about what content *is* quite open. We might debate whether content is conceptual, and indeed what it is for content to be conceptual.[29]

One significant debate is about whether the content of perception is just a matter of how things are in experience (content internalism), or whether part of the content of experience is determined by how things are in the world (content externalism). If we accept content externalism (and I wish to remain neutral about this here), then how things seem to us in experience is not the same as the content of experience. We can still talk about the luminous content of experience (i.e. that aspect of experience which does just depend on how things seem to us in experience), but as suggested above, this might not be needed for justification if one is a reliabilist. If content externalism is true, that there is McTchange might well be part of the content of experience. If we are reliabilists about justification and externalists about mental content, then we might have exceptionally good evidence that there is McTchange because that there is McTchange might well be part of the contents of experience, and reliably so, since if there were no McTchange, there would be no experience. Schellenberg (2018), for example, defends a view on which we use perceptual capacities to perceive the world but distinguishes between factive perceptual evidence and phenomenal evidence. Factive evidence is not luminous, whereas phenomenal evidence is. But factive perceptual evidence is 'evidence that is determined by the environment to which we are perceptually related such that the evidence is guaranteed to be an accurate guide to the environment' (p.8). Not being luminous, this evidence could be a guide to the presence of McTchange without our being able to detect it. Similarly, for other accounts of perception that allow that we can have non-luminous factive perceptual relations to the world.

Of course, non-luminous factive perceptual evidence would not be persuasive evidence to a sceptic of McTchange, since we couldn't tell whether or not we had such evidence. Prosser thinks we can't so much as represent McTchange. That's because he thinks we can't *conceptually* distinguish it from other potential contents. But as we saw in §2.1.4, we can do so via a definite description.

Not only can we think carefully about the content of experience, but we might want to make a distinction between time *perception* and 'time consciousness'.[30] This is roughly a distinction between time as it occurs in the phenomenal content provided by our sense modalities and proprioception, and time as dealt with in lived experience more broadly. As noted in the previous chapter, the separation of 'observations' from other parts of one's theory cannot be done cleanly. But we can still draw such a distinction without insisting on a dichotomy.

We might notice that an object is moving, changing colour or changing other qualitative properties; these seem to be aptly thought of as perceptual. A sense of *weltschmerz*, ennui, nostalgia or wistfulness might be thought of as requiring resources besides the perceptual, since they depend on one's emotional attitudes towards past and/or future in complex ways and should be thought of as part of 'time consciousness'

[29] See, e.g., McDowell (1994:162); Brewer (1999), Crane (2009), Schellenberg (2011; 2018).
[30] Cf. McInerney (1991).

more generally.[31] The wider intellectual and social context of a person will be required to make sense of these emotions, the activity of planning or the act of apologising, for example. In some broad sense, we have experience of apologising or planning, but these aren't related in the right way to the sense modalities to count as *perceptual* experiences. It is plausible that time consciousness more generally plays a role in allowing us to represent McTchange as distinct from mere variation, and so to be *so much as able* to represent to ourselves that there is McTchange.

We might make the distinction between low-level and high-level views of perceptual content. Susanna Siegel (2009; 2010) has defended a high-level account of perceptual experience where causal relationships, for example, will count as amongst the contents of visual experience. So, on such a view, it's plausible that there might be some perceptual experience of some kind that counts as having content which is temporal. Alternatively, we might think only very low-level things count as perceptual: we might restrict ourselves to those aspects of experience that are immune to cognitive penetration (i.e. that are unaffected by the content of our beliefs more generally) because they are part of a sensory 'module' that is fast, informationally encapsulated (i.e. unaffected by the content of our beliefs more generally) and mandatory (i.e. unaffected by whether we want to perceive them that way).[32]

There are plenty of middle cases; we might wonder whether to include forms of recognition and anticipation as coming under the perceptual, though of course, which things are recognised and how quickly they are recognised will depend on one's social context.[33] Affordances, which Prosser counts as perceptual, as noted above, seem to be middle cases.

What work we think 'perceptual experience' can do in motivating belief in McTchange is going to depend on what kind of account of perceptual contents we offer. If we allow that perceptual experience is 'cognitively penetrated' by our beliefs more generally – that what we *perceive* represents things differently depending on what beliefs we have – then we are left with the question Siegel (2010:10) raises: 'If experience can be influenced by beliefs to begin with, then what happens to its status as a tribunal that can be used to test beliefs against the world?.'[34]

The claim that experience gives us evidence of McTchange can be taken to mean that we have some informationally encapsulated contact with McTchange, immune to cognitive penetration, that puts us in contact with the world. But, as I argued in §2.1.2, the feeling of time speeding up or slowing down seems a much more plausible candidate for such a perceptual experience and seems antithetical to what metaphysicians are interested in. When we appeal to experience as motivating belief in McTchange, it is not because experience provides an indubitable tribunal of the senses, or observation

31 I'll come back to affective features of time consciousness, like nostalgia, in Chapter 8, but it just won't be the same.

32 See Siegel and Byrne (2017) for a discussion.

33 See, e.g., Del Pinal and Spaulding (2018) for a discussion of recent work on implicit bias, which is a way (though not the only one) in which perceptual experience might be penetrated by beliefs.

34 Siegel (2016) acknowledges that this debate is related to the earlier debate about whether we can separate observations from theories, in terms of which I have framed things so far.

sentences that are individually confirmed as such. Rather, it invites us to treat some differences we find in experience as defeasibly guiding us to contours in the world.

The distinction between perception and time consciousness more broadly doesn't matter so much for this kind of hypothesis generation.[35] As Peirce (1903) says, 'abductive inference shades into perceptual judgement without any sharp line of demarcation between them' (CP5.181). Perception, successful strategies for navigating the world, apt emotional responses and presuppositions of inquiry are different sources of the same kind of justification. It is only if we were interested in perceptual experience, which provides us with directly confirmed observation sentences, that perception would play a distinct justificatory role. If we discovered that, in order to get our robots to navigate the world successfully, we needed to programme them to treat space and time quite differently, that would be equally good as a motivation for belief in McTchange as conscious experience of time differing from conscious experience of space. It is this kind of contrast, not an ability to detect whether time is passing, that in §2.1.4 I argued made McTchange intelligible.

Given that what counts as perceptual is contested, there's a question about how reliable we are at describing the content of our own experience. Hoerl (2014) and Miller, Holcombe and Latham (2020) consider that we might be either misdescribing or inferring that we have perceptual experience of passage, even though we don't have any such experience. Misdescription is a matter of using language suggestive of McTchange, and consequently coming to believe we have perceptions of McTchange, because we are relying on metaphysically loaded concepts. Inferentialism, by contrast, is believing we have perceptions of McTchange, because we are committed to McTchange and so believe that perceptions of change are *a fortiori* of McTchange. As Miller et al. argue, this looks to be an issue on which we can get empirical evidence. Consider the difficulties of getting a sense of what divides perceptual experience of time phenomena from passage experience more generally, and debates about cognitive penetration, and about the relationship between our language and the ability to describe our experience. Given all this, it seems possible that we might both find variation in people's experience (over time and between individuals) and worry about the unreliability of people's descriptions of whether something is *perceptual*.

There seem to be at least five ways in which perception may fail to be a good guide to McTchange.

1. It could be that the relevant content is externalist, and so we can't tell if we have it.
2. We might not be able to tell whether a state is perceptual or non-perceptual (just as we can't tell whether something is a memory or a confabulation by introspection).
3. Perceptual experience may be subject to cognitive penetration, so experience may be formed in response to metaphysical commitments, rather than the commitments formed in response to perception.
4. We may misdescribe our experience.

[35] Thanks to Zoe Drayson for pressing me on this point.

5. We may infer that our experience has content it doesn't because the language we use to describe the content adds metaphysical baggage.

As Akiko Frischhut (2015:143) says, 'all parties have conceded far too easily that there is [a perceptual] experience that needs explaining in the first place'.

Frischhut goes on to argue that 'only experiences in the strict sense, that is, perceptual experiences, can succeed in the role of securing the dynamic theory an intuitive advantage over the static one' (p.144). That is where I disagree. Indeed, the next chapter will discuss a non-perceptual sort of experience that does seem to give an intuitive advantage (albeit a defeasible one) to a dynamic view: our first-personal experience of persisting through time by enduring. Before we consider that case, we should consider what sorts of experience we might appeal to; for one way to motivate a difference in experience between time and space might be to look at all the various ways in which time matters in experience. Individually they might not do any more work than a feeling time is passing slowly when we're bored, but collectively they might suggest that there is some distinctively temporal aspect of the world we ought to be accounting for.

2.5 Varieties of temporal experience

In this section, I'll consider, in slightly more detail, what sorts of experience we should consider as providing the relevant temporal data. I'll consider qualitative change and movement, extension through time and duration, temporal order and succession.[36] I'll also consider purported feelings of 'nowness', 'flow' and 'jerk and whoosh'. Some of these features might seem to provide more impetus to believe in McTchange than others, but as Miller et al. (2020) warned us, the use of 'passage-friendly' language may cause us to misdescribe these experiences. We should no more read metaphysics off the terms that we happen to use in English to describe the world, than we should read it off the feeling of time passing slowly when we're bored. Nonetheless, that we find passage-friendly language so useful in describing the world may be another datum we need to account for in our explanations of the world.

2.5.1 Qualitative change and movement

Let's start with one of the most obvious temporal experiences: qualitative change. This presupposes some object of experience that possesses some qualities. We have experience, it seems, of change in what these qualities are. We notice that the trees are in bloom, and then we notice an abundance of fruit. We notice the lemons turn from green to yellow, then we notice the leaves turn from green to brown. We notice the sky turn from blue to grey. These examples are all, in some sense, *visual* examples, though they are unlikely to count as *perceptions* of change. They are merely perceptions of

[36] This list is taken from McInerney (1991) and Torrengo (2017).

difference; we notice the colour of the leaves being different from what we remember it being but don't have visual awareness of the colour changing. But we also have experiences of change, where we experience someone's face reddening when they blush.[37] The examples of qualitative change need not be visual; we may experience someone's voice as being lower than we remember, someone's voice changing in pitch or volume, or feel our own cheeks reddening as we realise that our yodelling was itself observed by bemused onlookers.

Movement might be thought of as a special case of qualitative change if you think of location as being a relation between a thing and other things (e.g. the observer) or a thing and space. Movement, on such an understanding, is change in location. Once more, movement can be inferred from noticing that something isn't located where you remember it being, or because you observe something moving. C. D. Broad (1923:351) uses the example of seeing a second hand on an analogue clock move, but noticing the hour hand has moved.

Qualitative change and movement are some of the most straightforward cases to explain in terms of temporal variation. Just as things vary in colour across their surface, they vary in colour across time. Movement is no more than varying across location, as we might represent in a graph plotting distance from a certain point over time. It is important to distinguish between noticing something has changed and observing a change. As William James (1890:629) pointed out, 'a succession of feelings, in and of itself, is not a feeling of succession'. The observation of change or movement seems to have something additional – a feeling of change or movement – in addition to successive noticing of difference. But although this difference matters for the analysis of experience, it doesn't matter so much for the metaphysics of time. Noticing that the old grey mare ain't what she used to be is enough to raise the question of whether this change is to be explained in terms of McTchange or temporal variation.

2.5.2 Extension through time and duration over time

We are aware of things changing over time, but we are also aware of changes in what things there are. We notice that the working day has finished, and the rush hour has begun. We make comparative judgements; noticing the days getting longer as summer gets near, or that the pain lasted much longer than my hand was on the stovetop. We can judge that two things were contemporaneous in experience, where one started before and continued after the other, or that both coincided. We can even consider two events that are presented to us at different times and note that they are different in duration. I remember at school a teacher took us outside to demonstrate that the speed of light was faster than the speed of sound. She took two cricket bats to the other side of the playing fields and hit them together with some violence. After half a second or so, we heard a long echoey *crack*. The sound came later than the sight of the impact (which, presumably, came slightly after the impact itself). Indeed, I can judge that the physics lesson happened much longer ago than either the time it took to relay the

[37] They might be blushing because you are scrutinising the colour of their face so closely. Please conduct phenomenological experiments considerately.

story or the time it took for the sound to travel from the cricket bat. Perception is certainly involved here, but so too is memory. Anticipation may also be involved in such judgements of duration and extension through time. I anticipate that reading this book will take less time than writing it has.

Extension and duration are likely to be cases the defender of a static view of time thinks of as relatively easy. The amount of time an event takes up is no different, from a mathematical point of view, than the amount of space an event takes up. These are just quantities along a dimension.

2.5.3 Temporal order and succession

'Succession' is definitely one of the passage-friendly words Miller et al. warned us about. There's a neutral reading on which succession is just a matter of temporal order, sometimes known as the 'arrow' of time.[38] We might judge that one event came before another, or in between two events and so on. Time order isn't simply a matter of arranging events in a sequence, but of explaining why one end of the sequence counts as being *before* the other, later, end. First the earlier events happen, *and then* the later events. Music, as Husserl (1900–1) noted, relies on our experience of temporal order. But, for all I have said, considerations of the arrow of time are just ones about how to label the axes on our graphical representations of distance and time.

'Succession', then, might be used to describe not merely the arrow of time, but the episodic nature of it. We have awareness of change, not as mere variation, but as a difference between what *was* the case and what *now* is the case. Perhaps also, we experience an anticipation that what is now the case is about to stop being the case in favour of some other eventuality. As J. D. Velleman (2005:198) puts it, 'I don't just anticipate experiencing the future; I anticipate experiencing it as the payoff of this anticipation, as the cadence resolving the present, anticipatory phase of thought.'

The use of 'present' or 'now' in such an explanation may be exactly the sort of passage-friendliness of which we should be cautious, lest it cause us to misdescribe our experience, and it may well infect the concept of succession. But defenders of a static view of time may be perfectly happy with the word 'now' provided it is understood as an indexical that picks out the time of usage in just the way that 'here' does. What would motivate McTchange is some reason to think that 'now' works differently from 'here'?

Maybe succession does this. Imagine looking at a mountain range, the Sierra Nevada, say. You see a series of peaks in panorama, and they are ordered from south to north. Imagine you are viewing them on a train, slowly heading northwards. The southern ones are nearer the start of your journey, and the northern ones are nearer the end. But nonetheless, you see the peaks in panorama; this is not to see them successively. Succession seems to be a matter of seeing episodes that replace the previous one, rather than merely abut it. Scanning the horizon seems to be a matter of attending to different bits of a space that is spread out before you; successively experiencing times seems to be very different. For myself, I have difficulty thinking of succession in this sense without

38 E.g. Price (1996). The expression was popularised by Eddington (1929:68).

appealing to McTchange, for even if I tell myself that this is merely my perspective on things, it seems my perspective is McTchanging. P. T. Geach articulates a similar worry:

> Even if a man's impressions as to which realities are past, present, and future are illusory, the fact that he has in that case different and uncombinable illusions shows that at least his illusions really are successive – that they are not all present together, but now one illusion is present and now another. (Geach 1969:92)

I'll revisit succession in the final section of this chapter, and again in the next chapter; for now, let's note it as a phenomenon of particular interest. Before then, I want to discuss apparent awareness of presence, flow, jerk and whoosh.

2.5.4 Presence, flow, jerk and whoosh

Sometimes people seem to report some additional feeling or sense of time in addition to those I've outlined. Paul (2010), quoted above, describes a special 'feel' of 'nowness'. D. C. Williams (1951:466), rather sarcastically, writes:

> we are immediately and poignantly involved in the jerk and whoosh of process, the felt flow of one moment into the next. Here is the focus of being. Here is the shore whence the youngster watches the golden mornings swing toward him like serried bright breakers from the ocean of the future. Here is the flood on which the oldster wakes in the night to shudder at its swollen black torrent cascading him into the abyss.

He continues: 'It would be futile to try to deny these experiences, but their correct description is another matter' (Ibid.). Is there a feeling of flow or passage in addition to an experience of succession or duration? I'm not sure. There are certainly times when one finds oneself with a sense of urgency, but it can be hard to separate this from the feeling of one's bladder being full or of physiological arousal more generally. I recognise the feeling of time passing slowly when one is bored. But when one separates this into a failure to measure duration and listlessness, I can't tell if there are any ingredients left over that we ought to consider. But in this section, I want to discuss what might have been left out by the previous sections, taking a lesson from Hans Reichenbach (1956:316): 'This is the scientific approach. Do not expect answers before you have found clear meanings. But do not throw away unclear questions.' It may be the case that there is something that the previous discussions have missed, and I want to be open to that.

One candidate is 'presence'. It strikes us that the world is presented to us 'now'. Consider J. J. Valberg (1992:5):

> We have a strong inclination to view objects that are present in experience as being, at that very time, existent: if something is *now* present, it *now* exists. This inclination to encompass the object within the temporal present does not extend to reference, or thought in general. There is no problem about referring to, or thinking of, objects which no longer exist. But our inclination is to say that

> such things cannot be present in experience, that they cannot *now* be objects of experience.

Presence in experience may seem to be an extra-temporal phenomenon that we need to consider. But things are spatially, as well as temporally present in experience.[39] Is the feeling of nowness distinct from a feeling of hereness? It is widely remarked that there is an immediacy to perception; that it 'transparently' presents its objects to us, and is thus 'diaphanous'.[40] That things are presented to us in experience, however, doesn't seem like it could provide a difference between space and time, at least not by itself.

Perhaps things are not merely presented to us, but presented as ephemeral, fleeting or evanescent. It is noteworthy that the metaphors we appeal to are apt for a range of sense modalities. 'Whoosh' seems to be onomatopoeia for something passing one suddenly, while 'jerk' seems to be proprioceptive, as of being pulled off-balance. Flow suggests a river of time, but the metaphor need not be visual; rivers, when one encounters them, often make distinctive sounds as the water babbles or thunders downstream, and one can immerse oneself in the river and feel the currents and eddies pulling at one.

One proposal that seems to capture this metaphorical appeal nicely is from Giuliano Torrengo (2017:166). He proposes the 'phenomenal modifier view', where, in addition to experiencing physical objects move and change, events of long or short duration come to be succeeded by other events of longer or shorter duration, and the like; every experience is had in a certain way. Rather than positing an additional feature of the contents of experience (perceptual or otherwise), we should consider a ubiquitous modification of the *way* that content is represented:

> the feeling of the passage of time should be understood as a modifier of the character and content of experience, just as the *blurred* or *vivid* nature of a visual experience can be seen as modifying the way the experience seems feels to us.

There is not, on Torrengo's view, some *whooshiness* that we detect in experience, or represent, but the things represented are, as it were, represented *whooshily*.

I'm intrigued by this suggestion (though given that phenomenal modifications aren't the appropriate objects of perception – they are precisely features of the vehicle – it is hard to attend to them); I find myself attending to some object of experience and trying to distinguish my judgements of it from the way it is presented. This is a case where I have no idea how to separate the perceptual from the non-perceptual.[41] Nonetheless, Torrengo presents an evocative idea:

[39] Though, just as we are presented with things in experience that are really far away, for example, mountains on the horizon, we are presented with events that took place a long time ago, for example, repeats of *Steptoe and Son*. And, of course, the night sky can present us with objects that are really far away, and events that took place a very long time ago. For all that, they are presented to us in experience. Presence to experience, in this sense, doesn't require us to take a view on whether there is any time lag between things happening and our being aware of things.

[40] G. E. Moore (1903) .

[41] I am reminded of Merleau-Ponty's (1962:313) observation that 'the blue of a carpet would never be the same blue were it not a woolly blue'. Am I judging the blue to be a woolly blue, or viewing the blue woollily? I don't know.

> we feel as if what we represent is related to something that exceeds reality as a whole – namely, the fact that reality as a whole is dynamic. It is as if all our representations whispered 'I am about to become older' in an aside to the subjects who are having them. (Torrengo 2017:187)

The idea of a whispered aside is poetic. Perhaps more is whispered, however: that this fleetingness is just the same as all the previous moments, and that there's nothing you can do about it.[42] 'This too shall pass away . . .'[43] I don't know. I don't want to commit myself here, but I admit that I'm intrigued by Torrengo's suggestion. But Torrengo isn't, I should be clear, advancing an argument that we should believe that time passes on the basis of this phenomenal modifier, merely explaining, as this section has tried to, why appeals to experience are so often made when defending the reality of McTchange. In the next section, I want to consider a puzzle about time experience, which was raised by Geach above. Does our experience of time, even if not perceptual evidence of time's passage, nonetheless presuppose time's passage?

2.6 Puzzles of experience

It might seem, from our experience, that the sorts of things we appeal to – observations of change, movement, duration and succession – seem to involve some period of time *during which* we experience them. This seems to be in tension with a splitting of reality into past, present and future, since the present, as experienced, seems too long not to spill over into the past, present and future. But if experiences are temporally extended, there seems to be a challenge in explaining the succession of experiences, since we seem to experience a 'stream of consciousness'.[44] I think we should accept that our temporal experience is more like the saccades of the eye than a continuous stream. Just as we have mechanisms that keep the page before us appearing steady when we move our eyes, even though the image on the back of our retina is suddenly, discontinuously different,[45] we have mechanisms to make the changes we experience seem to flow into one another.

Accounts are available of the psychological mechanisms whereby a feeling of unity is brought to the successive discontinuous images on our retinas, or to our perceptual experience more generally.[46] I don't think the problem is the appearance of *continuity*, or even the appearance of change. The problem is to explain how a change in appearances is possible at all. Indeed, there is a trap here for defenders of a static view: one can rid

42 Thanks to Miri Albahari for this suggestion. Torrengo, in conversation, was nonplussed by the suggested additions to the whisper.

43 Cf. Farid Al-Din Attar (1976:235).

44 James (1890).

45 Shebliske (1977).

46 See, e.g., Dainton (2000); Grush (2006); Hoerl (2013); Rashbrook (2013); Wittmann (2014/2016); Kon and Miller (2015); Power (2018).

McTchange from an account of the world, only to place McTchange in consciousness as Weyl (1949:116) did:

> Only to the gaze of my consciousness, crawling up the life-line of my body, does the world fleetingly come to life.

If we have consciousness's gaze crawling up a world-line through space-time, it seems that this requires McTchange, for the gaze isn't changing merely relative to space-time, but changing *simpliciter*. Contemporary defenders of a static view of time are less likely to get rid of McTchange from the world, only to attribute it to consciousness. Consider Frischhut (2015:149):

> 'movement' through time consists in our occupying various temporal perspectives, simply by virtue of existing at different times.

This is the sort of account we would expect from a defender of a static view, and indeed they seem committed to saying something like this. But this doesn't seem to be enough for *succession* of experience.

2.6.1 Extentionalism versus retentionalism

Attempts by defenders of the static view to account for experience of succession tend to appeal to accounts offered by William James (1890), on which experience of time involves a temporally extended present, or by Edmund Husserl (1893–1917), on which we have, in addition to experiences of a particular moment, retentions of how things just were and protensions (or anticipations) of what will shortly be.[47] James and Husserl were interested in the unity of experience: how, for example, listening to a melody felt like a melody rather than an experience of a sound followed by another experience of another sound (such that I was never aware of hearing more than one note). It is unclear to me that either James[48] or Husserl[49] would have had much sympathy for a static view of time, so let's call the view the James-esque view 'Extensionalism' and the pseudoHusserlian view 'Retentionalism'.[50] Hoerl (2013:375) explains the two views like this:

> for the [re]tentionalist, temporal experience is to be explained in terms of the idea that perceptual experience can be intentionally directed not just towards the present, but also towards a small portion of the past (as well as, perhaps, a small

47 E.g. Dainton (2011), Power (2018).

48 Cf. James (1909:213): 'I myself find no good warrant for even suspecting the existence of any reality of a higher denomination than that distributed and strung-along and flowing sort of reality which we finite beings swim in.'

49 See Zahavi (2007) for a discussion of Husserl's views.

50 This usage follows Dainton (2011), who surveys the relevant literature on this debate much more thoroughly than I shall. Hoerl (2013) used 'intentionalism' rather than 'retentionalism', but this risks confusion with a term used in discussions about mental content more generally.

> portion of the future). For the extensionalist, by contrast, the explanation of why we can have, say, experiences of succession, is to be sought in the fact that episodes of perceptual experience themselves unfold over a period of clock time.

As stated, the two views offered are not obviously incompatible – it's not clear that belief in intentional directedness towards the past or future is an *alternative* to the passage of time (indeed McInerney (1991) explicitly appeals to both). The extensionalist is normally thought to say that we experience some duration in perception *as a whole*. This is to be understood that we have some temporally extended period of time before our consciousness. Valtteri Arstila (2016:166) describes the extensionalist view as follows:

> the experience of one flash succeeding another can be explained as follows: At the time we experience the latter flash, the first flash lingers in our consciousness as past or preceding content. Because we are conscious of both flashes during the same specious present, we also experience the succession.

It's important to note that the cognitive mechanisms that explain the experienced unity of a melody or movement aren't likely to play much of a role in settling the issue of whether there is McTchange. As Sean Power (2018) acknowledges, many other sorts of experience are going to be very important in explaining time on a static view. And the cognitive mechanisms are going to be neutral on the issue at hand; cognitive scientists may not have an opinion either way on whether the changes they note in a lobe of the brain are mere variations over time rather than McTchanges.

The debate in this area has, I think, been hampered by a preoccupation with temporal ontology. Power (2018) explores two options: one on which only a single moment exists, and one on which all moments equally exist ('Presentism' and 'Eternalism' respectively).[51] Dainton (2011) does somewhat better, exploring more views, including a version of the view I will be defending later in the book.[52] I have nothing against temporal ontology, although I think it fails to explain key features of the passage of time by itself.[53] In this context, my worry is that bringing in temporal ontology so centrally distracts from the question about competing accounts of change. As we saw in §1.3, Fine (2005) has argued that no temporal ontology, by itself, gets us an account of McTchange, so if McTchange is the thing that we're supposed to find a natural commitment given our experience, temporal ontology won't help.

Showing that a particular temporal ontology, such as Presentism, fails to provide an account of experience, as Dainton (2011), Dorato (2015) and Frischhut (2017) try to do, won't help either. The interaction between McTchange and considerations of existence is going to get quite complicated, as we shall see in Chapter 9, but having the wrong or *inadequately explained* temporal ontology won't give the defender of a

51 More on these in Chapter 9.

52 Dainton calls this the 'Grow+Glow' view. I will not follow him in doing so because it suggests that the view is arrived at by bodging together the worst bits of other views and so promotes a misunderstanding of the view I wish to defend.

53 I've argued for this in Forbes (2015).

static view reasons to reject McTchange, nor give defenders of a dynamic view reasons to accept it. It is for this reason that questions of temporal ontology will be shelved as far as possible until Part III of this book. This section has aimed to show that the debate between extensionalists and retentionalists doesn't help us get at the question of whether we need McTchange to explain experience.

2.6.2 Parmenidean Presentism

Imagine there is no McTchange, and we are trying to account for the change in what we experience. We seem compelled to take something like Frichhut's view, quoted above, that we have different perspectives on the world at different times, but there is no McTchange in what perspective we occupy. The extensionalist and retentionalist provide two options for why we experience changes as flowing or continual; a single perspective is temporally extended, or a single perspective contains protentions and retentions of future and past (respectively). But, whichever of those options I accept, it nonetheless feels like I successively experience such feelings of succession. As Velleman said (quoted above), we anticipate the future as a payoff to this anticipation. This seems to have no analogue with space; there is something special – *episodic* – about time which makes it different from space. Even if the world doesn't come in episodes, we have to explain why our experience does. And this leads to a puzzle about experience.

If there is no McTchange, why think there is more than one time? Let's accept, for *reductio*, that one of the various accounts offered by defenders of a static view of time captures the feeling of succession. This seems to undermine the claim that we experience more than one time because it gives us an account of experience that doesn't need to have an objective change from one felt episode to another. *This experience* feels like it's about to become another, as if we're hearing a whispered aside. But this experience is on par with all the others that feel like they're about to become another. This is how things feel from *this* perspective on the world, and from other perspectives on the world things feel differently. But our consciousness is not crawling up a 'world-line'. We are 'trapped in the amber of this moment'.[54] If there is no McTchange, I only have this perspective to go on for what the world is like. Sure, it *feels* like I have had others, and will have others, but that is just to say that there are other perspectives that *feel like* they follow this one. But on this view, my experience *right now* provides no evidence for there ever having been or ever going to be other times. If we explain the feeling of succession without already being committed to something successively experiencing change, we have an argument for a static view we might associate with Parmenides, where there is no past, no future and no change, and our experiences provide no reason to think there is. All they provide evidence of is an occurrent state of *seeming* to be in the midst of change.[55]

[54] Cf. Vonnegut (1991:63).

[55] This argument has an affinity with a similar argument used by Torrengo (2018) which is used for a somewhat different purpose.

Even if we think that we don't have perceptual evidence of McTchange in experience, we experience time in ways that have no spatial analogue that seem to presuppose us as changing subjects of experience. If persisting through change requires McTchange, that would provide some basis in experience for thinking that there is McTchange. I shall address this in the next chapter.

3

Enduring McTchange

We step and do not step into the same rivers; we are and are not.

—Heraclitus, fragment 49a

Our first-personal experience of persisting through time is some reason for accepting that there is McTchange. In this chapter, I will consider what our experience of persisting through time is, and why that might be reason to think that there is McTchange.

Let's begin with an observation from L. A. Paul (2017:266):

> At subjectively near times, I naturally represent my first-personal perspective as a continuously enduring agential perspective from the near past, through the specious present, into the near future. The nature and character of my experience is as of being the very same enduring self, that is, as of qualitative continuity (with respect to sameness of agential perspective), from the inside, through my temporal window.

Paul argues that we have subjective experiences as of persisting through time by being the same individual that changes in what properties it has.[1] We have a temporal point of view on the world – a temporal window – that seems to survive the change in what the world is like. We can remember having been the same person only recently and anticipate being that person in the future. Paul is particularly interested in the limits of that anticipation and memory, but I'm interested in the idea that it first personally seems to us that we are persisting in this way. I think this first-personal experience of persistence might be one of the central reasons to believe in McTchange.

In the next section, I will describe some different ways of thinking about persistence so we can see what is distinctive about our first-personal awareness of enduring over time. I will adopt a specific account that we are proper subjects of change. I will then think about the role of such first-personal awareness of endurance in making sense of our lives, and the relation between that first-personal awareness and the metaphysics of persistence we might adopt. This will allow me to argue that we need to accept that we first personally persist by enduring and that endurance requires McTchange.

[1] Paul doesn't think we, in fact, endure, merely that we have experiences *as of* enduring.

3.1 Rival accounts of persistence

Persisting through time is a matter of being the same thing over time, even when there is change during that time. The philosophical problem, then, is one of change; how can things be the same through some change? There are a few popular answers: endurantism, perdurantism and exdurantism being the most widely defended. I am going to defend endurantism. In particular, I am going to defend endurantism understood in terms of Haslanger's (2003) *Proper Subject of Change* condition: the change in properties involved in the persistence of an entity is a genuine change in the entity that persists, rather than variation in its parts or unchanging relations to counterparts. That is, according to endurantism, a persisting entity is the proper subject of change.

Perdurantism, defended by, for example, Quine (1950) and Armstrong (1980), is the view that objects persist by having a variety of numerically distinct temporal parts, each of which has a specific set of unchanging properties. For example, a shoe might have a temporal part at time t_1 which is polished, and a temporal part at time t_2 which is scuffed. The shoe persists through change in virtue of having parts at different times which have different properties. Note that on this view, it is not the shoe which is scuffed, but merely a temporal part of the shoe. There is nothing which was once polished and is now scuffed. We might talk as if there were, but strictly speaking, there is not.

Exdurantism, defended by, for example, Sider (1997) and Hawley (2001), holds that what persists are temporal stages. They persist by standing in counterpart relations to other stages. On such a view, the scuffed shoe-at-t_2 has persisted through change in virtue of standing in the right causal, historical and similarity relations to a shoe-at-t_1 that was polished. The shoe is genuinely scuffed, on this view, rather than merely a part of it being scuffed, as the perdurantist would have it. Nonetheless, the shoe is not an object that has changed. Strictly speaking, the shoe is a momentary object that is related in certain ways to another momentary object with different properties. To talk of these two momentary objects as being the same is strictly inaccurate; they merely stand in counterpart relations.

Endurantism holds that, when an object persists from time t_1 to time t_2, it is the numerically same object at t_1 and t_2, and, if there is any change, it is the properties possessed by the object that differ. That is to say that the shoe is the (numerically) *same* shoe that was once polished and is now scuffed, and it was *the shoe* that was polished and is now scuffed, not merely a temporal part of it that was polished and a different temporal part that was scuffed. So, when Paul, above, describes an experience as of subjectively enduring, it is an experience as of persisting in this sense she has in mind.

Endurantists differ on how exactly to account for this change in the properties had by persisting objects – some explain such changes by appealing to tense changes (Zimmerman (1998); Markosian (2004)), while others do so by treating possession or instantiation of properties as being relative to times (Johnston 1987; Haslanger 1989). Often you will see endurantism described as the view that objects are 'wholly present' at each time they exist.[2] I find this formulation unhelpful.[3] I defend an alternative, and somewhat heterodox, approach to endurantism. What unites all versions of endurantism is that, unlike perdurantism or exdurantism, endurantism meets Haslanger's (2003) *Proper Subject of Change* condition: the change in properties is a genuine change in the entity that persists. That is, the persisting entity is the proper subject of change. The exdurantist's stages don't change, and the perduring object, extended over space-time, does not change its parts; it simply is those parts spread over space-time. Endurantism, uniquely, gives us a view on which the thing that our theory attributes the persistence to and the thing changing are the same. It may be that there are occurrents that meet the proper subject of change condition, as well as continuants. But, as I'm using the word 'enduring', an entity persists by enduring *if* it persists through change and is the proper subject of change.

If I am giving a somewhat heterodox formulation of endurantism, and talk of being 'wholly present' is murky, what is the endurantist getting at? The thought behind this talk of being 'wholly present' can be explained with this example: Tenzing Norgay climbed Everest. If I ask whether *all* of him climbed Everest, or merely part of him did, the answer will come that all of him did. The claim that all of Sherpa Tenzing climbed Everest contrasts with the claim that he left significant bits of himself at base on the route up. This is an intuitive distinction we make with, for example, events. If we say 'the climb was dangerous' we might talk of parts of the climb that were dangerous, and parts that were not especially dangerous. Crucially, these might be *temporal parts*. The bit of the climb ascending the South Col might be more dangerous than the bit before, for example.

The intuitive distinction between the climb and Tenzing Norgay is one between *continuants* and *occurrents*.[4] Continuants are what we tend to think of as being individual things that change over time. If anything endures, continuants are the intuitive candidates to do so.[5] According to Peter Simons (2000:59–60), amongst the continuants are

[2] Consider the definition from Miller (2016): 'If x endures from t to t* then there is something that is wholly present at t, and something that is wholly present at t*, and the thing wholly present at t is identical to the thing wholly present at t*. Indeed, endurantism is also sometimes known as three-dimensionalism since it is taken to be the view that persisting objects are three-dimensional.'

[3] Mark Johnston (1984) was responsible for this formulation in terms of being 'wholly present', which was then popularised by David Lewis (1986). Johnston (2011) has recanted this formulation, now preferring to talk of a persisting object 'exhibiting its essence' at each time it exists. Johnston (2011:52 fn36) says this of talk of enduring entities as being 'wholly present': 'That should stop; for given this account of endurance, it is quite unclear how a variably constituted entity, one that is capable of gaining and losing parts, can even be a candidate to be an enduring entity. (Are all of the parts it has over time to be present at each time? Of course not! But then which parts are to be wholly present at each time?) Yet many enduring substances are variably constituted.'

[4] This distinction from W. E. Johnson (1921:199; 1924:78–101), cf. P. Simons (2000).

[5] I will be using 'continuants' to pick out the category that *prima facie* endures, and 'endurance' to pick out the metaphysical way of persisting. So, I'm stipulating that you and I are continuants, but making it a substantive question whether we do in fact endure. This may differ from usage elsewhere, though separation of the folk concept from the substantive metaphysical claim is not always cleanly marked.

> ourselves and other organisms such as cats, corncrakes, cowslips and cocci; geo- and astrophysical objects like cliffs, cols, cordilleras, continents, comets, and clusters of galaxies; artifacts like cufflinks, cups, cars, computers and cruise liners; and social entities like committees, clubs, companies, colleges, and countries like Croatia.

It is not, then, merely organisms like us that are continuants; *things* generally are. Ordinarily, we talk about various different types of entities. Events, like births, marriages and deaths;[6] things like shoes, ships, cabbages and kings; stuffs, like ceiling wax and CO_2;[7] states of affairs, like the water's being exactly 20°C, or Burt Bacharach's being twenty-four hours from Tulsa;[8] and processes like digestion, respiration and inflation;[9] and activities like reading, writing, speaking and listening.[10]

'Things' is being used to pick out a common-sense ontological category, contrasted with activities, processes, events, states of affairs or stuff. Stuffs are continuants, and states of affairs plausibly are too. Events are occurrents; they occur in phases, unless they are very short events. It is debated whether processes should be thought of as continuants or occurrents.[11] But things are our paradigm case of continuants, so let's not worry about the other categories too much, except to acknowledge that we ordinarily talk as if other categories are available.

Note that perdurantists, Quine (1950) explicitly, treat continuants as a special case of occurrents. The endurantist thinks that continuants persist by enduring, but needn't make any claim about how occurrent categories persist. They may be no occurrents in existence, or they may vary across their temporal parts; the endurantist is committed to at least some things persisting by enduring through change. The perdurantist, by contrast, treats everything as being analogous to an occurrent; they have phases given by temporal parts.

Having introduced the distinction between continuants and occurrents, which is meant to track our intuitive application of ways of persisting to different ontological categories, we might make a similar but slightly different distinction between three- and four-dimensionalism. The four-dimensional has temporal parts, and the three-dimensional lacks temporal parts. We typically think of persistence by enduring as being related to being a continuant, and to treating an object as having spatial but not temporal parts, at least in ordinary language. But note that exdurantism also gives us three-dimensionalism, since the 'things' to which the exdurantist appeals – stages or time-slices – lack any temporal parts. I'll discuss in the next section how endurantism

6 Cf. Davidson (1980).

7 Cf. Markosian (2015).

8 Cf. Armstrong (1996).

9 Cf. Steward (2013); Stout (1997). Crowther (2018) argues that processes should be treated as stuff-like.

10 Kenny (1963) and Vendler (1967) use 'activities', and Mourelatos (1978) talks of processes to mean roughly the same thing, though Mourelatos (1993) acknowledges that this may have been a poor choice of wording. Nonetheless, the terms have often since been used to pick out more or less the same category.

11 See e.g. Stout (1997); (2015); Steward (2013) Processes will receive a much fuller discussion in Chapter 4.

is compatible with having temporal parts; for now, I'll simply observe that nothing we have said rules it out. One confusing feature of the debate is that we have three slightly overlapping distinctions: enduring/perduring/exduring; continuing/occurring; 3D/4D, and different parts of the literature make use of different distinctions without mentioning the others or treating them as equivalent. I mention these multiple distinctions so as to be clear how they differ from the central distinction that matters to me: being the proper subject of change or not being a proper subject of change.

One might try to make enduring entities three-dimensional *and* temporally extended entities that are wholly present. Miller (2016), for example, considers endurance to be a matter of being multiply located in the manner of universals, or, say, photographs are said to be multiply located. So, just as redness is located everywhere something red is (and not partially located, but wholly located), and the photograph of Tenzing at Everest's summit is located everywhere prints of it are (and not partially located, but wholly located), Sherpa Tenzing is wholly located everywhere Sherpa Tenzing's stages are. Notwithstanding that the person who introduced the term 'wholly present' recanted it, this doesn't seem to capture endurance well. Universal redness doesn't persist through change, and a photograph of Tenzing doesn't change by having another print made, but Sherpa Tenzing *did* persist through change. So, treating endurance as a matter of being multiply located three-dimensional object doesn't seem to help explain persistence by enduring; it only makes endurance murkier.

If exdurantism gives us three-dimensional entities, does it also give us continuants? Are stages continuants? It seems *stages* aren't continuants because although they exhibit all their essence at each time they exist, stages do not *continue* to exist. The exdurantist aims to give an account of continuants that treats talk of continuation as not being literally identical over time, but of stages standing in the appropriate similarity-cum-causal counterpart relations. The exdurantist, then, will certainly *talk* of continuation, but central to their dispute with the endurantist is whether they are entitled to do so, given that there is no thing such that it continues.

We have three popular accounts of persistence, then.[12] I've summarised these in Table 3.1.

Persisting by enduring is a matter of being the proper subject of change. To anticipate §3.4, reasons to accept the claim that you, dear reader, are a proper subject of McTchange, would be reasons to accept the claim that there is McTchange. But there

Table 3.1 Accounts of Persistence

	Perdurantism	**Exdurantism**	**Endurantism**
Proper Subject of Change	No – variation amongst temporal parts	No – relations to counterpart stages	Yes – the persisting thing changes
Continuants or Occurrents	Continuants treated like occurrents, but continuant talk is fine	Nothing strictly a continuant but continuant talk is fine	At least some continuants; compatible with there also being occurrents
3D or 4D	4D	3D	3D or 4D

[12] Other, more esoteric accounts are available.

is much to be done before then. In the next section, we will look at two objections to thinking of ourselves as enduring, before looking at what way being a proper subject of change might matter in §3.3.

3.2 Problems for endurance

I am arguing that our first-personal awareness of our own persistence through time is what motivates McTchange. But this argument can only work if we can make sense of persisting by enduring. In this section, I will consider some problems for endurance. First, I will consider the problem of change: that endurance can't allow us to make sense of things persisting and changing. Second, I will consider some arguments to the effect that the very idea of enduring through change is *incoherent*.

The 'problem of change', as Katherine Hawley (2001:16) describes it, is of how enduring entities can have incompatible properties. We can easily make sense of a banana being yellow in some places and green in others. Appeal to temporal parts, counterpart stages, etc. can explain how a banana can be green at one time and yellow at another, by appealing to the variation of the banana over time. For the perdurantist, the banana has some temporal parts at one time and other temporal parts at another time. For the exdurantist, the green banana stands in similarity-cum-causal relations to a yellow banana at a later time. Explaining endurance, by contrast, is trickier; a banana cannot be green all over and yellow all over without qualification. It can't persist through change by being the proper subject of change without having different properties on either side of the change. But it also can't have inconsistent properties. The problem, then, is how one thing can have different properties from itself.

3.2.1 Responses to the problem of change

Two strategies that are advocated are as follows: you can relativise the having of properties to times (such that the banana isn't green, it's green-at-t), or one can have the properties in a temporalised way, that is, have instantiation as relative to times or tenses (e.g. the banana is green *pastly*). The former of these doesn't seem to involve any commitment to McTchange. The latter might do, if instantiation is relative to tenses, provided we think what is present changes.

The former of these views seems to involve making properties more complex than we thought. Where we had thought that there were monadic properties, we now take properties that looked to be monadic to be relational, and properties that we took to be relational are relational to more things than we thought. A lingering concern about the view that properties are relations to times is a familiar one. As Sally Haslanger (2003:331) puts it: 'One might reasonably ask, however, in what sense does the relationalist account really capture *change*?.' I will come back to this thought in §3.5. For now, I want to say that if our first-personal experience of persisting through change is meant to motivate a commitment to McTchange, it can hardly be an objection to an account of endurance that it gives us endurance without a commitment to McTchange. To the extent that we are appealing to endurance as a reason to think time passes in a way that has no spatial analogue, we must have a reason to think that the right account of endurance is one that requires McTchange.

The strategy of having the *instantiation* of properties relative to times keeps the properties roughly as we thought they were, but the way the properties are instantiated removes the incompatibility. One can do this by treating instantiation as relative to times, such that the banana is t_1 green and is t_2 yellow. On such a view, yellowness would only be incompatible with greenness if they were instantiated in the same way (e.g. a temporally unqualified way). The context, one might argue, provides a restriction of the relevant temporal qualification to the instantiation, such that, in ordinary speech, we don't pronounce the subscript indices (i.e. t_1, t_2, etc.). Alternatively, we can appeal to tenses, rather than dates, in our temporal qualifications: to say that the banana is green *pastly* – or better, the banana *was* green – means that the greenness and yellowness of the banana are no longer incompatible. A banana that was green and is now yellow doesn't have incompatible properties. If one wants to avoid commitment to McTchange, but nonetheless be an endurantist, one has the option of indexing instantiation to times or making properties themselves time-relative, then. Alternatively, exdurantism gives one talk as if things persist by enduring but sacrifices the proper subject of change condition.

Occasionally, one sees defences of endurantism that insist that things can only persist by enduring if only the present exists, such as Hinchliff (1996); Zimmerman (1998); Merricks (1999). I will postpone discussion of this view until §9.2, though insofar as one has a motivation for thinking that only the present exists, and that what's present changes, one has a motivation for accepting McTchange.

The problem of change, then, is usually a problem for the logic of enduring entities: they either have incompatible properties, or the properties things have are more complex than we expect, or the manner of instantiation is more complex than we expect. In the next section, I will consider a more serious objection: that the very idea of enduring is *incoherent*. That is, rather than finding ways to explain what it is for an object to endure, we must abandon the idea altogether.

3.2.2 The incoherence of endurance

J. David Velleman (2006) and A. S. Meincke (2019) think that endurance is incoherent. Given its incoherence, we shouldn't hold it as a view of our own persistence. I will argue that it is not, at least in the way I understand it as being a proper subject of change.

Endurance is incoherent, according to Velleman (2006:8), because it involves thinking that there are multiple complete people at different times:

> I am tempted to say that all of my temporal parts are present at a single point in time because I tend to think of myself as my present self – a momentary subject whose existence is indeed complete in the here-and-now. I am tempted to say that I nevertheless persist through time because I tend to think of this self, complete in the moment, as nevertheless existing at other moments. And because I therefore conceive of each moment in my temporal extension as containing my complete self, I am tempted to deny that it contains a mere temporal part of me. There I am,

> all of me, at my fifth birthday party; here I am, all of me, remembering that party; there I will be, all of me, on my seventy-fifth birthday – as if one and the same momentary subject can play the several parts of my five-year-old, 53-year-old, and 75-year-old selves.

Note here that it is assumed that enduring entities cannot have temporal parts, because they must, definitionally, be wholly present. Velleman then argues that I am both complete at this moment and that I exist at other moments. This, he argues, is incoherent. But it is not clear how much of this is required to capture the common-sense commitments that Velleman does such a good job of articulating. The metaphysics seems to be getting in the way.

Must we say there are no (non-present) temporal parts?[13] We set up the example of endurantism with a claim that Sherpa Tenzing, rather than merely part of him, climbed Everest. It is not clear that the claim that Sherpa Tenzing climbed Everest requires us to say that Sherpa Tenzing, strictly speaking, has no temporal parts. To see this, let us consider the following view:

> Suppose what properties I *currently* have depend on my current states, my past states, and some causal laws that constrain my future states. How old I am, for example, might depend on the sequence of past states I've had, such that there are such states for each year I've been alive. We can imagine a view on which I acquire temporal parts as I get older; there are all the temporal parts for my first decade, all the ones for my second decade (which combine to form the temporal parts for my first twenty years) and so on. But given that my current age depends not merely on some temporal parts, on this view, but on my improper temporal part (i.e. my entire history), it would be odd to say my most recent temporal part was forty years old. It is not a temporal part, but I, myself, that am forty years old!

Saying that Sherpa Tenzing *simpliciter* climbed Everest is compatible with accepting that Tenzing has temporal parts, if we don't attribute the climbing to the temporal parts but to Tenzing himself. The key is not to identify me with the mereological sum of my temporal parts. A commitment to something enduring might require that a person isn't identical to an aggregate of temporal parts, but that doesn't rule out a person being constituted by temporal parts, or having temporal parts. Once we, as Johnston (2011) suggests, dispense with endurance as being a matter of being 'wholly present' and focus instead on being the proper subject of change, the purported incoherence disappears. An event may just be a succession of temporal parts, but the relationship between a thing and its temporal parts may be one of constitution, or coincidence or something else, without our ruling out that there are such temporal parts – it is implausible to suppose that persons are mere aggregates of parts.[14] If one starts by assuming that a thing is

13 In §9.2.2 we will discuss a view that there simply are no non-present times. I will postpone discussion of that view until then.

14 Constitution is normally concerned with how spatial parts combine to create an object that is distinct from the mere collection of those parts. I am using 'constitution' in an extended sense where temporal parts can constitute an enduring object without that object being identical to the

simply an aggregate of its parts, then the suggestion that a person can coincide with their temporal parts but not be identical to that mereological sum won't get very far. Endurance, on such a view, really would be incoherent. But temporal parts may indeed be useful, given that many of a thing's properties supervene on states over a significant temporal interval, in order to make sense of what properties something currently has.[15]

Hofweber and Velleman's (2011) proposed 'reconfiguring' of the persistence debate, where endurance is a 'notion of an object whose identity is fully determined within the context of each moment of its existence' (p. 50), seems to insist that we cannot rely on other temporal parts to work out what an enduring object's current properties are. But this, as they concede it might, has the consequence that paradigm cases of enduring entities don't count as enduring on their view.

There's a confusion that sometimes arises hereanent, which it would be good to pre-empt. When I ask, 'Did all of Sherpa Tenzing climb Everest?', as we considered in §1, the thought was that all of Sherpa Tenzing went up Everest, rather than merely part of him. I claimed, just now, that what current properties something has depends on non-present temporal parts. Am I not thereby committed to claiming that the temporal part of him from his first birthday to his second birthday climbed Everest? I might claim that the sentence 'Tenzing Norgay climbed Everest' depends on that temporal part, amongst others: perhaps something in his childhood influenced his agreeing to the expedition. It being he who climbed Everest depends on him having the history he does, and not another history. But it would be a category mistake to claim that a temporal part climbed Everest, as much as it would be to say that his foot climbed Everest. He used his foot to climb Everest, and no doubt used his years of experience, but it was he who did the climbing, not the foot or the years of experience. Sherpa Tenzing is the proper subject of the climbing, not his temporal (nor, indeed, his spatial) parts. What's going on here is something like the fallacy of division: Sherpa Tenzing has some properties, but it doesn't follow that those properties can be attributed to his temporal parts.[16]

One could object here that the idea that an object is not identical to its history doesn't resolve the incompatibility that we were concerned with in the previous section. For all I have said, it's still true that the banana has, over the course of its history, different colours, and that Sherpa Tenzing has, over the course of his history, different properties. If it were a question of parts of the banana being one way, and parts being another, these colours wouldn't be incompatible, but the banana is one and the same thing throughout the relevant period. Saying that the banana has *changed*

collection of temporal parts that make it up. But nothing in my argument requires accepting that it is this relationship that temporal parts have to people, just that it is some suitable dependence relationship that is not identity.

15 I'm indebted to R. A. Briggs for discussion of these issues. For those interested in a more technical discussion of these issues, the semantics given in Briggs and Forbes (2012) shows how you can get present-tense claims about things that depend on a significant temporal interval that are derived from sequences of sets of statements that don't involve such a temporal interval. Although we don't show how this works for claims about things or people specifically, the same procedure should apply.

16 There's a similar fallacy of division worry for presentness. If I say 'this year is present', I am not thereby committed to claiming that each temporal part of the year is present. I am not thereby committed to there being some smallest temporal part that is present either.

is simply unhelpful, the objector might insist.[17] It amounts to saying that the banana is both the same thing and different.[18] This may seem reminiscent of the argument that McTchange is incoherent, which we discussed and dismissed in Chapter 1. It *is* the same argument, at heart; Velleman explicitly draws this comparison. There is purported to be a problem with things having incompatible properties to themselves, but the properties are only incompatible if one does not allow there to be McTchange in what properties something has.

We saw, in our earlier discussion, that practical problems of measuring change are distinct from giving an account of the coherence of McTchange. Just as there are practical problems about measuring change, there are practical problems of re-identifying the same thing before and after change. The problems of giving identity conditions for particular things are legion,[19] and require careful work for the different kinds that we are interested in. But the problem of change for enduring entities is meant to be a problem for the *very possibility* of things persisting through change. This problem doesn't get going any more than McTaggart's argument for the incoherence of McTchange does. How can there be a problem for the very possibility of things persisting through change when *you've made it this far?*!

A. S. Meincke (2019) has a rather different challenge to the coherence of endurantism: that it is incompatible with McTchange! Whereas Velleman tries to show that the endurantist is both committed to temporal parts and to denying them, Meincke aims to show that the endurantist is committed to an unchanging thing that changes its properties. It is not the change that we must get rid of, but the substance that changes.

We have already seen ways of thinking of endurantism on which properties are indexed to times, or their instantiation is indexed to times. These are specifically the target of Meincke's attack. The worry here is that the properties an object has don't change, but are merely indexed to different times in a different way. That is, it is counted as an objection to endurantism that it fails to provide things which meet the proper subject of change condition because the things we end up with are unchanging bundles of temporally indexed properties or temporally/tensedly indexed instantiations of properties. If we think that there is some *essence* to an object that does not change, or some *substratum* that unchangingly possesses the properties, that doesn't help, since that doesn't change either. Meincke (2019:22) is damning:

> Endurantism is as far from delivering the only coherent account of change as anything could be, and just to the extent that it fails to accommodate change it also fails to account for persistence.

17 Lewis (1986) is just such an objector. 'It is not a solution to say how commonplace and indubitable it is that we have different shapes at different times. To say that is only to insist – rightly – that it must be possible somehow.' Equally, to insist that things *can't* have different properties at different times, when they indubitably do, isn't to create a problem needing a solution.

18 Heraclitus, quoted above, clearly saw the difficulty of being both the same and different.

19 Haslanger (2003) lists paradoxes of motion, material constitution, growth and the replacement of parts as relevant puzzles. Working out, for example in what way different properties of Sherpa Tenzing depend on temporal parts is going to be very difficult, and likely different for different properties.

The culprit, it seems, is 'thing ontology' or 'substance ontology'. This approach to ontology, which belongs to a tradition often associated with Aristotle, views substances as unchanging entities that possess changing properties. Meincke (2019:23) says:

> Substance ontology crucially rests upon the idea that change – if there is any – must have a non-changing basis. Whatever appears to change is believed to have an unchanging core. Change, if it exists at all, is taken to be parasitic upon stasis; it does not have an independent ontological basis.

Aristotle exegesis is notoriously difficult, but I can't help but note that he's going to turn up in §4.1.1 as one of the inspirations for contemporary discussions of activity and process, the chief rival to substance ontology. Aristotle seems to be both hero and villain in this story, perhaps suggesting that he had a rather rich set of views that are not easily reduced to philosophical melodrama. So let's set Aristotle's views to one side. Must the endurantist be committed to 'substance ontology'?

Substances, in the most basic form, are things of which you can predicate properties, but can't be predicated of anything else. So, I can say of a horse that it is grey, but I can't say of a cough that it is horse.[20] To the extent that I am defining endurantism in terms of the proper subject of change condition, I am open to the claim that the proper subjects of change are not substances or things but processes. But I probably am committed to at least one thing that can't be predicated of anything else. I think there has to be something doing the changing, which isn't itself a mere property of something else; it can't be properties all the way down.

In fairness, in this weak sense of substance, processes may well count as substances. This sense of substance isn't committed to the ontological priority of stasis, but to the claim that there are property-bearing ontological categories that are not themselves properties. There is a richer sense of substance on offer, where we divide substances into unchanging essences and changing accidental properties. The essential properties are the ones that make something the thing it is, and the accidents are any other properties. But there's one sense in which essences must be unchanging and a sense in which they needn't. Different things have different persistence conditions: not every change is one that something can persist through. Essences are invoked as circumscribing those persistence conditions. So, in some sense essences must be unchanging in a case of persistence: something can't persist through change from being the thing that it is to becoming a different thing. But in some sense, essences may perfectly reasonably change. If the history of something is essential to it, and it can persist, that history must continually change. But it being the thing that it is requires having *that* history. So, the substance may depend on a processual essence which is continually changing but is the *same process* of history unfolding, for example. It may be that we cannot non-trivially *define* an essence in terms of necessary and sufficient conditions, though it doesn't seem like that's generally how we have been judging cases of persistence in most cases anyway. The version of endurantism I want to defend simply lacks commitment to the claim that substances, or their essences, cannot change. It *is* committed to the claim that if

[20] It might be hoarse, of course.

we have reason to believe anything persists, we have first-personal reason to believe we, ourselves, persist.

Meincke's arguments do, however, give us reason to think that if there are enduring substances, they must be parasitic upon McTchange. Insofar as you, dear reader, are a proper subject of change and cannot be predicated of some other thing, I'm happy to call you a substance, but I have no claims that you are fundamentally unchanging; merely that you are the same person over time. You are not a bundle of properties or an aggregate of parts that must never change to maintain identity, but precisely something that can persist through McTchange. Indeed, far from being an objection to my view, Meincke's argument is grist for my mill. There are views of endurantism that are not dynamic, and Meincke's argument is effectively that they fail, in the end, to offer proper subjects of change.

The dispute between substance ontology and process ontology has more to it, of course. Process philosophers are keen to emphasise the interdependence of different processes, treating substances as things that lack that interdependence on other substances. But I want to distance myself from a conception of substance that goes so far. While talk of substances may be attractive to those who think of substances as the sort of distinct existences that are independent of one another, nothing about our first-personal sense of our own persistence through time – the case that motivated this discussion of endurantism – compels us to accept that we aren't interdependently bound up with our environments and communities. Indeed, the negative health effects of solitary confinement are well known and include heightened levels of anxiety and panic; paranoia; cognitive dysfunction, hypersensitivity to stimuli and hallucinations; depression, mood swings, and loss of emotional control (Craig 2018:372). This dependence on culture drives the plot in classic fictional stories such as J. M. Barrie's (1902) play *The Admirable Crichton* or William Golding's (1954) *Lord of the Flies*. It isn't much of a stretch to think that we are dependent in important ways on our environments and communities, despite being paradigm cases of things which endure.

I want to finish this section with a brief aside on the *relevance* of this debate. In addition to claiming endurance is incoherent, Velleman (2006) argues that there are *moral* advantages to rejecting the enduring self as an illusion; we can rid ourselves of suffering caused by egoism and the fear of death, of the loneliness of the proprietorship of a single life, and of pain and unpleasantness being transferred into suffering or panic.[21] At least for Velleman, then, the dispute about whether we endure is taken to have consequences for how we live.[22] Velleman links these purported moral consequences with Buddhist attempts to overcome 'suffering' [Pāli: *dukka*; Sanskrit: *duhkha*]. Velleman aligns himself with this Buddhist soteriological project, beginning his paper with a claim about what Buddhists believe:

> Buddhists believe that the existence of an enduring self is an illusion and that this illusion is the root of the suffering inherent in the human conditions.

[21] Cf. Velleman (2006:18) and Partfit (1971:27); (1984:280).

[22] Indicating that there may be an explanatory difference between the views under discussion. Cf. §1.5.

I agree that endurance has consequences for how we live but want to suggest that the Buddhists may not be the allies Velleman hopes. Velleman also draws attention to similarities Derek Parfit (1984) noted between his [Parfit's] views and those of Buddhists, in particular a link between a lack of egoism and a rejection of an enduring self. As Velleman observes, Parfit 'claims that the consolations of his view can be obtained by attending to the philosophical arguments for it, whereas the Buddhists believe that they can be obtained only through long and arduous meditational practice'. However, not all Buddhists reject an enduring self in the relevant sense, and even amongst those that do, many do so precisely because of a commitment to change, perhaps of the kind that motivates Meincke.[23]

[23] All Buddhists are committed to the 'no-self' thesis; much Buddhist philosophy is aimed at working towards acceptance of the no-self thesis, culminating in a moment of 'extinguishing' [nibbāna (Pāli) or nirvana (Sanskrit)] whereupon one escapes from the cycle of suffering and achieves enlightenment. These Buddhist views arise out of a dialogue within classical Indian philosophy about the nature of the self with what we would now think of as Hindu philosophers, who typically defended a view where there was a transcendent self [Ātman (Sanskrit)]. Cf. Perrett (2016:170). It was this 'true self' that one discovered only in meditative practice (and is perhaps a view of endurantism parasitic upon stasis in the way Meincke criticised above). While Buddhists followed the Hindu traditions in the importance of meditative practice, they denied the existence of such a thickly conceived self to be discovered through meditation. In the context of debate with these Hindu traditions, then, a better translation might be the 'no true-self' thesis.

It's not clear that what the Buddhists are arguing against, when they deny the existence of a self, amounts to the denial of the coherence of anything enduring through change as I have defended it. There is some room between the denial of an immortal soul and denying that the person who orders a cup of coffee and the person who picks it up can be the same. Indeed, this debate is one that has taken place within Buddhism. The 'Pudgalavāda' view was the view that there is some minimal sense in which a person endures. Amber Carpenter (2014:32) describes the view like this:

> It [the self] is, perhaps, a bit of metaphysical glue, making it possible for us to re-identify the same person over time and through change; it is perhaps a bit of psychological glue, preventing us falling off the steep precipice into nihilism. If the continuity between different psychological events is real . . . then the self must be equally real, for it just is this continuity.

As Carpenter notes, however, this view, in the Classical Indian context, is a minority view within Buddhism. Philosophers of the Nyāya Hindu school, against the majority of Buddhists who deny the self even in this sense of 'metaphysical glue', defend what was the Classical Indian common sense position. As Dasti and Bryant (2013:5–6) put it:

> Nyāya philosophers agree that the enduring individual self must be the locus of agency and moral responsibility, and correspondingly attack Buddhist no-self theories. They argue that rejection of an enduring self makes it impossible to explain moral responsibility over time: if there is no enduring self, I am not the same individual that I was last week, or, for that matter, in my previous birth. This disassociation, Nyāya argues, would make me free of moral responsibility for what 'he' did back then as much as I am currently not responsible for what my neighbo[u]r does right now.

It is striking that common sense, in this case at least, seems to be common between West and East. The need to explain reincarnation, which, with the exception of the Cārvākas tradition, is ubiquitous in classical Indian philosophy, no longer has the pull of common sense, however. The conditional claim that if there were reincarnation, there would need to be something that persisted through it plausibly does still have a claim to common sense.

What should we say to those many Buddhists who reject the Pudgalavāda view? I have argued above that Velleman is wrong to think that there is something incoherent about enduring through change, and I will happily concede that there need be no thickly conceived transcendent self that is eternal and not subject to suffering. Where Buddhist philosophers claim there is something incoherent about enduring through change, I will say much the same to them as I did to Velleman.

Even if we are sympathetic to Buddhist arguments in favour of the no-self thesis, their equal emphasis on the transience, evanescence and inconstancy of all things[24] suggests that Buddhists may often not be opponents to McTchange like Velleman, but rather proponents of processes like Meincke. Buddhists' objections to the eternal self are often appeals to the assumption that selves, were there to be any, wouldn't be able to change in the ways they would have to do the work that is required of them.[25] The alternative to selves they often present – where we are an 'imposition of a unity and coherence on a complex, multifaceted stream of events and processes'[26] – may well make a good case for belief in McTchange. The next chapter will argue that the need to appeal to processes in our explanations gives us a reason to accept that there is McTchange; those views, like Meincke's and various Buddhist views that reject an enduring self *in favour of a view of people as processes*, will perhaps be more sympathetic to that chapter than this one.[27]

I have shown why I disagree with Velleman's and Meincke's arguments that endurance is incoherent, at least as I understand it here as being the proper subject of change. I agree with Velleman, however, that this debate *does* have consequences for how we live. The aim of the next section is to think about some of the ways in which it makes a difference to our understanding of agency and how we think about persistence.[28]

3.3 Am I the proper subject for change?

We started the chapter with the thought that we have a first-personal awareness of ourselves as continuants, that is, that we have experiences *as of* enduring through time. We considered some worries about endurance and moved on to the idea that there was a problem, one with significant moral consequences, for thinking of ourselves as enduring *selves*.

What positive is there to be said in favour of the self? One obvious reason to care about the self (particularly ourself) is the special concern we have for ourself over time.

Where Buddhists' reasons are tied to a soteriological need to attain extinguishing of the self to escape from suffering, these may not be persuasive unless one accepts escape from suffering as an achievable and desirable goal. It is not clear, for all the wisdom we might habitually attribute to Buddhism, that an extinguishing of the self is or should be our primary ethical goal.

24 anicca (Pāli) or anitya (Sanskrit).

25 Cf. Carpenter (2014).

26 Garfield (2013:174). Garfield is specifically discussing Mādhyamika Buddhism.

27 As Miller (2017) argues, we see a diverse range of views about time in Buddhist philosophy. She claims it is straightforward that the Sautrāntikas and Theravādins are committed to McTchange; they hold that the world is genuinely McTchanging as new events and objects come into, and pass out of, existence. Others, like Nāgārjuna, Miller suggests, may be thought of, on different readings, as error-theorists, fictionalists or deflationists about time. It is far from obvious that any side in the debate will be able to appeal to Buddhists *en bloc* as allies. This may, however, be good evidence that these debates are not merely confections of a peculiarly Western sensibility, or of misuse or abuse of terms in English, but of issues with relevance to how we live that can arise independently of a particular philosophical tradition or language.

28 These themes of agency and making sense of our temporality as agents will resonate through Chapters 7 and 8.

It is not mere narcissism to do so, but tied to the very possibility of agency. As Susan Wolf (1986:719) argues:

> Being a rational agent involves recognising one's ability to make decisions, form intentions, and plan for the future – recognizing one's ability to make one's *own* decisions, form one's *own* intentions, and plan for one's *own* future, that is (emphasis original).

Wolf argues that the metaphysics of personal identity is irrelevant to the question of justifying a special concern for ourselves. But I will argue that we need to accept that we, first personally, are the proper subjects of change in order to justify the special concern we have for ourselves.

The central thought, which I take from Wolf, is that the incoherence is not in an enduring self, but in a self built out of momentary stages. Thinking of ourselves as momentary subjects of experience, from which are built durations is 'ruled out':

> If the reason we care about persons is that persons are able to live interesting, admirable, and rewarding lives, we may answer that time slices of persons, much less experiences of time slices, are incapable of living lives at all. (p. 709)

The idea that we have, distributed across space-time, a series of experiences, from which we build up persons, seems, instead, to dispense with them. It is not merely that we cannot make sense of why we should care about other perspectives in a temporal ordering of experiences containing the perspective you adopt *right now.* If that were the only concern, answers to the effect that those experiences stand in similarity-cum-causal relations to you might have some mileage.[29] As exdurantists like Baber (2021:133) argue, 'we care about how things go for earlier and later stages that are [appropriately] related to us because the way things are for them is the way things were and will be for us.' But the objection is not simply one about how we go from momentary stages to other momentary stages, but rather that momentary stages aren't going to be the right kind of building blocks in the first place. Christine Korsgaard (1989:113–14) puts the point forcefully:

> To ask why the present self should cooperate with the future ones is to assume that the present self has reasons with which it already identifies, and which are independent of those later selves . . . to the extent that you regulate your choices by identifying yourself as the one who is implementing something like a particular plan of life, you need to identify with your future in order to *be what you are even now*. (emphasis original)

Recall Hofweber and Velleman's (2011) thought, in §2.2, that we cannot rely on other temporal parts to work out what an enduring object's current properties are. Korsgaard's argument suggests that the problem is the other way around. We cannot

[29] Cf. Baber (2021).

work out what the current properties of an enduring *subject* are unless we think of them as being things with histories and futures. The argument in favour of you and me being proper subjects of change goes as follows: as a matter of practical necessity, we have a special concern for ourselves in the future. That special concern is a requirement of the unity of agency – in particular, diachronic unity of agency. Diachronic unity of agency requires the 'metaphysical glue' that there are proper subjects of change. I will work through those different claims individually to make clear how the argument is supposed to go.

3.3.1 The practical necessity of a special concern for the future

That there is a practical necessity that we have a special concern for our own future is meant to contrast with a theoretical, metaphysical necessity. It is not that a special concern for our own future follows from some theory about selves, or persons. Rather, it is that, in order to act, we must presuppose certain things. As such, those things are practically necessary; that is, necessary for practical matters – for action.[30]

Why is a special concern for the future practically necessary?

> It may be impossible to *be* a self, or at any rate a rational agent, without caring especially *about* oneself. Being a rational agent involves recognising one's ability to make decisions, form intentions, and plan for the future – recognising one's ability to make one's *own* decisions, form one's *own* intentions, and plan for one's *own* future, that is. (Wolf 1986:719)

The central thought is that there's a bunch of things tied up with agency *as such* where we respond to reasons, carry out intentions and monitor whether our actions achieve what we intend them to. The special concern we have for the future is a concern with being the author of our own actions; being the sort of reasons-responsive creature that can carry out intentions, and monitor whether one is achieving one's aims.[31] This conception of agency as taking place over time may rule out instantaneous time-slices as having agency – and *a fortiori* exdurantist selves – because intentions require a duration of something whose intentions they are.

Reasons wouldn't count as reasons unless we took them to matter long enough to respond to; intentions wouldn't count as intentions unless they could sometimes last

[30] Emmet (1992:50) makes a similar argument, in response to the Buddhist/Parfitian 'no ownership' view of the self:

> 'No ownership' may advocate the absence of moral possessiveness, but it also may dissolve moral responsibility. I can be concerned about the consequences of my actions in future states, whosesoever they are, but I cannot acknowledge them as *mine*. If I am to take responsibility for my past actions I shall need to think of myself as enduring . . .

[31] Intentions, as I'm thinking of them, aren't merely mental states that precede actions, but things agents carry out by acting. The talk of self-monitoring is meant to emphasise that I am not concerned with the mere issuing of an instruction to one's muscles (as one might to a servant or robot) that is never spoken of again. Usually, with intentions, we think carrying out an intention involves some interest in whether one does so successfully.

long enough to be carried out. In order to monitor our actions, we need to care about the actions we are monitoring, and about what they aim towards. To make sense of agency, we have to think of agency as lasting long enough for reasons, intentions and self-monitoring to get into the picture. If agency is to be *coherent*, we need to think of it as involving not merely a relationship of special concern between two time-slices, but as a changing unity across time.

Perhaps you think the reason I insist on *agents* as the proper subject of change is that I am thinking of agency in a mysterious way. List and Pettit (2011) think of agency primarily in terms of 'states' of a 'system'; representational states, motivational states and capacities to process these states to produce intervention. The system represents how things are, how things are required to be, and a capacity to change them. On List and Pettit's functionalist view of agency, it looks like being an agent is more or less just a matter of the right relations holding between different states. The unity of agency, then, might just be a matter of the right similarity-cum-causal relations holding between the various exdurantist counterpart states. I'm not quite clear on how a *capacity* to intervene becomes a relationship between various states, but if something like this picture were to work, the diachronic unity of agency would not provide any reason to think you or I were proper subjects of change, rather than single person-stages standing in the appropriate similarity-cum-causal relations to counterpart person-stages. My contention is, however, that you cannot build up persons from states and relations between them. The states must be, at least sometimes, understood as the limits of processes or activities that take time, rather than processes or activities being built up from collections of states. This will be the subject of the next chapter, so I will postpone further discussion of this until then. As it is, even if we accept List and Pettit's view of agency (to which I do not want to commit myself), the idea of a *capacity* to intervene seems to require something with that capacity that persists throughout the intervention it brings about.

3.3.2 The unity of agency

There seem to be two elements to the unity of agency. One is unity at a time: there needs to be some kind of resultant effect produced from amongst one's various desires, intentions and so on. Agency requires that one does not get paralysed by conflict. This does not seem to require anything quite so strong as consistency amongst one's motivations, just that any conflict amongst them doesn't prevent the possibility of any action whatever. As Peter Godfrey-Smith (2018) observes, the octopus has arms that seem to act intelligently somewhat independently of one another. Nagel (1971) has also influentially discussed the case of so-called 'split-brain' patients, where the *corpus callosum* has been severed, and some experiments reveal phenomena where the two hemispheres of the brain seem to act independently of one another. A lack of unity can be found even more commonly in group agency, such as (I'm informed) marriages and in parliaments. But even taking into account these examples of disunity, acting requires that there is some action that results. To act and then to retract the action requires unity at a time on both occasions, even if it shows an inconsistency across time. Two independent actions don't involve a problematic disunity, provided they don't conflict. Where there is conflict amongst intentions, attempted actions tend not to succeed or become different actions, but one *cannot* (practical necessity again) carry out an inconsistent plan.

This unity at a time – what we might call 'synchronic unity' – is not of primary interest to us here. It is the unity over time that is key to our being proper subjects of change. Let's call this 'diachronic unity'. How does the diachronic unity of agency get us a connection between change and agency, though? The idea is that the unity of agency requires an agent being the proper subject of change, since agency is related to differences between before and after acting.[32] In lots of cases, one acts in order to bring about a state of affairs that differs from the current one. On those occasions where one acts in order to maintain the current state of affairs, one does so to avoid a change. Acting requires a proper subject of change, even in those cases where the subject seeks to avoid change.

One potential counter-example to this claim might be, as it were, suicidal acts, where the point of the act is to end the agent. These are, it seems, the exception amongst actions. And even these acts require a proper subject of change. Consider cases where one attempts to strangle oneself and, upon passing out, releases one's grip, thus failing to strangle oneself. Although, by definition, one does not survive one's own suicide, one has to complete the act by which one brings it about, and that requires an agent to persist through the performance of *that* act.

Perhaps you worry that the argument one, oneself, is a proper subject of change, requires understanding a unity of agency (and so a special concern for our future selves) as a matter of *identity* over time. Why can't unity of agency (and so a special concern for our future selves) be achieved without a subject which is *identical* over time? The argument, let me be clear, is specifically about agency here. Agents need to be around for some time, and to bring about changes which they survive *in order to be* agents. Momentary time-slices don't last long enough to act, and an attempt to introduce agency via counterpart relations between static stages seems to force us to locate agency *in* some states or to equate agency with some mere variation in what states something is in. The agent seems to drop out unless it is a subject of change to which it can be appropriate to ascribe certain sorts of *responsibility* for change.

The question of *identity* depends on what it is that is taken to be identical. Wolf (1986) and Korsgaard (1989) have argued that the unity of agency, rather than identity of persons, is what is doing the work here. I need not claim that we need an account of the identity of persons, rather than some other psychophysical entities, in order to make sense of the unity of agency, and so of one's special concern for one's future. Amy Kind (2004:548) has responded to Wolf:

> On my [Kind's] account, our special concern for our present selves is viewed as the foundation upon which our special concern for our futures is built. In order for someone's concern for her present self to be fundamental in this way, we must be

32 This will be revisited in Chapter 8, when the difference between before and after is more in focus. Here the concern is with change.

> able to understand her concern for her present self as independent of her concern for the future.

Kind, then, seems to disagree with the claim I have endorsed that there needs to be diachronic unity. But Kind immediately goes on to consider the objection (attributed to Robert Adams):

> that someone's self-concern at any given moment is so permeated by concern for herself as an enduring entity that there is no way to make sense of her concern for her present self in isolation from it.

Kind then goes on to admit that this objection is incredibly plausible if we take a sufficiently short time span. Indeed, the observation from Paul (2017), with which we started the chapter, was in the context of Paul's investigation of how long this subjective sense of endurance lasts. As long as we think there is some short duration for which we cannot make sense of present self-concern without making sense of oneself as having diachronic unity of agency, we have reason to accept that we are the proper subjects of change, and so enduring entities.

3.4 Am I a proper subject of McTchange?

Let's assume I have persuaded you that, as a matter of practical necessity, if we are to have a special concern for our future (or indeed our present), we must have unity of agency, which requires us being the proper subject of change. Why should practical necessities tell us anything about metaphysics?

One could respond to this by denying the possibility of agency. If it is practically necessary to have unity of agency, and I'm right this requires enduring subjects, then an easy solution would be to deny that we have agency. But I won't take this response seriously.[33] Rather, the point of this section is to consider the response that the change that we persist through by having unity of agency is not McTchange. Paul (2017), herself, is keen to emphasise that she doesn't think the continuously enduring agential perspective she describes commits us to anything like McTchange. Nonetheless, this is the challenge I face: to explain why we should think the appearance of enduring through change gives us reason to believe in McTchange.

3.4.1 Endurance needs McTchange

Why do we need to think of endurantism as being a matter of enduring McTchange? As I argued in §3.2.2, if one allows that things can McTchange their properties, the worry about endurance being incoherent goes away, so having a view on which endurance is endurance through McTchange can help avoid objections to the very

[33] See Forbes (2023b) for a response to concerns of this kind.

coherence of endurance. Nonetheless, you might object that one need not accept that endurance *requires* McTchange. Haslanger (2003) surveys views on which endurantism does not go with McTchange. The options are roughly as follows:

Relational properties: Rather than my ball being yellow *simpliciter* it would be yellow-at-time-*t*.

Tensing the copula: Rather than my ball being *simpliciter* yellow it would be-at-time-*t* yellow. We keep the same property of yellowness, but having of the property is tensed.

Adverbialism: Like tensing the copula, adverbialism changes the manner in which things have properties. But rather than positing lots of different copulae for each time, one modifies the standard ones in lots of ways. On this view, my ball would be *t*-ly yellow, rather than being *simpliciter* yellow.

SOFism:[34] This option is slightly less easy to explain than some of the alternatives, so I will quote Haslanger (2003:346) directly:

> SOFism differs from the accounts I've so far considered by suggesting that the principal job of an ordinary subject-predicate sentence is not to express a primitive predication between an object and a property. . . . Instead, it takes an ordinary statement to be making a claim about a state of affairs type obtaining, or alternatively, about a token state of affairs being of a certain type.

The idea here is that we make the matter of whether the state of affairs obtains relative to times, but the state of affairs itself does not change and need not involve any temporally qualified copulae or properties. Alternatively, whether a state of affairs is of a certain type may be relative to times, but the state of affairs itself need not contain any temporally qualified copulae or properties.

There are two further options here, suggested by Fine (2005). We could accept that truth itself is relative to times (*a la* Fine's relativist realism about tense), or we could abandon the requirement that what's true at different times is compatible (*a la* Fine's fragmentalist realism about tense). So, provided we are prepared to modify properties, copulae, the obtaining of states of affairs, the types to which tokens belong, truth or we are prepared to reject the requirement that the properties something has are logically consistent, we need not accept that McTchange is required by endurantism. But those alternatives all strike me as highly unattractive. In fact, I'll admit it now, I have no argument which compels anyone to accept McTchange. But the challenge of Chapters 1 and 2 was to show that it was coherent, and, in this chapter, that there was some reason to believe in it.

The purpose of this section is to give reasons why one might prefer accepting McTchange over any of these alternatives. I don't wish to devote a lot of time to surveying the faults with these different approaches to endurantism, which are variations on the theme that they make something temporally qualified which isn't usually thought to be temporally qualified. Instead, I wish to address the issue in general. Why think that our first-personal experience is evidence of a changing world, rather than one which is

[34] SOF stands for 'states of affairs', but the acronym allows for a pun on 'sophism'.

bound to a partial, parochial perspective, which is in no way privileged, besides being ours?[35]

3.4.2 Changing perspectives

The question of whether there is McTchange is one about whether cases of change can be reduced to temporal variation. In a remark of Geach's, quoted in Chapter 2, we considered the thought that even if experience is illusory, we have experience of succession of such illusions. To describe our experience as of enduring through change, we need to appeal to succession of experience. But succession-to-us may be explained without any genuine succession. For it to be McTchange, there must not be some more encompassing perspective that we can adopt from which the succession we experience is mere variation over time.

Kurt Vonnegut (1969) invites us to imagine an alien race who doesn't have our first-personal experience as of endurance. He imagines a human – Billy Pilgrim – in a zoo run by these aliens, with a tour guide who explains what things are like for humans like you, me and Billy Pilgrim:

> The guide invited the crowd to imagine that they were looking across a desert at a mountain range on a day that was bright and clear. They could look at a peak or a bird or a cloud, at a stone right in front of them, or even down into a canyon behind them. But among them was this poor thing, and his head was encased in a steel sphere which he could never take off. There was only one eyehole through which he could look, and welded to that eye hole were six feet of pipe.
>
> This was only the beginning of Billy's miseries in the metaphor. He was also strapped to a steel lattice which was bolted to a flatcar on rails, and there was no way he could turn his head or touch the pipe. The far end of the pipe rested on a bi-pod which was also bolted to the flatcar. All Billy could see was the little dot at the end of the pipe. He didn't know he was on a flatcar, didn't even know there was anything peculiar about his situation. (p.94)

I will have more to say about these aliens in §8.3. What concerns us here is the thought that such aliens would be able to make sense of our perspective-bound situation. Of course, it would seem to us as though this perspective fitted how the world really is, but that's only because of our inability to escape how things seem to us.

We can think of some descriptions of the world as being from a perspective. Every experience Billy Pilgrim has is from a particular perspective in time (and space, and on him), but what seems to Billy Pilgrim as change in what time it is, is mere variation in perspective (according to the fiction). Why should we think that our first-personal experience changing is evidence of our first-personal experience McTchanging? As I

[35] I recall Hugh Mellor at an Australian conference putting the point to Heather Dyke as follows: 'It's like thinking your kids are special. Everyone *thinks* their kids are special, but no-one's kids really *are* special.' Dyke demurred, accepting Mellor's argument in the temporal, but not the filial case. Cf. Dyke (2012:177 fn. 6). Incidentally, Mellor (1998) is an example of someone who defended endurantism about persistence but not McTchange.

argued in the previous chapter, we shouldn't think that the content of our experience is in a position to distinguish between these two hypotheses. The question is one of how to extrapolate from our individual case to what is the case. Is there some description we can give of the world where the change in perspective disappears, or do we always have to have a world in which we take up a McTchanging temporal perspective?

I fear I will not settle this question here. Nothing I say will provide a decisive reason in favour of believing in McTchange. But I think that the contrast between time and space, which seems to occur across history and cultures, gives us some reason for thinking that our temporal perspective on the world is inescapable in a way that our location in space, or even our first-personal consideration of our own interests, is not.

The contrast between time and space, and the contrast between 'now' and 'I', give us some reason to think time is special. We are quite happy to acknowledge that not everyone with whom we might have dialogue or collaboration is in the same place as us or the same person as us (as we might when we yell out of a window 'what are you doing over there?!'). But there is something odd about treating people at other times as interlocutors/collaborators. (Think of a small child writing a letter to their future self and eagerly awaiting the response.) Consider, for example, Brian Barry (1977:269–70):

> A truistic fundamental difference between our relations with our successors and our relations with our contemporaries, then, is the absolute difference in power. The present inhabitants of Britain may believe that, although they have some discretion in the amount of aid they give to the people of Bangladesh, they have little to hope or fear from the present inhabitants of Bangladesh in return. But they cannot be sure that later geopolitical events may not change this in their own lifetime. We can be quite certain, however, that people alive in several centuries time will not be able to do anything that will make us better off or worse off now, although we can to some degree make them better off or worse off.

Even many critics of McTchange acknowledge that there is something striking about the universality with which our perspectives on the world seem to be constrained by time (and change in what perspective we adopt). Ismael (2012) and Callender (2017) both acknowledge that there is something about our first-personal experience of time that requires an explanation that has no analogue to space (or the first/third-personal distinction). For them, it is a matter of reconciling our embeddedness in time with a non-perspectival description of the world given by science. But the very fact that Ismael, Callender and I agree about the pervasiveness of our embeddedness in time gives us some reason to accept the reality of McTchange.

If we have the relatively unambitious goal of explaining why people should think there is McTchange – why it has the claim to be the common-sense position – then the disanalogy between our communal changing temporal perspective and our first-personal, but not communal, spatial or subjective perspective can provide this. It is this contrast, rather than responses to vignettes by roughly 70 per cent of participants in empirical experiments, that gives McTchange the claim to be the abductive hypothesis worth investigating. This will not be enough, however, to *conclude* that there is

McTchange. As I will discuss in the next subsection, opponents to McTchange offer a number of debunking arguments, to the effect that our collective experience of viewing the world from a temporal perspective cannot give evidence for believing in a world containing McTchange, because our experience can be explained without appeal to McTchange. Indeed, it the contention of these philosophers that their alternative explanations are better. Consider Callender (2017:227–8):

> The sketch appeals throughout to known and independently posited physical constrains and psychological mechanisms. The net effect of these mechanism and constraints is sufficient, I submit, to undermine confidence that such a theory cannot be right and that that there must instead be genuine and fundamental temporal flow [i.e. McTchange].

3.4.3 Debunking arguments

In having a debate between dynamic and static views, one of the central (and most difficult to respond to) tools is the debunking argument. Debunking arguments work like this: take some apparent feature of the world (e.g. moral properties, causation, the direction of time), and agree completely with a description of the appearance of such a feature. Then come up with an explanation for the appearance of that feature that makes no appeal to the reality of any such feature. Given that we can explain an apparent feature of the world without accepting the reality of such a feature, it is unnecessary to commit ourselves to such a feature of the world. We have thereby 'debunked' such a feature.

Debunking arguments seem to work in some contexts. Experiences as of *déjà vu*, ghosts, out-of-body experiences and alien abductions all look like cases where we might wish to explain the experience without accepting that we have really been here before/seen dead people/hovered above ourselves/been probed by little green men. But crucial to the debunking argument is that it aims to undermine our confidence in some claim, rather than to provide evidence that it is false; this is why it is so often used to argue against commitment to the existence of some controversial entities. Providing positive evidence for negative existence claims is notoriously difficult.

Heather Dyke (2012:179), for example, offers an evolutionary story about why we should not accept that time passes: 'there are compelling reasons for thinking that the fundamental tensed belief system is an AIR [adaptive imaginary representation]' rather than a MOR (a model of reality). The terminology is from David Sloan Wilson (1990; 1995; 2002). The central thought is that it can be incredibly useful to have a false picture of reality if it motivates people to do things that are in their interests. Confidence in oneself, even when not supported by the evidence, can make success more likely. Believing one's enemies to be despicable monsters makes one more motivated to fight them. William James (1896:24) gives a number of examples:

> Who gains promotions, boons, appointments, but the man in whose life they are seen to play the part of live hypotheses, who discounts them, sacrifices other

> things for their sake before they have come, and takes risks for them in advance? His faith acts on the powers above him as a claim, and creates its own verification.

Believing that one is a proper subject of McTchange (if we grant that this is the common sense view) may be another example of a distorted take on reality – an AIR.

Dyke gives a number of reasons for believing this. One is a more-or-less computational one. 'Once they have adopted this tensed AIR, individuals do not have to make complex calculations in order to decide how best to act' (p.179). Another is that, once adopted, such representations become extremely tenacious. That we (as a human species, as mammals, as '*dasein*'/'creatures like us') find it compelling to think of ourselves as having an inescapably temporal perspective on a McTchanging world is *exactly what we should expect*, Dyke argues, if our tensed system of beliefs is an AIR. So the appeal to our experience (even if this experience is common between you, me, Confucius and philosophers of the Nyāya Hindu school) doesn't give us any reason to believe in McTchange, she claims.

It is, of course, open to us to respond that the reason believing in a dynamic world is so tenacious is because it *is* a dynamic world. In fact, the contrast between a MOR and an AIR might be somewhat unhelpful. It is plausible that we rarely have MORs for lots of things, given how cognitively costly they would be. Our visual system and our memory are largely reliant on noticing changes in particular salient features, rather than tracking how everything looks/looked constantly. But to describe this as a 'distorted' picture of reality is premature; it suggests some alternative picture of reality with which we might compare such things, a perspective that was veridical. We can make sense of distortion of vision by comparison with other sense modalities, and distortion of memory by comparison with other records, but what is it that we are to compare our purportedly distorted tensed belief system with? The very philosophers who attempt to debunk it rely on it as much as anyone.

This is not to deny that there are specific pathologies whereby we lose track of time, prioritise the short term over the long term to our own cost, or assume that trends which have been stable for a few years are permanent and have held since time immemorial. But these pathologies can be noted by comparing our experience with others, or by finding that our own recalcitrant experience corrects mistakes (e.g. by the alarm clock going off).

Dismissing a belief in McTchange as a distortion of reality only works if (in some broad sense) it is contradicted by experience. Now, Dyke will argue that our experience *is* contradicted by experience, in the form of our best scientific theories. In particular, the Special Theory of Relativity clearly poses challenges to common assumptions about time (though not, I shall argue, McTchange as such). In Chapter 5, I will address this particular worry at length. For now, we should note that the question of whether we are proper subjects of McTchange comes down to whether our best explanation of the world appeals to some changeless perspective from which change is mere variation across a temporal dimension, or whether the embedded perspective in time is really inescapable. Whether we accept debunking arguments, here as generally, comes down to what the best explanation we have is.

Given we only have limited evidence available, some of this may come down to philosophical temperament. I find myself of the temperament of Thomas Reid (1863:6.170):

> Now, I confess I would rather make one of those credulous fools whom nature imposes upon, than one of those wise and rational philosophers who resolve to withhold assent [to common sense] at all this expense.

In this particular case, we are not being asked to disbelieve our senses but to refuse to infer from first-personal and interpersonal experience to the general case. I am prepared to claim that all normally functioning creatures like us experience an asymmetry between time and space manifested in our experience as of enduring through time because there *is* McTchange through which we endure, at the risk of being a credulous fool. Our next job shall be to make clear what the costs – especially the costs to science – are in withholding such assent.

4

The science of McTchange

But every one of these words, like every equation in physics, presupposes our *pre-scientific experience of the world, and this reference to the world in which we* live *goes to make up the proposition's valid meaning.*

—Merleau-Ponty (1962:432)

In the first chapter, we set up the idea that time might be unlike space in some interesting and important ways: time passes, and space does not. This led to the subsequent two chapters focused on what reasons we might have for believing in the change characteristic of time passing – McTchange. In Chapter 2, we saw that we couldn't be said to detect the passage of time, although there are some reasons in time-experience to think there might be McTchange. Chapter 3 focused on one of these reasons – our personal experience as of enduring through change – and argued that this gives us a *prima facie* reason to think that there is McTchange. This chapter, and the next, tries to move beyond appeals to common sense and makes the case that our best theory of the world should include McTchange. Chapter 5 will deal with concerns about time as they come out of relativistic physics, but the aim of this chapter is to argue that McTchange is plausibly *required by* scientific explanations, rather than hostile to it. There will be two parts to this argument. One is to make the case that the content of scientific theories involves a commitment to an irreducibly dynamic category: processes. The other is that the practice of science presupposes a commitment to treating the phenomena to be explained as undergoing McTchange. The impression that science does not involve such commitments, I shall argue, comes out of a deliberately timeless conception of scientific explanation. Such a conception leaves much of what we take to be valuable about science not counting as 'proper science'.

When we choose an explanation of the world, we choose amongst competing theories. Sometimes, we can build into our criteria of theory choice that McTchange is ruled out of a good theory. W. V. Quine (1981:26) remarked, 'observation sentences are occasion sentences, whereas theory is formulated in *eternal* sentences, true or false once for all. What logical connection can there be between the two? Evidently, we first need to eternalize the observation sentence' (emphasis original). The assumption that a properly formulated theory *must* appear in eternal sentences that are true once and for all, I suspect, might explain why the debate about McTchange is often set up as one of scientific respectability against common sense. This is despite the fact that Quine took science to be a more rigorous version of common sense.

What is the scientific case for McTchange then? The argument for processes in our role of processes in our best theory goes something like this:

1. Processes of decomposition, metabolism, respiring, etc. are commitments of biology.
2. Biology is part of our best theory of the world.
3. Therefore, processes are commitments of our best theory.
4. Processes require McTchange.
5. Therefore McTchange is a commitment of our best theory.

The most prominent example of this has come from the philosophy of biology, as illustrated by Nicholson and Dupré (2018). The idea is that processes, an irreducibly dynamic category (and therefore one requiring McTchange), are required to explain organisms. One reason that processes are particularly significant is because, as we saw in the previous chapter, some of the most plausible instances of McTchange – changes in things – can be explained away as variations between parts of events. Processes, as contrasted with events, undergo change themselves, and as such do not admit of such change being explained away as temporal variation.[1]

The argument for McTchange in the practice of science, by contrast, doesn't rely on processes, but on opportunities. That science often requires making interventions when they are timely, and not when they are inopportune, not just to gain knowledge, but to secure desirable outcomes in the world, is a reason to think that the world we are attempting to give a scientific account of is a changing world. The practice of science *presupposes* an embedded perspective and so is at odds with a conception of objectivity that treats change as mere temporal variation. The Quinean idea that the work of science is concerned only or primarily with formulating eternal sentences of a theory ignores huge amounts of what makes science of interest. The contrast between an 'objective' scientific image of the world and 'subjective' human experience of the world doesn't give us grounds to treat McTchange as merely a feature of experience and not of reality as described by science. Or so I will argue.

4.1 Processes

In the previous chapter, we considered our first-personal awareness as of enduring. In §3.2.2, we discussed arguments suggesting that we should not treat ourselves as enduring things but as a temporary convergence of processes. Although I don't think

1 A. N. Prior (1962) argues that changes in things are often mistakenly treated as requiring changes in events, but claims events themselves do not change – rather they *are* changes (i.e. temporal variations). One instance of such a mistake is thinking of 'temporal flow' as a change in events. Accepting processes means that the case for McTchange doesn't depend entirely on changes in things, which, for example Quine (1950) treats as temporal variation across events.

we need to claim that we ourselves are not enduring things, the need to appeal to processes in our theories provides another reason for accepting McTchange. According to Helen Steward (2013:809) these are '*goers-on in time*' (emphasis original). Processes (and activities, which I'll treat as processes that, as it were, *belong* to something), as these terms shall be used here, are *irreducibly dynamic*. Other categories, such as states or events, need not be dynamic. If there are processes, there is McTchange.

This chapter will set out the scientific case for McTchange. Part of that case is a scientific commitment to processes. I'll argue that, even if we don't think enduring entities are evidence of McTchange, we should think processes are.

Processes are importantly related to events, on the view I shall defend. Following Rowland Stout (1997; 2018), I'll base my understanding of processes around the progressive/perfective distinction.[2] While events will happ*en*/have happen*ed*, processes are happen*ing*, will be happen*ing*, were happen*ing*. Stout is sensitive to concerns that this linguistic difference may not amount to much:

> I can say: 'An apple is decaying', or I can say: 'The decaying of an apple is happening.' If the second does not add anything to the first, then it looks as though talking about processes is quite spurious. (1997:20)

But Stout thinks he has an explanation of why events have to differ from processes; they have different modal properties. Although we might have a variation of things over time that coincides with a process, processes are not identical with variations over time:

> An ongoing process manifests itself in a sequence of outcomes over time – a sequence that satisfies a pattern characteristic of that type of process. . . . This sequence of outcomes happened (perfective). It is a completed event, not an ongoing process. (Stout 2018:3–4)

This idea that processes have something in addition to a sequence of events is naturally appealing for a defender of McTchange. Crucial here is the thought that events are simply temporal extensions, whose identity conditions are given by the temporal parts they contain. Processes have distinct identity conditions from their temporal parts (they could have had more or different temporal parts and have nonetheless been the same process).[3] As Steward (2013:805) puts it:

> When we think of a process as something delimited [from events understood in this way], we conceive of ourselves as having delimited a thing which is both a genuine *individual* (and not a mere mass) but also one that is conceived of

2 Forbes (2023a) discusses the role of the perfective-imperfective distinction in relation to tense in more detail.

3 As we saw earlier in §3.2.2, I am committed to rejecting mereological essentialism about things and processes (the claim that such entities have their parts essentially), the claim that things or processes are identical to their parts, and that sameness of parts entails sameness of identity for things and processes.

> only as *accidentally* thus delimited – a thing which might, for instance, have gone on longer, or in somewhat different ways – or might indeed *yet* go on for longer. But there are not ways in which we can think about events, as characterised by the assumption of mereological essentialism.

Steward is trying to make sense of the respect in which processes are continuants, while also being occurrents. As discussed in the previous chapter, things are paradigmatic continuants, and events are paradigmatic occurrents. Processes seem to be thing-like and event-like. Processes most definitely are extended over time. They seem to spread out in a way people do not; all of Edmund Hillary may have climbed Everest, but the process of climbing was spread out through time.

Processes can be interrupted, and events – understood as perduring mereological sums of times – cannot. The thought seems to be this: events represent spans of time over which there is some change. Processes are spans of time over which there is some change, but are themselves proper subjects of change. So, although they have different parts at different times, they are nonetheless not mere variations over time, as events are. As Steward (2015:114) says: 'A process, unlike an event, *is* capable of change. It can have a property at one time, but a different property at another' (emphasis original). In a slogan: processes are changes which change.

Processes are thought of as having a *modal* element – capacities or capabilities to change are paradigmatically *modal* properties. We have moved, then, from *categorical* properties, which capture how things *manifestly are*, to modal properties, which capture what *will be, can be,* or *must be and so on*. It is in virtue of their modal character that processes are different from mere events.

A theme we will see emerge over the book is that McTchange and modal realism seem bound together.[4] Insofar as processes give us a reason to accept McTchange, they require us to accept that McTchange is constrained by some modal element in our story of the world.

The category of process is, as Stout points out, especially useful in providing an explanation that is within a temporal perspective. If we want to explain our first-personal experience as of enduring, discussed in the last chapter, the idea that we are *doing* things rather than merely nodes in a world-line straggled across space-time seems indispensable. But it doesn't yet get us to the claim we need in this chapter, which is that our best scientific theories *require* appeals to processes, and that processes involve McTchange. From *our perspective,* it seems like processes are distinct from events, but we need the claim that processes are (*simplicter*) distinct from events.

Stout (2018:4) goes on to observe that Newtonian physics very much relies on reducing processes to events:

> The moral of the success of Newtonian physics and mathematics might seem to be that descriptions of completed events – sequences of stages – are all we need

[4] I use 'modal realism' to refer to a commitment to the irreducibly modal, not to be confused with Genuine Modal Realism *a la* Lewis (1986), which claims modality is entirely reducible to the non-modal through a commitment to a plurality of possible worlds.

> for a scientific account of the occurrences in nature. From these, by applying a bit of differential calculus, it looks as though we can describe change in nature. These descriptions can employ the progressive aspect, but such descriptions are grounded in perfective aspect descriptions. So it seems that the laws of nature apply to completed events not to ongoing processes, and we can derive our talk of ongoing processes from them.

The success of this Newtonian picture suggests the possibility of a reduction of process to mere variations in properties over time. And if that is how processes are to be treated in science, then processes provide no reason to accept that there is McTchange. I will argue that there is a good case that science is committed to McTchange, both in biology and in physics. The contrast between a subjective view of time and a scientific one, I shall claim, causes problems for our understanding of what science is.

This is not the only way of thinking of processes. Some distinguish between processes and events in terms of whether they have some terminus. We might, for example, distinguish *walking* (an activity or process) from *walking to the shops* (an event). This distinction is based on the possibility of a certain type of inference from φ-ing to having φ-ed. This is because, at any point during a walk, it is true that you have walked, but it is not until you reach the shops (and the journey is complete) that it is true that you have walked-to-the-shops.[5]

The distinction between things that are defined in terms of boundary conditions and those that are not may well be a useful distinction, but it is difficult to see why it would be a distinction of metaphysical category that would be useful to us here. Insofar as the distinction between events and processes is supposed to give us a distinction relevant for McTchange, it is unclear how disagreements about boundary conditions would do it. Walking to the shops seems to have no better or worse a claim to provide an example of McTchange than walking aimlessly; particularly when walking aimlessly often ends up placing one outside a shop.

As Kathleen Gill (1993:383) puts it:

> we experience the physical world as an ongoing causal flow through time. In describing the occurrences we pick out segments of that flow. . . . But in principle that causal flow could be segmented at any point. Any state which obtains at a moment is the culmination of the causal processes leading up to it, and so could be used as an end-point which ties together preceding changes.

The idea that events and processes differ in virtue of the former having a terminus and the latter lacking one seems to make very little difference to the questions that

[5] This distinction is adapted from Aristotle's *Metaphysics* Θ, 6. Something like this test is used to distinguish between *kinêsis* and *energeia*. Anthony Kenny (1963) translates *energeia* as 'activity' and *kinêsis* as 'event', though translating Aristotle here is rather tricky. As Stevie Makin (2006) notes, *energeia* has a wider application than 'activity' suggests, and Makin translates *kinêsis* as 'change', rather than event. Sticking with the Greek origins, we may describe this distinction as between events as *telic* and processes as *atelic*; *telos* is Greek for end, or aim, and the idea is that events are defined by their boundaries in a way which processes are not.

concern us. This is largely because we seem to be able to characterise exactly the same phenomenon as bounded or not, depending on our whim. Particular decisions about what description some occurrence falls under may well be useful for explaining actions, intentions and responsibility,[6] but if our ambition is to claim scientific respectability for the distinction between events and processes, this does not look to be a useful way of framing that distinction.[7]

There is some dispute about whether processes can be individual, or one can merely have some quantity of a process, as Thomas Crowther (2018) argues. It might be that processes, by analogy with stuffs, don't form countable individuals. Perhaps one can have *some* digestion, but not *two* digestions. Alternatively, processes might come both as mass terms and count terms. But such a discussion would distract us here from a distinction between events and processes that looks tractable: events as mereological sums of temporal parts, and processes as temporally extended entities that need not have their temporal parts essentially.

I want to avoid many of these disputes if I can. The aim of this section was to clarify how I will be using the words 'process' and 'activity'. I will be reserving them for an irreducibly dynamic category of goers-on in time, related to events but with different modal properties. It is to this irreducibly dynamic category, and *a fortiori* McTchange, that I shall argue science is committed. In the next section, I will discuss what talk of 'commitment' to McTchange means.

4.2 Commitments of our best theory

What would a scientific case for McTchange look like in terms of the content of scientific theories? Roughly, I want to say that processes are important to some important scientific explanations, and that provides a reason to think there are processes. Processes, I will then argue, are irreducibly dynamic and so commit us to McTchange.

4.2.1 Getting committed

How does one show that processes are 'important' to scientific explanations? One option is to show that processes are indispensable to our best theory. There are various ways this might go. Following Michael Resnik (1995:166) we might think of an 'indispensability' argument flowing from three premises: Indispensability, Confirmational Holism and Naturalism.[8] Naturalism, here, means that we take what is

6 I have in mind the role that acting under a description plays in, for example, Anscombe (1963), Foot (1967) and Davidson (1980).

7 Insofar as the progressive/perfective distinction is linked with the *energeia/kinesis* distinction, it is quite tenuous. But one further Greek term that is often brought in is *dunamis*, variously translated as potentiality or capacity. See Makin (2006:xxii) cf. §7.1.2 when potentiality will play a significant role. The idea that *energeia* is the exercise of some *dunamis* ends up with the idea that process is thought of 'in an Aristotelian spirit as ways of exercising capacities' (Stout 2018: 13). As we have seen in §3.2.2, Aristotle is appealed to for many, often conflicting, purposes.

8 One might think of this as a 'Quine-Putnam indispensability argument'. Liggins (2008) argues that neither Quine nor Putnam straightforwardly use indispensability arguments for mathematical

real to be a matter of what science commits us to. Confirmational holism claims that evidence does not directly confirm individual hypotheses in the absence of a broader theory and auxiliary hypotheses,[9] and indispensability claims that some ontological category *c* is an indispensable component of our best theory. If we accept that science is what gives our ontological commitments, and that it does so at the level of theory, then if *c* is a commitment of our best theory, we are ontologically committed to *c*.

Naturally, a question follows: what is it to be our *best* theory? You might think that our best theory is whatever scientists tell us it is. But of course, this faces the problem that scientists are not normally in the business of providing ontological commitments; they are in the business of doing science. They may well demur that the theory they provide requires an interpretation, and it is the *best* interpretation that gives us our ontological commitments. For Quine (1969) it was a matter of putting the theory in a rigorous, cleaned-up language for science: first-order logic. We might call this the 'canonical notation' or, following Sider (2011), 'ontologese'. The idea that we have a special language from which we can directly read ontological commitments is quite attractive; the commitments of our best theory are those the best theory commits us to when expressed in ontologese. This may not, however, help us settle which of our competing theories is the best.

What makes a theory the best? Peter Lipton (1991:59) suggests two different ways of understanding what theory is the best:

> We may characterize it [the best of two competing explanations] as the explanation that is most warranted: the 'likeliest' or most probable explanation. On the other hand, we may characterize the best explanation as the one which would, if correct, be the most explanatory or provide the most understanding: the 'loveliest' explanation. The criteria of likeliness and loveliness may well pick out the same explanation in a particular competition, but they are clearly different sorts of standard. Likeliness speaks of truth; loveliness of potential understanding.

One way of thinking about this distinction (which oversimplifies things somewhat) is that science is well placed to provide grounds for thinking some explanation is likelier than another because it appeals to empirical confirmation, but in choosing between two competing explanations that equally account for the data, we must choose the loveliest. This oversimplifies things because it ignores the holism of confirmation – there is no sharp separation between what the data confirms and what is an auxiliary hypothesis – but it also oversimplifies things because we only settle *what counts as data* when we interpret a theory. All observation of data is to some extent theory-laden. A further complication is that Lipton believes that loveliness is a guide to likeliness. If we think there is no sharp distinction between observation and background theory, assessing the relationship of loveliness and likeliness is difficult. In §2.2.4, we discussed whether we can separate out hypotheses for which

Platonism in the way often attributed to them, and Misak (2011) traces the development of appeals to indispensability from C. S. Peirce, Chauncy Wright and William James.

[9] Cf. Quine (1953), which we discussed in §1.5

we have pragmatic support and those for which we have evidence. Lipton is assuming that loveliness does offer *evidence* that a theory is true, but as we discussed then, the relation between empirical confirmation and the abductive process of hypothesis selection may be complicated. In any case, the best theory will be one that combines likeliness and loveliness in the best way.[10]

4.2.2 What counts as science?

For the purposes of this chapter, we're assuming 'naturalism' – the view that what is real is a matter of what science commits us to.[11] But what do we mean by 'science'? It doesn't seem like we need to adopt any theoretically driven view here. We can accept the various uncontroversial examples of sciences. As Hookway (1988:69) explains:

> Granted Quine's insistence that we start inquiry in the middle of things, equipped with a body of beliefs and prejudices, concerned to rebuild our boat at sea, perhaps we rely on a list of those disciplines which we would all agree are sciences: physics, chemistry, biology, and so on. Then we identify the scientific method by observing which the methods which those disciplines employ. Since there is much consensus about which disciplines are scientific, this seems a sensible Quinean starting point.

I agree with Quine's insistence on starting inquiry in the middle of things. Neurath's metaphor of repairing a ship at sea advocates the same policy of reflective equilibrium Aristotle recommended in the *Nichomachean Ethics*.[12] There is, as it were, no dry dock for us to set up a new system of beliefs, so we must improve upon the ones we have. But it is worth noting that Quine himself opts for a view in which fundamental physics is the science that counts as providing our best theory. The various other uncontroversial sciences, such as biology and chemistry, are taken to have no commitments beyond those of fundamental physics. It is often the case, therefore, that other sciences are ignored in favour of fundamental physics.

It is not clear to me that we should ignore the other sciences. Our understanding of how biology relates to quantum gravity is worse than our understanding of quantum gravity. To use physics as a conceptual foundation, we would need a secure understanding of the most fundamental aspects of the physical world, and a secure understanding of the method of moving from more fundamental explanations to less

[10] You may simply disagree with me over what matters to our theory being the best. The aim of this section, at the very least, is to suggest that we disagree about which interpretation of science best describes the world, rather than some contest between science and a non-scientific alternative.

[11] Papineau (2016) has a discussion of various senses of 'naturalism'. As he points out, the great majority of philosophers accept naturalism in the sense that 'they would both reject "supernatural" entities, and allow that science is a possible route (if not necessarily the only one) to important truths about the "human spirit"'. In this sense I am most certainly a naturalist, and I assume you, dear reader, likewise are. I assume, for this chapter only, that scientific methods are the only methods of generating ontological commitments. In general, however, I am open to the possibility that, for example, the humanities can also generate ontological commitments.

[12] 1145b1, quoted in the preface.

fundamental explanations. But both of these are the subject of considerable debate.[13] Cartwright (1999), for example, has argued that our understanding of fundamental physics is limited to laboratory settings, and that not only do we not *have* a unified theory of how the various sciences fit together, but that pursuing one would be a dangerous waste of time.

We are better to proceed from what we do understand of science than from promissory notes about what science will say in the end. In the next section, we will explore the case for processes in biology, and then consider what case there is for processes in physics if we do take the more austere conception of science.

4.3 Processes in biology

Processes are useful for biology and other life sciences for a number of reasons. Nicholson and Dupré (2018) argue for a 'thoroughgoing processualism'. That is, they think an account of biology can be given by appealing *only* to processes, where things and stuffs are characterised as regions of stability amongst processes. Processes, as they understand them, are irreducibly dynamic and distinct from a sequence of events. One might be more pluralist than Nicholson and Dupré while still accepting the central claim that biology requires *some* processes that are irreducibly dynamic. As Nicholson and Dupré (2018:13) put it: 'To conceive of processes as series of discrete temporal episodes is to overlook the very dynamicity that process philosophy is intended to emphasize.' The *point* of bringing processes is to give an alternative to thinking of things as properties had at times. Dupré (2013:31) contrasts things and processes as follows: 'One can imagine a rock undergoing no changes at all without thereby ceasing to be a rock; a mouse in the same state of stasis is an ex-mouse.'

Dupré and others[14] argue that processes afford a different kind of explanation: thinking of things in terms of states and properties requires one to explain how one goes from one state to another. On a view based around things, one needs to explain why things change (and, indeed, how it is possible for them to survive such changes). Processes, by contrast, are changes, such that one needs to explain stability. A process changing requires no explanation, because that's what processes do. But if changes are prevented from taking place, this would need some explanation. Dupré and Bertolaso (2018:323–4) use the example of cancer, which comes in many different forms with many different causal factors in different cases. Attempts to explain the causes for change from no cancer to cancer seem to have no common characterisation, they claim:

[13] E.g. Ladyman (2008) defends the reduction of the special sciences to physics, while Menzies and List (2010) defend the special sciences as emergent from physics. Shaffer (2010) argues that the whole universe is more fundamental than any individual entities, and Bernstein (2020) rejects Shaffer's view and the view that the smallest things are the most fundamental.

[14] E.g. Dupré (2013), Dupré and Bertolaso (2018), Nicholson and Dupré (2018) and Dupré (2020).

> There is, however, one general characterization that applies to all forms of cancer: it is a failure of the processes that regulate the production and destruction of cells. . . . The proper balance of cell types is not something that is achieved once and then maintained by inertia; its maintenance requires a continuous and dynamic set of activities.

The claim for the role of processes rests on the explanatory resources they provide. Because change is part of the nature of a process, stability needs explanation, rather than change. Continuing the cancer example Dupré (2020:109) says:

> The question we should ask first is why we ever *don't* get cancer. From this point of view it is wholly unsurprising that the causes of cancer are distributed across the genetic, epigenetic, cellular, tissue, and environmental levels. (emphasis original)

Taking change not to need grounding in states and properties of those states means that explanations of the change from one state to another are no longer required. If states are grounded in processes, then stability needs explanation, rather than change. Dupré argues that this change of explanatory priorities is needed to make sense of the life sciences.

If the prevention of change is sustained, and relatively counterfactually robust, further processes may well be what explain this stability. A mouse's living is to be explained by various processes that keep the mouse from decomposing. These might include metabolising, respiring, excreting and maintaining homeostasis.

Dupré's arguments extend beyond this change of explanatory priority. The attacks on substance ontology involve the substance ontologists' need for sharp boundaries between substances and other things, and the lack of dependence of substances on other things. These claims depend more on our account of what it is to be a substance, rather than whether there is McTchange, so I won't address them here. Processes are intended to be much more interdependent than substances are normally conceived to be, but I suspect this is better thought of as a distinction, rather than a dichotomy. I don't have any commitments to recombinable distinct existences of the kind defended by Armstrong (1989).[15] Nonetheless, despite their vague boundaries, and interdependence with their environment, it is possible to transplant organisms to new environments without destroying them; in this sense, they are saliently not dependent for their persistence on the particular context from which they were transplanted. Emmet (1992) discusses 'things-in-process' which combine aspects of substance-based accounts and process-based accounts – living organisms being paradigm cases. These things-in-process would serve the role of substances without denying that processes are required to explain them.[16]

15 See Thomas (1995) for criticism.

16 Wiggins (2016) similarly argues for the need to allow both things and processes. However, I am not arguing for a view in which there are *only* processes.

Another source of McTchange in the sciences is appeal to mechanistic explanations, in biology, the health sciences and elsewhere. Mechanisms are understood by Machamer, Darden and Craver (2000:3) as involving activities:

> Mechanisms are entities and activities organized such that they are productive of regular changes from start or set-up to finish or termination conditions. . . . Mechanisms are composed of both *entities* (with their properties) and *activities*. Activities are the producers of change. . . . Entities are the things that engage in activities.

As I remarked above, I'm not distinguishing between processes and activities here, though it should be clear that Machamer, Darden and Craver aren't thinking of an account where *everything* is to be explained in terms of processes. Activities are goers-on in time, and seem to involve changes that aren't merely entities varying in the states they have relative to times; it would be odd to list activities separately if they were just entities having states successively.[17] The idea that an appeal to mechanisms involves an appeal to organised entities and activities means that mechanistic explanations provide another reason to think McTchange is part of our best theory. If we were to replace 1 with 1*, we would have a very similar argument for processes in biology:

1*. Activities are commitments of mechanistic explanations in biology.

These arguments are straightforward, and it seems clear that biology does discuss processes/activities a lot. But perhaps they don't make it into our best theory. There are two reasons why they might not. One is that when we put the theory into the canonical notation/ontologese, there will be no McTchange. The other is that some other theory, such as physics, might be more fundamental, and so only the commitments of the more fundamental theory count as part of our best theory in the relevant sense. In the next section, I will consider whether physics is committed to processes, so if you think the case for processes in biology is inconclusive for either of these reasons, you may be more persuaded by that case. For the remainder of this section, I want to respond to one of the worries I have just raised: do processes make it into the theory when expressed in the language of the ontology room?

Someone who accepts biology talks about 'processes' may deny that biology is committed to processes as Dupré (2013) and Nicholson and Dupré (2018) describe. As we saw above, there is a significant dispute about how processes differ from other categories of entities. What's the case for thinking that the processes biology is committed to are *irreducibly* dynamic? The answer is bound up with the types of explanation we can offer. As we saw above, processes require us to explain changes and allow changes to be explanatorily primitive. Processes will carry on unless interrupted, but changes in the states of things need explanation. Irreducibly dynamic processes

[17] Processes count as proper subjects of change, and so count as enduring by the standards of the previous chapter. Nonetheless, there may be reasons I haven't discussed to distinguish between enduring things and enduring processes, based on criteria for counting as a thing rather than a process.

allow us the resources to explain the stability of properties in spite of change. But, so process philosophers argue, building up dynamism from complexes of states at times is much harder. Nicholas Rescher (2000:23–4) compares the irreducibility of dynamism with the Aristotelian view that Nature abhors a vaccuum:[18]

> Process philosophy has replaced *horror vacui* with a *horror separationis*; being impelled by the paradoxes of Zeno into the conviction that once reality falls apart into disjointed discreteness, not all the king's horses and all the king's men can put it together again.

The comparison with a now long-discredited physical theory may not do Rescher's *horror separationis* any favours; the belief that there can be no vaccua in nature shows a failure in imagination that is a dangerous sin of omission in a metaphysician. To confidently claim that we cannot countenance explanations of processes that reduce processes to states and properties seems somewhat hubristic. But the appeal to the extra explanatory resources that processes can provide seems to be a (defeasible) scientific reason to commit ourselves to processes.

Because processes are *goers-on* in time, and are not mere hunks or regions of space-time, they have, as it were, 'natures' (though natures deeply interdependent on other processes). One can appeal to properties of a process in order to explain some biological phenomenon (or the operation of a mechanism), and these properties include not merely some manifest state the process is now in, but properties that constrain what the process can or will do (e.g. carry on unless interrupted). The argument that processes change the explanatory priorities depends on thinking that processes have a modal profile that scientists can appeal to. As it happens, I think there is nothing alarming in this, but we will begin to see a package emerge where commitment to McTchange goes hand in glove with commitment to irreducible modality, which shapes the course McTchange takes. Those hostile to modal properties may find McTchange harder to accept. But we would do well to locate such a dispute in metaphysics, not in science. Although it may be possible to construct a scientific account of the world without appeal to modality or McTchange, if one insists on excluding them from the ontologese, the claim that this makes one's account *more scientific* is dubious. Biological explanations, at least, *seem* to require explanations that appeal to modal properties and processes.

It is certainly possible, though, that some story can be told in which processes are explained in terms of something else, and the change that processes are committed to isn't McTchange after all. As much as I would like to be able to argue that processes *irreducibly* commit one to McTchange with the confidence of process theorists such as Rescher (2000) and Nicholson and Dupré (2018), I can't see that I have evidence enough to rule such a reduction out entirely.

Am I claiming that practising biologists, if surveyed, would report being committed to McTchange? I'm not confident we would see any difference between biologists and the general population in terms of their reports of their views or their descriptions of their commitments. The claim is that McTchange offers better conceptual/explanatory

[18] Cf. Aristotle *Physics* 4.8.

resources with which to do biology. Accepting that there are processes, and that they do not need to be understood as constructions of sequences of states because they are irreducibly dynamic, allows for more straightforward explanations of phenomena with which biology is particularly concerned. Biology isn't particularly special here. Computer Science, for example, has a particular concern with processes, given that computation is something a computer processor *does*. Biology happens to be an example where the defence of processes has already been made, and an example that is paradigmatically a science.

That our best theory of the Universe requires McTchange is looking more plausible now than it was before; however, the life sciences seem to be committed to processes, and it is up for debate whether this commitment to processes can do without a commitment to McTchange. Accepting McTchange seems to open up explanatory resources of the kind Callender (2012) was sceptical.[19]

4.4 Processes in physics

Even if we find the case that biology requires processes plausible, it is open for someone who denies that there is McTchange to claim that physics doesn't involve processes or McTchange. One could assume that biology will reduce to chemistry and chemistry to physics, such that the appearance of McTchange in biology is to be explained as temporal variation in our canonical physical theory. If one assumes that (and it is not clear that this is an assumption that one can simply help oneself to![20]) then the scientific case for McTchange depends much more centrally on physics.

Process philosophers, such as Rescher (2000:12), think the case for processes in physics is just as strong as in biology:

> Modern physics teaches us that at the level of the very small there are no ongoing *things* (substances, objects) at all in nature – no particulars with a continuing descriptive identity of their own. There are only patterns of process that exhibit stabilities. (The orbit-jump of an 'electron' is not the mysterious transit of a well-defined physical object at all).

Rescher thinks that we aren't going to get fundamental substances – contemporary, tinier versions of the tiny corpuscles envisioned by early modern science – because quantum phenomena are going to be very small processes that combine to produce macroscopic processes. Indeed, he thinks the language of science suggests the canonical language of first-order logic, with subjects and predicates, is replaced in physics by 'the language of process'[21] – differential equations.

[19] Cf. the question **ED** in §1.5.

[20] Neurath (1935) and Cartwright (1999) would both resist the idea that there is a system that all boils down to physics, for example.

[21] Rescher (2000:13). He goes on to say that Leibniz, one of the developers of differential equations, was himself a process philosopher.

But differential equations look like they give us a textbook example of how processes can be reduced to variations across a dimension. We plot time on the x-axis, and acceleration on the y-axis, and we have variation in the process relative to time, but no McTchange. Rescher (2000:24) has a response: 'of course, such process representations are not themselves processes as such.' He goes on (p.25) to respond to a rejoinder to this rather quick move:

> 'But surely processes can be contemplated, thought of, described, and so on, without being exemplified.' Quite right. But process descriptions (a conceptualisation in general) do not *create* processes, any more than people descriptions create people. The principle 'to be is to be describable' holds for process *conceptions* all right, but not for processes as such.

To represent the world with some formula – a Hamiltonian or whathaveyou – either represents a possible change, or a past change, but not a change itself. Or so Rescher would claim.

I find myself thoroughly sympathetic to Rescher's position. I fear, however, that arguments are rather thin on the ground here. It *may* be a deficit of ontologese that it cannot exemplify McTchange, merely describe possible or past McTchange through giving examples of temporal variation. To insist that processes *require* McTchange in our best theory, and that descriptions of processes are insufficient to capture what makes processes important begins to look like table thumping.[22] In Chapter 1, I distinguished between McTchange and temporal variation, and said I would argue that we needed McTchange rather than mere temporal variation. Quoting Rescher saying the same thing doesn't constitute an argument.

What puts more pressure on the claim that physics requires McTchange is that 'process theories' of causation, such as those defended by Wesley Salmon (1998) and Phil Dowe (2000) don't involve any appeal to McTchange at all. On their views, we are to think of two 'worldlines' – paths through space-time – that interact in such a way that there are conserved quantities. These paths through space-time spatio-temporally vary but don't McTchange, however. Those sceptical of 'process ontology' may have room in their theories for things called 'processes' but which lack the irreducibly dynamic character of processes as I have been using the term.

If we notice that a theory of physics or of biology is committed to process talk, we may be able to interpret it as a commitment to worldlines with certain properties that describe their relations to other worldlines. It looks as though *any* commitment to processes can be understood in this way. We cannot reduce properties to sequences of states and properties of those states, but we might reduce them to worldlines. We are left with the worry that processes seem to have different modal properties from events, and the 'worldlines' account seems to treat processes as events. The talk of behaviour and interaction of worldlines suggests that these processes have some

[22] 'Table thumping' is a term used to describe the act of elaborating or defending a point by repeating it with more emphasis. The worry is that such emphasis does not put the audience in any better position than they previously were.

modal properties that mere paths through space-time lack, but there are strategies for explaining these modal properties away in non-modal terms.[23] As so often in these arguments, the reasons I have presented are resistible, but nonetheless, there is some reason to think that processes add important elements to our scientific theories.

The dispute about whether 'change' commits us to McTchange or mere temporal variation reoccurs for other terms like 'process'. There is more to say about how we should think of time at the most fundamental level of physics, but this will get extensive treatment in the next chapter. What we should conclude from this section is that physics coming with a commitment to McTchange has some defenders; Rescher's arguments are unlikely to force those committed to theories framed in terms of 'eternal sentences' of the canonical notation to give up. Nonetheless, we have a case for thinking that talk of processes is important in sciences, including the life sciences and physics, and that there are reasons to think that processes are not the same as worldlines that undergo temporal variation.

4.5 Presuppositions of the practice of science

Science, as has often been observed, is not merely our best theory of the world, but an ongoing inquiry into it. Science is itself a practice.[24] I want to suggest a relationship between the practice of science and the question of whether time involves McTchange. In simple terms, the suggestion is this: doing science involves changing things and seeing what happens, so there is something odd about thinking the goal of science is to provide an unchanging description of the world. This sounds like a *terrible* argument. Compare: 'science involves people making decisions based on implicit biases, so there is something odd about thinking the goal of science is to provide an unbiased description of the world.' It seems to be true that scientists in fact have various biases.[25] But it also seems right that the goal of science is to remove these biases. I will try to argue that a *sub specie aeternitatis* view of the world, where change has been carefully removed, is not part of the goal of science in the way impartiality more plausibly is.[26]

To consider the goals of science, let us consider the views of an early writer on the scientific method: Hippocrates.[27] Hippocrates has two words for time: '*chronos*' and '*kairos*'. The first refers to time roughly as clocks (i.e. 'chronometers') measure it. The second refers to an appropriate, proper, opportune or crucial moment.[28] That is, a

[23] E.g. Lewis (1986).

[24] That science is to be interpreted in light of its being a practice carried out over time has been advocated by many authors, including Kuhn (1962), Hull (1988), Caws (1993), Marcolis (1993), Pickering (1995) and Chang (2022).

[25] Scientists are, in this respect, no worse than philosophers!

[26] I certainly don't want to claim that science is *value*-free, merely that it aims to be free of bias.

[27] I shall be considering works in the corpus traditionally attributed to Hippocrates. The question of whether a man from Cos wrote them is not important in this context.

[28] *Kairos* is also an important concept for other ancient Greek writers on topics such as the timely moment for rhetorical persuasion or a military raid. *Kairos* is also an important concept in Heideggerian approaches to temporality, though used in a somewhat different way from my usage. In this tradition, it is a profound moment of change in one's own position in the world from which

timely moment. Hippocrates tells his readers that they must be concerned both with *chronos* and *kairos*. *The Precepts* begin with the following observations:

> Time [*chronos*] is that wherein there is opportunity [*kairos*], and opportunity [*kairos*] is that wherein there is no great time [*chronos*]. Healing is a matter of time [*chronos*] but it is sometimes a matter of opportunity [*kairos*]. However, knowing this, one must attend in medical practice not primarily to plausible theories, but to experience combined with reason. (Trans. Jones, I.313)

As medical practitioners, Hippocrates' readers should be interested not merely in facts about time, but in identifying *when* to act. Medicine is often considered a science, but the knowledge of medicine is often (as in Hippocrates' case) thought to be a skill (i.e. know-how) rather than pure theory (know-that). One can imagine the novice surgeon who understands well the theory of the circulatory system but is simply not quick enough at clamping the artery when the time comes. A surgeon checking the textbook to what the next step is, or fumbling around for the right instrument, may find the opportunity to clamp the artery in time to save the patient has passed. Similarly, knowing that it is important to contain a pandemic before it enters general circulation is much less useful than knowing that *this* is the time to introduce drastic prophylactic measures. Medical skill involves identifying and acting at the right moment, not merely knowing that the right moment succeeds those moments that are too early and precedes those moments that are too late.

One can respond to Hippocrates by setting up various dualisms: between contexts of discovery and of justification, between the practical and theoretical, between the embedded perspective and the objective view-from-nowhere. My aim in this section is to put these dualisms under pressure. Science involves the study of changing situations and involves and enables practical interventions in the world (indeed in identifying the right moments for so doing). Objectivity doesn't involve escaping from our embedded perspective, but being accurate in describing the contribution of the world to it. Indeed, the very idea that objectivity could be something so distant from our experience can be subjected to serious pressure.

4.5.1 The context of discovery and the context of justification

Let's switch for a moment from medicine to palaeontology; the study of ancient life through their fossilised remains. Much palaeontology is destructive. Studying some fossils usually involves destroying some evidence; one removes some rock to get access to other layers. In addition to developing theories about ancient life, palaeontology involves locating, preparing and examining the evidential basis for such theory. Alison Wylie (2019) considers the relationship amongst technicians and scientists preparing

new options in life become conceivable. Thanks to Polt (MS) for enlightening me on this usage. In my usage, that is just one way *kairos* may manifest, with quotidian opportunities to exercise one's skills equally counting as examples. See Smith (1969) for a discussion of *kairos* in the sense in which I use it.

fossils. The practice of technicians involves a number of considerations about how much time to expend on preparing fossils for scientists to access, how to ensure stability of the fossils over time (particularly in light of possibilities for improved conservations technique in future). Viewed as a practice, the science of palaeontology involves many, sometimes competing, considerations about the timeliness of action. These considerations of the timeliness of preparation, destruction and/or conservation seem to be considerations of objective features of the world. By destroying a piece of evidence, we remove the possibility of investigating it further. By neglecting an artefact, it may deteriorate and cease to be a source of evidence in future. One can see the parallel with medicine here; medical interventions likewise often affect your ability to investigate what is going on because they change the situation in undesirable ways.[29] Indeed, medicine and palaeontology are not exceptional cases here. These features are common to many sciences.

It's not just that sciences involve a practical element as well as a theoretical one. Sabina Leonelli (2016; 2018) argues that paying attention to the time involved in data generation, processing and dissemination brings into relief the ways these processes can affect the evidence we get. She points out that data from experiments can become unusable because the person who recorded it changes jobs, or because a hard disk breaks, and urges us to think of data as something that needs active curation in the life sciences, as much as in historical sciences like palaeontology. Curation of data involves considerations of the timeliness of interventions.

Leonelli (2018:744) considers processes of data generation, retention, preparation, communication and so on to be part of the context of discovery 'rather than to the [realm of] knowledge derived from such processes'. The distinction between the context of discovery and the context of justification comes from Hans Reichenbach (1938; 1951). The context of discovery is the context in which new hypotheses are formed or new data is gathered, and the context of justification is the context in which the evidence is synthesised and justified. The orthodox view, due to Reichenbach, is that the context of justification is what matters for science, and the act of discovery 'escapes logical analysis' (1951:231).

One can imagine, then, a situation in which one concedes that the practice of science involves all sorts of processes and the identification of various opportunities, but all that matters to science is the synchronic relations of justification between evidence and theory. As a matter of *practice*, understanding how data is gathered and disseminated is vital, though, conceptually, at least, this can be separated off from the business of justifying the theory in terms of evidence. Someone may argue that justification of theory by evidence can be done without any commitment to McTchange.

Even if we accept, with Leonelli, that there is a conceptual distinction between the context of justification and the context of discovery, we should agree wholeheartedly with her view that the context of discovery has been under-considered. The diachronic and dynamic situation we see in gathering, testing and recording data suggests that even idealised science will involve a diachronic process of inquiry and discovery. But I

29 Cf. Hippocrates, *Aphorisms*: '*ho dè kairòs oxús, hē dè peîra sphalerḗ, hē dè krísis khalepḗ*' ('opportunity [kairos] is fleeting, experimentation perilous, and judgement difficult' trans. Jones, IV.99).

think some pressure can be put on the distinction between the context of discovery and the context of justification. For one thing, inquiry is an *iterated* process of discovering and justifying. Scientists go from generating and ranking hypotheses for testing, to getting new data, to revising hypotheses and getting more data. Not only do I think the world that scientists describe changes and the processes by which they discover facts change, but the more purely epistemic aspect of science is itself tied to a history of inquiry. The current best theory is in part justified because it is the result of a history of fruitful inquiry.[30] Just as we might be holists about confirmation in a synchronic sense – we take theories rather than hypotheses to be the units of assessment – we might be holists diachronically too. Research programmes or paradigms, rather than mere theories, might be what we assess as justified or unjustified, and research programmes themselves are ongoing concerns, which have to respond to opportunities for testing, rather than the more abstract, synchronic, justificatory job of moving from data to theory.

If we think of science in this richer way, not merely as delivering an evidence base and theory, but as giving us an ongoing process of inquiry, it becomes harder to see science as giving us the output of a big theory of everything that is abstracted away from time. Science, in order to be good science, is concerned with tracking those objective features of the world – the *kairoi* – that make some interventions timely. Ian Hacking (1983:173), for example, places intervening quite centrally in our understanding of science: 'I shall insist on the truism that experimenting is not stating or reporting but doing', he says. This, I take it, is not merely a point that there is more to science than theory (though it is at least as much). Intervening on the world (a process of practical engagement with it) contributes directly to the *epistemic* value of science.

By intervening on the world – wiggling something and seeing what wiggles – we come to know about it. Indeed, we arrive at causal explanations because we are in a position, at least sometimes, to manipulate or intervene on the world. As James Woodward (2003:10–11) puts it:

> On this way of looking at matters, our interest in causal relationships and explanations grows out of a highly practical interest human beings have in manipulation and control; it is then extended to contexts in which manipulation is no longer a practical possibility. . . . If we had been unable to manipulate nature – if we had been, in Michael Dummett's (1964) example, intelligent trees capable only of passive observation – then it is a reasonable conjecture that we would never have developed the notions of causation and explanation and the practices associated with them that we presently possess.

Key to scientific explanations that concern the modal aspects of the world (including causation) is our practice of intervening on the world. Our ability to provide justification for some claim depends not merely on an abstract relationship between

[30] In Chapter 6, for example, I will argue that our understanding of the arrow of time depends on our taking our current knowledge to depend on past inquiry.

data and theory but on practical engagement in the world that allows us to distinguish between coincidence and causation, and between contingency and necessity.

One need not deny our practical engagement with the world in order to accept a distinction between contexts of justification and discovery, of course. One could locate the practical engagement with the world entirely on the discovery side. But I claim that the *justification* of modal claims comes from our practical engagement with the world. Our ability to shape inquiry in an iterated process of discovery and justification depends on our wiggling things and seeing what wiggles. This practical engagement involves *kairos* as well as *chronos*, since *kairos* is a timeliness *for intervention*. Thinking about the practice of science in this way hopefully undermines the dismissal of how we enquire about the world as being irrelevant to our best theory. It seems quite a leap to think that science viewed as a collection of *processes* – of preparation, curation, intervention, synthesis, hypothesis formation and the rest – is going to produce a world in which change is mere temporal variation. Even if I can't rule out such a view, the richer conception of science I have advocated means that the question of whether or not there is McTchange is not best framed as a competition between our subjectivity and respectable objective science. Rather, we must consider how best to make sense of practically engaged objective science that involves *kairos*, intervention and changing things to see what changes.

4.5.2 Pure and applied science

An understanding of the practice of science, then, plausibly involves an appeal to the timeliness of interventions, and investigating is plausibly a process (and as good a candidate to be irreducibly dynamic as any process). But why not think that the goal of science is fundamentally unlike the practice of science? For all I have said, the goal of science could be to describe something for which *kairos* is not relevant, even though the practice of science requires one to identify the timeliness of certain interventions.

One red herring to point out immediately is the idea that there is a single goal to science. One might follow Aboutorab Yaghmaie (2017) in making a distinction between pure sciences with cognitive goals, and applied sciences with design-based goals. Applied science is no less science for being applied. Applied science will certainly require some sense of the timeliness of interventions and need some account of how to carry out inquiry in the world. Nancy Cartwright (1999:18) has been one of the most vocal proponents of an engaged, applied, science:

> To me this is the great challenge that now faces philosophy of science: to develop methodologies, not for life in the laboratory where conditions can be set as one likes, but methodologies for life in the messy world we invariably inhabit. (1999:18)

The separation between pure science and applied science is useful for some purposes. When one applies to fund one's project, no doubt it is useful to be able to specify how one will measure success. But this division is not one between science and non-science, or between proper science and engineering, for example. For one thing, as Yaghmaie (2017) makes clear, science often needs to be applied in the manufacture of the very equipment used in doing the theoretical work. Viewing science diachronically, as I have been suggesting, gives us pure and applied science within the same research programme.

Cartwright's embrace of applied science isn't simply that we should design more stuff that can be useful in the world. Built into her view is the idea that the cognitive goals we get from science reflect the messy world in which applications take place:

> Rather I want to consider what image of the material world is most consistent with our experiences of it, including our impressive successes at understanding, predicting and manipulating it. (1999:9)

The push for 'pure' science that deals with *chronos* at the expense of *kairos* pushes us into thinking there is some problem of our subjective experience of time that is at odds with objective time in science. But, I have tried to argue, that time as opportunity – *kairos* – should not be excluded from the goals of our best description of the objective world.

4.5.3 The embedded perspective and the view from nowhere

Despite my appeal to the life sciences and medicine, and to the practice of carrying out science, I can imagine my interlocutor remaining unconvinced. If what they hanker after is a pure Platonic picture of the world, what I have said will be unpalatable. Isn't everything I have offered just the embedded perspective discussed in §3.5.2? I have only got science to involve *kairos* by contaminating the real nature of the world with subjectivity, my interlocutor may claim. As physicist Rafael Sorkin (2007:158) puts it:

> But the 'opponents of becoming' seem not to content themselves with the experience of a 'situated observer'. They want to imagine themselves as a 'super observer', who would take in all of existence at a glance.

I will conclude the chapter by arguing that we have independent reason to be suspicious of a view of science as aiming towards an unsituated, value-neutral superobserver. Not because science can never reach it – I have no objections to regulative ideals – but because there is something incoherent about such a project.

One way of framing this problem comes from Thomas Nagel (1986:27):

> to insist in every case that the most objective and detatched account of a phenomenon is the correct one is likely to lead to reductive conclusions . . . . Sometimes, in the philosophy of mind but also elsewhere, the truth is not to be found by travelling as far away from one's personal perspective as possible.

The urge to escape an embedded perspective for a more detached one doesn't necessarily lead to truth, because it may end up disposing of the very phenomena that we intended to capture. It is not that physics has no place for time – it abounds with processes – but that philosophers insist on interpreting *representations* of change as representing mere variations across a dimension. In this chapter, I have argued that this move is at best optional, and now I'm making a plea for the unsatisfactoriness of that particular option.

Hilary Putnam (1990; 1994) has been particularly strident in pushing against a dichotomy between a detached, 'absolute' world of facts and an embedded 'perspectival' world with values:

> It cannot be the case that scientific knowledge (future fundamental physics) is absolute and nothing else is; for fundamental physics cannot explain the possibility of *referring to* or *stating* anything, including fundamental physics itself. So, if everything that is *not* physics is 'perspectival,' then the notion of the 'absolute' itself is hopelessly perspectival! (1990:176, emphasis original)

Putnam is worried that a world that isn't encountered, investigated, interrogated and intervened upon by *creatures like us*[31] is simply too divorced from the world that our concepts have a grip on to be meaningful:

> we must accept the position that we are fated to occupy in any case, the position of beings who cannot have a view of the world that does not reflect interest and values, but who are, for all that, committed to regarding some views of the world – and, for that matter, some interests and values – as better than others. This may mean giving up a certain metaphysical picture of objectivity, but it does not mean giving up the idea that there are what Dewey called 'objective resolutions of problematical situations' – objective resolutions to problems that are *situated* in a place, a time, as opposed to an 'absolute' answer to 'perspective-independent' questions. And that is objectivity enough. (1990:178, emphasis original)

Putnam's views may seem strikingly anti-realist. In fact, Putnam's critiques of science as an absolute description of the world may seem at odds with my own metaphysical ambitions; am I not wanting to be a realist about McTchange? As I adverted in §2.2.4, the debate about whether there is McTchange is one about whether change is relative to something else. To be a realist about McTchange doesn't require saying that there is some *view from nowhere* that changes, but to deny that there is some *view from nowhere* where change is mere temporal variation. As I have tried to argue in this chapter, this isn't to supplant science with the subjective, but to acknowledge that science is a practice used to investigate the world we – you and I, dear reader – share. And this is entirely consonant with the point Putnam (1994:489) presses:

[31] i.e. ones that have a perspective on things, interests and values.

> treating physical science as a description of reality 'in itself' is a profound error. Physical science is a product of some of the same interests that underlie our conceptual construction of the *Lebenswelt* [lit. 'lived world']; it is no more a description of reality 'in itself' than is our everyday talk of colors or of meanings or of desires or of values or anything else.

The discussion of *kairos* was an attempt to acknowledge that science has, for as long as there has been science, been engaged with navigating and intervening upon a changing, unstable world in which some moments are more opportune for intervention than others. That is, science, for as long as there has been science, has been concerned with a world in which its practitioners are embedded. One might push back on Putnam's use of 'constructing'. If it is a construction, the building blocks are hewn from a world that we found ourselves in, and not just any way of piling them up will enable us to make sense of things.

Crucially, and Putnam (1994:490–1) is still my ally here, seeing science as being part of an embedded, lived world does not mean sacrificing a commitment to a distinction between my perspective and the truth, or between subjectivity and objectivity:

> But creating standards of objectivity is not, for all that, a matter of proceeding at hazard; whenever we stand within one of these practices, instead of looking for an Olympus from which to look down on them, we know perfectly well that there is a difference between competent and incompetent, informed and uninformed, fruitful and fruitless, reasonable and unreasonable ways of proceeding . . . . The standpoint of agents in an uncertain world and at an uncertain moment in history is not perhaps the one we would have chosen, but it is the one that has been given us to occupy.

If science aims to develop epistemic goods (truth, understanding, knowledge etc.), accepting McTchange is no barrier to doing so. But that is weak sauce. Where's the argument that *compels* you to accept McTchange? Both in this chapter and the previous chapter, I can see my opponents remaining unconvinced. Not through pig-headedness or bad faith, but because they don't see any reason that I have presented that makes them change their mind. Kuhn (1962) noted that in big scientific changes, there isn't a moment when the alternative view gets decisively refuted. There's always a way for a response to be formulated. But at some point, it starts to look like they are working very hard to maintain a view that doesn't have much going for it. There's a contrast between space and time that gets explained away. There's a first-personal awareness of unity of agency that has to be built up from sequences of states and properties. Processes in biology and physics similarly have to be built up from sequences of states and properties. Applied science, social science, the practice of science, a description of the world we actually live in all have to acquire a derivative status so that a timeless, soulless description of reality can be given. Maybe it can be done, but what's the point?

Suppose you think the aim of science, or metaphysics, is to 'carve nature at its joints'.[32] If you think that the best butchery will not cleave time from space (such that change is mere temporal variation), I would ask that you consider the butchery metaphor carefully. Butchery is most certainly an interest-relative process, where the separation of meat from bone is to aid cooking and eating, and to avoid blunting one's knife. The very metaphor appealed to by those who yearn for an 'objective' account of the world is one that supports the idea of science as a practice carried out for certain purposes.

Perhaps the best butchery of nature will avoid cleaving time from space, but this then becomes a matter determined by the results of physical inquiry. Putnam (1967), in much earlier work, presses the point that the Special Theory of Relativity looks to be in tension with the passage of time.[33] The job of the next chapter will be to examine whether there is room for McTchange in relativistic physics.

To sum up this chapter, however, we have been concerned with the scientific case for McTchange. What I have said has been far from conclusive – even in a monograph, there are bound to be loose ends. But I have tried throughout the chapter to put pressure on a view that may have appealed to the reader at the end of the previous chapter, viz. that I am only describing the *appearance* of time (its 'manifest image'), and the static view correctly represents the scientific, or *objective*, image. I began by highlighting ways in which both the life sciences and the physical sciences get explanatory resources by taking change to be irreducible to sequences of properties and states. I then moved on to suggest that science is concerned with time as ripe for intervention (i.e. *kairos*), and this shouldn't be dismissed merely as a feature of the context of discovery, or of applied science or of the embedded perspective. To insist upon those distinctions, I argued, is to force oneself into a very narrow conception of what science (and objectivity) is, and what it is for.

For all I have said, one can still insist on a Platonic conception of science, but the contrast between the manifest image and the scientific one that would result starts to look odd. Most of the sciences, except for physics, turn out to be describing the manifest image of the world, not the scientific one! What's more, there is a risk that the resulting scientific image is so detached from the interests and purposes of the people studying it that it may not help to make sense of the world in which we actually live. For all I am interested in how things seem in the world I find myself embedded in, we shouldn't take that to suggest that accepting McTchange is antithetical to science because I claim the world I am embedded in and the world science investigates are the very same world.

[32] The metaphor comes from Plato's *Phaedrus*, and has been taken up widely in contemporary work by, for example, Sider (2009) Campbell, O'Rourke and Slater (2011). A similar butchery metaphor appears in a story in the *Zhaungzi*, in which Butcher Ding explains why his knife stays so sharp when carving an ox: 'I follow upon its natural order' (*yi hu tian*). This point about the metaphor relying on our interests in carving returns later in §9.5.3.

[33] Putnam changed his mind over his career, so many of Putnam's views of the 1960s may be at odds with his views of the 1990s. The tension between relativity and the passage of time is still widely pushed as a reason to reject McTchange, as in Price (1996) and Callender (2017).

The scientific case for McTchange, then, rests on the role of processes in science affording different types of explanation – explaining the surprising constancy of some phenomena against a background of change – and on the role of fleeting opportunities for action to intervene on the world in a timely manner. Perhaps there are some conceptions of science as a description of local and particular matters of fact spread across space-time in which all there is to do is describe the entities and their properties at each location. But that image of science looks to be at odds with science as it is practised and the range of activities that we call science. If we are interested in our metaphysics doing justice to the world as investigated by science, we find it is a McTchanging world.

5

Relativity and McTchange

> *We common readers, to adapt a phrase from Samuel Johnson, are fitted neither to criticize physical theories nor to decide what precisely are their philosophical implications. We are dependent upon the scientists for an exposition of those developments which – so we find them proclaiming – have important and far-reaching consequences for philosophy. Unfortunately, however, our popular expositors do not always serve us very well.*
>
> —Stebbing (1937:ix)

It would be hopelessly inadequate to write a defence of McTchange without discussing the theory of relativity. But it would be a gross misrepresentation to present myself as one of the *cognoscenti* when it comes to theoretical physics. As such, it would be inappropriate for me to adjudicate on questions about which I am unqualified, and which are still the subject of live debate within science.[1]

I aim, in this chapter, to explain why relativistic physics is commonly thought to be in tension with McTchange (though I will argue that the tension is actually with 'absolute simultaneity', which is logically independent of a commitment to McTchange). This will involve a survey of some key features of the Special Theory of Relativity, the General Theory of Relativity (which, as the name suggests, has the Special Theory as a special case) and then current work in quantum gravity. The aim of this survey is twofold. First, I wish to defend McTchange against the charge that it is incompatible with our best science because empirical results have *refuted* dynamic views. Callender (2017:2), for example, describes 'manifest time' – a rigorous, cleaned-up version of common sense – as 'more or less rubbish'. Putnam (1967:240) is perhaps the classic argument against '"the man on the street's" view of the nature of time'. I argue that these attacks on our common sense depend on what we build into common sense. While relativity requires us to adjust how we think about what is at the same time as what at very small and very large scales, it is no threat to accepting McTchange.

Second, I wish to provide the lay reader with a sense of why metaphysicians interested in time should be *in dialogue* with physicists, rather than conceding that, in Stephen Hawking and Leonard Mlodinow's (2010:13) words, 'philosophy is dead'.

[1] I should acknowledge that the view I favour is heterodox amongst philosophers of physics – a source of some concern for me and any reasonable defenders of a dynamic view. But it is a heterodox view that nonetheless has the support of some eminent and well-qualified physicists.

Philosophy is not dead, and the report of its death was an exaggeration. Indeed, a respectable, if heterodox view, in quantum gravity – causal set theory – offers hope for an account of McTchange within our best physical theories.

Why is common sense under such pressure? Often, it is because we treat it as being largely the same as the Newtonian picture of space and time. So we'll start off by looking at this picture and how it has more than our common-sense commitments built into it.

5.1 The Newtonian picture of space and time

There are two important assumptions we need to get the Newtonian picture of space and time going. Imagine I'm on a train, and I see another train pass me. I can, from my window seat, contemplate the laws of motion that apply to the passing train, and the laws will be the same – 'invariant' – whatever speed my train is going, as long as it isn't subject to changing external forces, for example speeding up or slowing down. That is, provided that I'm not accelerating in any direction, I can treat my window seat as providing an 'inertial frame of reference', and the laws of motion are the same in any such inertial frame. It doesn't matter whether I'm heading north at 60 mph on the east coast of Scotland, heading south at 200 mph through the channel tunnel, or sitting at a signal waiting for a platform to become available. We'll call this assumption – that the laws of motion are invariant across inertial reference frames – 'frame relativity'.

We can add to this a further assumption, which we will call 'absolute rest'. This assumption is that there is some privileged inertial frame of reference, which is objectively not moving. Intuitively, you might think of my train sat at the signal waiting for a platform, and having another train come past. My train is at rest, and the other train is moving. But this isn't *absolute* rest; this is merely rest relative to the signal, the track and (frustratingly) the platform, which is getting no nearer. The signal, track and platform provide a convenient inertial frame here, but this isn't the privileged inertial frame of absolute rest, because the tectonic plate on which the track stands seems to be moving, the planet seems to be revolving on its axis and orbiting the sun, and the sun seems to be orbiting the centre of the galaxy and so on.[2] Our ordinary experiences of the world (including sitting in stationary trains) do not provide plausible candidates for *absolute* rest.

Absolute rest requires more discussion because it was a controversial assumption, even in Newton's day. We should distinguish between two versions of the claim that there is absolute rest. On one version, from amongst all the inertial frames available to us, we have some physical properties that make that frame, in preference to all others, suitable for being the one we hold as the basis for comparing others against. There used to be a metre stick in Paris, against which other distances were to be compared. We

2 It, for all we know, *might* be that your local habitual reference frame is *the* privileged inertial frame. But this would be a staggeringly fortuitous turn of events if the unique cosmically privileged inertial frame were the one that was most convenient for you. It would, quite literally, be the case that the world revolved around you.

might think of absolute rest as being likewise the inertial frame against which other inertial frames are to be compared.[3]

Such a view of 'absolute rest' is compatible with the claim that spatial and temporal properties are no more than relations between events. Once one has the physical data about such relationships, on such a view, there can be no further question about which the privileged resting frame is. Why is this? If one compares the various ways of treating things, one may be uniquely the loveliest. And if this is true, that loveliest of reference frames should be privileged.[4]

One could take a more 'substantial' view of space and time, where spatial and temporal properties are properties of a substance that objects and events are contained within.[5] On such a view, it could be the case, at least in principle, that there is some inertial frame that gets it right, *even though there is no empirical measure that could identify* this. If there is no physical interaction between the spacetime that events take place within and the events themselves, then the objective facts about the privileged resting frame would not be empirically detectable.

Although one could frame this as a debate with no empirical content, physicists who defend the substantial view of space and time assume that the nature of the container *is* empirically detectable. Newton, the most famous person to be associated with the view that space and time are a substance, thought there were physical consequences that could be tested.[6] Various experiments have subsequently been proposed to test this, including the Michelson-Morley experiment (Michelson and Morley 1887). The thought was that if you shone a light in the direction you were travelling in, such that it was reflected back at you, and shone a light off to one side of your direction of travel, also reflected back at you, you would expect to *catch up* with the light in your direction of travel, but not with the light orthogonal to you. But, in fact, we don't find that. The speed of light in such 'round trip' experiments isn't affected by which direction the light travels, and so can't be used to pick out one inertial frame as the absolute one. Though Newton might have hoped for an empirical test of whether there is absolute space and time, round-trip experiments with light aren't going to provide it. Frame relativity applies as much to light (and electromagnetic radiation more generally) as it does to mechanics. But note that this doesn't tell us anything about whether space and time are absolute or relational, merely that we can't use light as evidence that space and time are absolute. The debate between absolute spacetime and relational space and time is back to being one that we can't obviously empirically test.

The Newtonian picture involved these commitments to frame relativity and absolute rest. But some consideration of the historical framing of this picture will show how it can hardly be taken to be a neutral, scientific and common-sense background against which relativistic developments occurred. Emily Thomas (2018:208) speculates that

3 It is notable that the metre stick in Paris no longer has this function, precisely because it seemed an arbitrary choice and didn't seem to be unchanging in length in the way required.

4 Cf. the discussion of Lipton (1991) in §4.2. Also see Swinburne (1981) for an argument specifically about spacetime along these lines.

5 This is the sort of gloss one finds of Newton's view, for example, in Sklar (1974:162), though see Thomas (2018) for a fuller discussion of how we should understand the commitments of Newton, and his contemporaries, to the idea of time as absolute.

6 Though there are some interpretative challenges here. See. Ryanasiewicz (2011).

contemporary anti-realism about the passage of time (i.e. rejection of McTchange) may have its origins in the early modern debates about whether space and time are absolute and relational. But, as she argues, those debates were bound up with theological considerations about the relation of space and time to an omnipresent and eternal God. A rejection of McTchange on scientific grounds is one thing, but a rejection of McTchange on the grounds of seventeenth-century theology is altogether different. Similarly, if we treat the Newtonian picture as providing the view of the man on the street, we potentially attribute theological convictions that they may lack.

One of the key commitments of the Newtonian picture is that time and space have a certain geometrical structure: that points in space persist through time; that we can make sense of a plane of space at a single moment of time. Indeed, following Maudlin (2012), I find it useful to think about what sort of geometrical transformations the Newtonian picture of time affords. In Figure 5.1, borrowed from Maudlin (2012:9), we can see three different ways of mapping the central shape (a circle in a triangle).

Topological transformation preserves continuity. If something is transformed topologically, any continuous line will still be continuous, and any discontinuity will

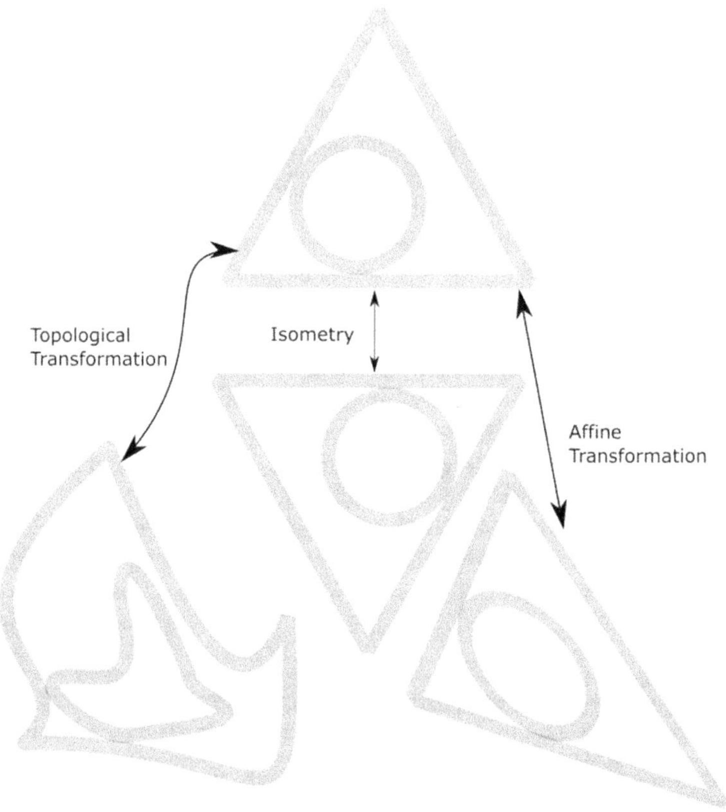

Figure 5.1 Transformations which preserve different features.

remain discontinuous. This is nonetheless compatible with significant changes. It is a renowned witticism amongst mathematicians, I am informed, that they cannot distinguish between a doughnut and a mug of coffee because one can perform a topological transformation from one to the other.[7] In fact, as Maudlin notes, the topological transformation depicted is also a diffeomorphic transformation, given that it preserves the corners of the triangles.

Affine transformation preserves continuity, smoothness and straight lines. Nonetheless, it doesn't preserve angles and distances. One question for us to consider is to what extent we take the world to come with things that can be transformed, preserving these features. Depending on what sorts of commitments to affine transformations we build into common-sense, relativistic physics will overturn it to a greater or lesser extent. When we draw lines with a straight edge, we don't tend to think of the shape of the straight edge as getting changed depending on its orientation (though with a long chalk line, we may be concerned that it's difficult to get it sufficiently taut when working against gravity at some height above the ground). It seems we do rely on straightness being stable through transformations at the scales at which we do, for example carpentry or interior design.

Isometry is a mapping of a shape that preserves continuity, smoothness, straight lines, angles and distances. In some sense, you might not allow that this is a *transformation* of the shape, because it preserves the shape. Again, as with affine transformations, we must think carefully about how much we are committed to transformations preserving distances in common sense. The first mile and the final mile of a marathon don't necessarily feel like the same distance when we run them, and different runners may experience them very differently. But we do tend to treat a marathon in Greece and a marathon in London as being the same length, even if we accept that the experience of running them may differ.

Newton gives us a picture of time and space with a rich but straightforward geometry; the space and time we get has affine structure and a metric. So not only do we preserve continuity and smoothness but we also have a structure that allows us to ask whether two distinct distances are the same length. From here on, we'll be specifically focused on time: if time passes from t_1 to t_2, and from t_2 to t_3, if time has a metric, there is a fact about what the ratio is between t_1-t_2 and t_2-t_3. As we shall see, relativity will not bear such richness out. But it would be a mistake to view these geometrical possibilities as core to common sense. Even if they are useful for the scales of carpentry and interior design, common sense doesn't commit to the same assumptions when designing Satellite Navigation. That space and time form a manifold spacetime that has a geometry with affine and metrical structure, more than just an approximation of those features at human scales, is a mathematical imposition on our experience to render it suitable to more precise scientific investigation. The Newtonian picture is an understandable extension of common sense, but although we might treat the ground beneath our feet as flat, we can readily understand that our common-sense horizons are limited because this flatness is merely an approximation that is useful for certain purposes.

[7] How we all laughed . . .

One of the dangers here is that 'common-sense' involves much that is a result of early mathematical education. If we imbibe Euclidean geometry and Newtonian theology with our nursery milk, we need to be careful about separating the presuppositions of our getting by in the world from cultural prejudices.[8]

Working out our common-sense commitments is tricky, and their relationship to evidence is far from straightforward. As we saw in Chapter 2, we should not look for a distinction *within* experience from which we can read off our commitments. Experience will not distinguish between a commitment to affine structure and metric applying at human scales, and a commitment to those applying universally. We do, however, find an apparent problem for my view. In Chapters 3 and 4, I argued for treating our embedded conception of the world as reflecting a genuinely McTchanging world; one in which we extend our treatment of environments and scales with which we are familiar to provide a story of a world which genuinely changes, rather than merely appears to do so from our perspective. Here, I seem content to accept that spacetime merely seems to have an affine and metric structure from our local perspective, but that we *should not* extend our understanding of these local scales to the Universe more generally.

The difference of approach between my attitude to McTchange, and my attitude to affine and metric structure is not an inconsistency of principle. The Euclidean geometry that had been presupposed as applying to space is motivated by appeal to axioms that are broadly plausible across time and culture.[9] It's just that we have evidence that they don't *work* when we try to do things across sufficiently large (or small) scales, for example when it comes to designing satellite navigation or landing rovers on Mars. I don't think that we have similar evidence that belief in McTchange fails to make sense of our practice.[10]

5.2 Special relativity

We have done much of the work to set up for the Special Theory of Relativity. This theory comes from accepting three facts. Here is Sklar's (1974:277) summary of them:

8 The nursery milk scheme entitles children under five who attend approved day care facilities to one-third of a pint of milk each day free of charge. The relevant geometrical concepts are taught two to three years after that, rather than contemporaneously, as I suggested. https://www.nurserymilk.co.uk/; https://www.gov.uk/government/publications/national-curriculum-in-england-mathematics-programmes-of-study/national-curriculum-in-england-mathematics-programmes-of-study.

9 See, for example, cross-cultural studies of geometric intuitions Dehaene, Izard, Pica and Spelke (2006); van der Ham, Hamami and Mumma (2017).

10 Price (1996) thinks that quantum phenomena provide just such evidence. I will not discuss his arguments in detail, but I will discuss approaches in quantum gravity that appeal to McTchange in §5.4. If these approaches work, quantum phenomena don't provide a reason to reject McTchange.

FACT ONE: All round-trip light propagation experiments give equal times for the propagation of the light in either direction around the closed course, so long as the laboratory is at rest in some inertial frame.

FACT TWO: There are no round-trip signals of arbitrarily high velocity, that is, it is not the case that for each round-trip course and for each time interval, no matter how small, a signal can be found which traverses the course in less than the specified time interval.

FACT THREE: If two distinct events cannot be connected by a causal signal that travels no faster than a light signal, then they cannot be connected by a signal at all.

FACT ONE amounts to the extension of frame relativity to cover electromagnetic radiation. FACTS TWO and THREE are what lead to the immense philosophical significance of *light* being subject to such relativity.

It is widely known that sound takes a while to travel across distances. If you were going to see a band perform in a large stadium, for example the Last Place Champs at Wembley, their sound engineers would have to set up speakers at spatial intervals throughout the stadium so people at the back could hear. Such sound engineers would routinely arrange for a delay on the sound coming out of the speakers at the far side of the stadium, so the sounds coming out of those speakers are in time with those coming from the speakers by the stage. The signal cable going to the speakers travels faster than the sound carries, and as such, the delay must be added.

Imagine if an aeroplane flew overhead during the concert. The people at the back would hear the plane overhead during some particular bass riff, say, but the bassist might hear the plane overhead shortly before the bass riff. That is to say, what seems to be *simultaneous* to these distinct listeners is different in different parts of the stadium. This is because it takes different amounts of time for these distinct signals to reach them. *No one has ever suggested to me that this is a reason to think that time cannot pass!*

Things work in an analogous way for electromagnetic radiation. Things can look simultaneous to one observer but appear in different orders to another. In the case of sound, we might appeal to visual signals to get a sense of what is taking place at the same time as what. Light travels around 882,000 times as fast as sound and so allows us to use it as a guide for shared ordering of phenomena, because the differences between the distinct signals reaching our observers will be much shorter. The problem is that once we work over short enough time scales, or long enough spatial distances, we will get similar effects where light signals take different amounts of time to reach distinct observers. We could establish a privileged resting frame by having *even faster* causal signals – infinitely fast ones – that would get rid of these effects where signals arrive in different orders for different people. But we can't defer to some faster signal to give us some natural, privileged inertial frame from which to pick out which observations of things co-present in experience really are happening at the same time, because no causal signals go faster than light in a vacuum. That is, we can't get the kind of coordinate system that Newton had assumed.

So, how can we make sense of what is at the same time as us? Let's introduce a term-of-art – 'contemporaneity':

> Two entities are contemporaneous iff$_{df}$ each entity is such that a causal signal could in principle be sent from one entity to the other and back to the original entity.

So if two writers, call them Leibniz and Clarke, are each able to send letters to one another, and receive replies from the other, they must be contemporaneous. I could read letters Leibniz wrote, but he cannot read letters I write in reply. I could send letters to Leibniz, but I will not get a reply; we are not contemporaneous. We can make sense of the contemporaneity of two enduring entities because they are around for long enough that each can interact with the other. For most uses that non-physicists will have experience of, when we mean things are at the same time, we mean that they are contemporaneous in this sense.

Note that contemporaneity is invariant, in the sense that, regardless of which inertial frame we adopt, iff *x* is contemporaneous with *y*, *y* is contemporaneous with *x*. I think that, for most senses in which we folk talk about things being 'at the same time as' other things, contemporaneity is good enough for our purposes. We don't tend to have much day-to-day experience at intervening upon scales at which contemporaneity isn't relevant.[11] Common sense has a sense of 'at the same time as' that is useful at the scales we mostly work at, and it is unthreatened by the developments of relativistic physics. In spite of what Callender (2017) says, we don't need to dismiss the manifest image as 'mostly rubbish', but merely as of limited application to those situations we are used to dealing with in ordinary life. That contemporaneity doesn't apply to very short events or entities at large spatial distances is no more an indictment of the usefulness of the concept than the fact that aeroplanes don't work in space is a sign that aeroplanes are useless.

If there were no upper limit to the speed at which causal signals could be sent, contemporaneity would be enough to give us a frame-invariant sense in which things were at the same time as each other at any scale, even when dealing with momentary events, as in the classic physics examples. But, as FACTS TWO and THREE show, there is an upper limit to the speed at which causal signals can be sent. So we need a sense of 'at the same time as' that can be used for the time between the causal signal being sent and received. 'Simultaneity', as used in relativistic physics, is a technical term, especially useful in cases where contemporaneity doesn't apply. Here's a definition of it from Sklar (1974:256):

> To give a Reichenbachian coordinative definition for simultaneity for distinct events, pick a particular inertial frame *S*. One says that *e* and *e*', two distinct [instantaneous] events, are simultaneous relative to *S* if and only if: (1) a clock at rest in *S* at the place (relative to *S*) of *e*' records the time of *e*' as $(t_1+t_2/2)$, and (2) a light signal sent from the place of *e*' (relative to *S*, i.e. from the spatial point said by an observer at rest in *S* to be the place where *e*' occurs) at t_1, whose reflection at the place of *e* (relative to *S*) is an event coincident with *e*, arrives back at the place of *e*' (relative to *S*) at time t_2.

[11] That's not to say that we don't experience phenomena that rely on such scales for their explanation.

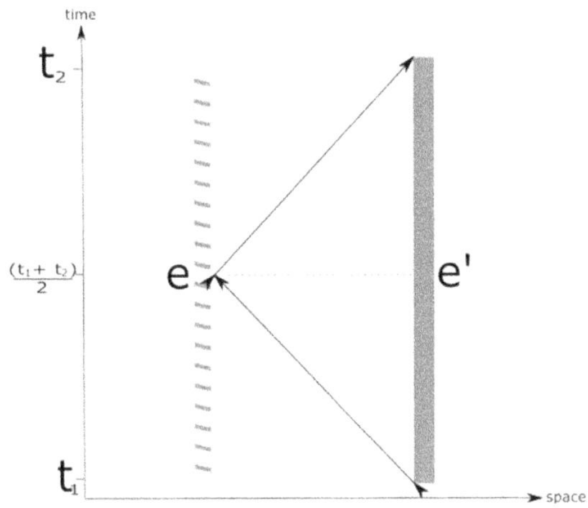

Figure 5.2 Reichenbachian simultaneity.

So, *e'* is halfway between the light signal being sent to *e* and a light signal being received from *e* in that inertial frame (see Figure 5.2). One might, intuitively think that halfway between sending the signal and receiving the signal, where the signal has a constant speed, will give you the time at which the reflection of the signal takes place. So at *e'* you can report, 'the signal is on its way back now!' to the assembled company. If you are sending a signal to Mars and back, it's going to take about half an hour for the round trip, so there might be an awkward pause in which you wish to announce this.

This definition of 'simultaneity' – also referred to as 'Einstein's simultaneity convention' – is relative to your inertial frame *S* that you started off by picking. It's not so much that if you pick a different frame, $S^\star$, that you are going to get a different answer about what is simultaneous – though of course you may well do – but that you are asking a different question. Event *e'* may be simultaneous$_S$ to *e*, and distinct event $e^\star$ may be simultaneous$_{S\star}$ with *e*, but 'simultaneity$_S$' and 'simultaneity$_{S\star}$' don't have the same content.[12] Both terms are relativised to a coordinate system given by different inertial frames. When we're dealing with scales (either large distances in space, or events lasting short amounts of time) at which contemporaneity is not helpful, all we have is a frame-relative notion of simultaneity.

Nothing about this relativistic notion of simultaneity says anything about an absolute sense of *being at the same time* for scales where contemporaneity isn't enough. That's the problem! Because we tend to characterise *being at the same time as* in terms of causal relations, the Special Theory of Relativity robs us of terms in which we could frame such a concept; the causal signals don't move fast enough.[13] As Reichenbach

[12] We might, following Kaplan (1989), say that 'I' or 'simultaneous' are indexicals, and have the same character in different contexts, but the combination of character and context yields different content.

[13] We'll revisit whether it is causal sequence that allows us to handle on temporal order in Chapter 6. One thing to note is that if we are discussing distinct instantaneous events, we will assume that they can only be causally related to events earlier or later than them, but two extended contemporaneous

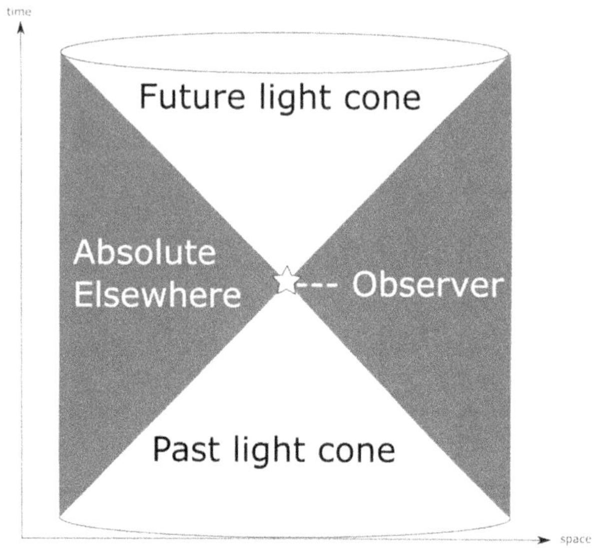

Figure 5.3 Past and future light cones.

(1949:306) says, 'Absolute simultaneity would hold in a world in which there exists no upper limit for the speed of signals, i.e., for causal transmission.' To recall the geometrical discussion around Figure 5.1, we have a situation where, assuming light always travels in straight lines, we can perform affine transformations between different observers' descriptions of the world, but have no method of preserving isometry (sameness of distances) when shifting from a description of the world from one perspective to a description of the world from another. That's what prevents us from picking out a privileged resting frame. No causal signal moves fast enough to give us a physically definable tool with which to do this.

Because there is a maximum speed of causal signals, any two observers in the same place and time – regardless of which inertial frame we use and their relative speeds – will be able to divide the world in the same ways as follows: all the points from which they could possibly receive a causal signal will be the same, and all the points to which they could possibly send a causal signal will be the same. We can call the region of the Universe from which the two observers could possibly receive causal signals their 'past light-cone', and the region to which they could possibly send causal signals their 'future light-cone'. So two observers in the same place and time will (invariantly) have the same past light cones and the same future light cones. There will be a region neither in their past light cones nor in their future light cones, from which they can neither receive causal signals nor to which they can possibly send them. Let's call this region 'absolute elsewhere'[14] (see Figure 5.3)

events, for example, the Lancashire Cotton famine and the American Civil War, can be causally related to each other.

[14] Things in Absolute Elsewhere are also known as being 'space-like' separated from us.

One of the big challenges for the philosopher of time dealing with special relativity is to say something about absolute elsewhere. Which bits of it, if any, are in the past, and which, if any, are in the future? Is there an objective inertial frame against which all others should be compared, or is the idea of an objective inertial reference frame meaningless? It may be that philosophers of physics have interesting answers to these questions based on a detailed engagement with the geometries in play, and it may be that certain scientific discoveries mean that some inertial frames start to look like more plausible candidates for being ones against which others are to be compared. Armchair philosophers like myself might be well advised to take the advice of Ludwig Wittgenstein (1922:§7) on many of these questions: 'Whereof we cannot speak, thereof we must be silent.'

Even while applying Wittgenstein's advice to absolute elsewhere, we can see that Special Relativity has presented no barrier to McTchange. From my armchair, I notice that what's in my past light cone keeps changing, such that there is more in it now than there was. Is this change in the content of my past light cone McTchange, or mere temporal variation? The answer rests on the discussion of the previous two chapters. Nothing about the Special Theory of Relativity suggests that change in the content of my past light cone is only true relative to an inertial frame.

Defenders of a static view may balk at Wittgensteinian quietism because McTchange is meant to be a commitment not merely to change-from-my-embedded-perspective or change-relative-to-my-local-inertial-frame, but change not *relative* to anything else. The fact that I can't provide an objective sense of universal change that applies to absolute elsewhere might be thought to be an objection to thinking change is best characterised as being McTchange.

Much literature tends to be against a *privileged present*, rather than McTchange. Putnam (1967:241) provides an argument against the 'Absolute Present', for example.[15] Putnam argues:

> If it is the case that all and only the things that stand in a certain relation *R* to me-now are real, and you-now are also real, then it is also the case that all and only the things that stand in the relation *R* to you-now are real.

If we take Einstein's simultaneity convention as our relation R, we end up with things that are in our own past or future light-cones counting as real because there is a chain of relations of what's simultaneous relative to each individual's respective frame of reference that leads to our own past light-cone or our own future light cone. What this shows is that we can't use a succession of relative simultaneities in the Reichenbachian/Einsteinian sense to establish a privileged present moment across the Universe. But, as Putnam points out, we have no grounds to provide any single frame of reference with privileged status either.

[15] We will revisit different ways in which the present may be privileged in Chapter 9, but my position is that we should avoid any commitments to what, if anything, is absolutely present besides the cases we have evidence for, viz. ourselves and those contemporaneous with us. We may retain a scholarly neutrality towards the status of the absolute elsewhere.

More recently, philosophers have been careful to separate questions of the absolute present from questions of McTchange (e.g. Callender, 2017). Nothing about Putnam's argument shows that there is an objection to McTchange in the content of our past light cone as we age. But, in virtue of my commitment to McTchange, am I not committed to there being an objective, absolute, frame-invariant present that spans the Universe from side to side?

The thrust of this chapter so far has been to explain that we lack the tools to give meaning to such a frame-invariant present.[16] It's not so much that such a hypothesis would contradict Relativity, but that we can't make physical sense of the claim with the tools available. This rejection of the meaningfulness of a claim that there is an absolute present doesn't amount to a view that has been advanced by Stein (1968), Dorato (2002) and Savitt (2002) that there are *independent* questions to be asked about each spacetime point as to whether it is present, such that we deny the transitivity of simultaneity, or temporal priority (i.e. being earlier than). If event A is frame-invariantly present, and distinct event B is (according to the privileged frame) at the same time as A, then it follows that B is also frame-invariantly present. I do not wish to deny that. I'm just claiming that we're not in a position to specify what it would mean for B to be at the same time as A according to the privileged frame if they are not in the same place. This lack of ability doesn't amount to a rejection of transitivity and doesn't touch on the issue of whether there is McTchange.

As outlined in §2.2.4 and §4.5.3, the debate around McTchange is about the absence of some dimension against which things merely vary. Our inability to identify some objective inertial frame is just evidence of how difficult it is to escape an embedded perspective. We can equally read this discovery as evidence that our embedded perspective is inescapable – even for our fundamental scientific theories – as we could that there is some objective God's-eye-view of the world (as yet epistemically inaccessible to us) by comparison with which our embedded perspective is hopelessly parochial.

5.3 The general theory of relativity

The General Theory of Relativity adds consideration of mass to the principles of the Special Theory. So, what counts as an inertial frame depends on its relation to massive objects. Inertial frames are those, as we saw, where we are not speeding up or slowing down. We need to refine that slightly; inertial frames are ones where we are not subject to external forces. In the General Theory of Relativity, we need to consider whether gravity is an external force. If we, and the whole room, were in free fall, speeding up as we fell to earth, the force of gravity would be constantly acting upon us. So being in free fall will nonetheless count as an inertial frame. In this case, we could choose an inertial frame such that gravity doesn't count as a force acting on us; it gets built into the context. But in such a context, whatever isn't in free fall will be whooshing past us because some force is preventing it from falling. As we saw in Special Relativity, our inability

[16] Not to mention the problems with thinking of the Universe having sides.

to provide an objective inertial frame against which to compare others means that we can switch between inertial frames that represent the world differently. When it comes to representing General Relativity geometrically, massive objects affect what counts as an inertial frame by effectively deforming spacetime, so two objects in the same inertial frame, free falling towards earth, may not stay the same distance apart (see Figure 5.4).

If I drop two cannonballs from space, and they fall freely to the ground, they will get closer together as they near the ground. This is because they are both falling to the same point – the centre of gravity of the Earth – and they will get nearer to each other as they get near to the centre. Far from being a revolutionary overturning of common sense, this is a thought Aristotle could have had. But it means that the 'flat' geometry we had been assuming in Special Relativity and previously in the Newtonian picture is no longer helpful. Not only can we not preserve isometry in General Relativity, but affine transformations will be beyond us too. For we cannot assume light always travels in straight lines, given that it is affected by massive objects. Nothing will count as a straight edge against which we can define straightness.

Given that our definitions of what spatial and temporal relations *are* are all provided relative to inertial frames, physicists often say that, according to General Relativity *spacetime is curved*. We need to be cautious about committing ourselves to a particular position in the debate about whether spacetime *just is* the temporal spatio-temporal relations that we generate from taking up the perspective of an inertial frame, or whether spacetime is something independent of that which we are attempting to describe through such a geometric metaphor. So General Relativistic diagrams in books like this get much more visually stressful than Special Relativistic diagrams. But it is no more worrying than having a road that appears straight on a map but, on closer inspection, goes straight over a mountain. The straight road may, in fact, be longer than one that curves around the mountain because the change in altitude wasn't obvious from the map (see Figure 5.5). The challenge in representing the world helpfully doesn't mean that the terrain we live in is unknown to us, and that cartographers are the only ones who understand what is being mapped.

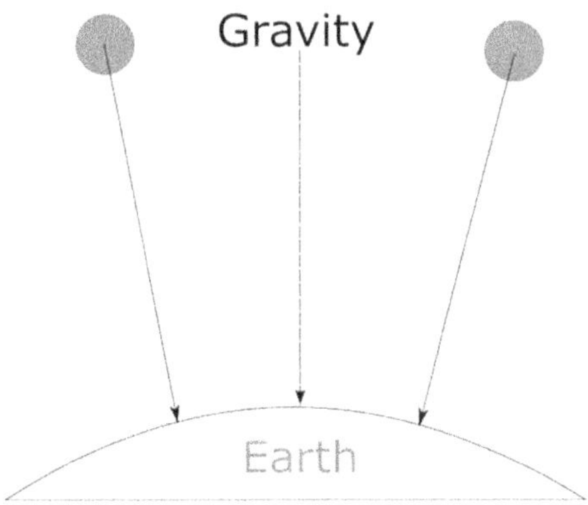

Figure 5.4 Two cannonballs getting closer as they fall.

Figure 5.5 A line represented from two perspectives.

Just as in the case of the Special Theory of Relativity, the issue is that we can't find a way to describe the world independent of an inertial frame, and it's tricky to single out any inertial frame as privileged. The fact that inertial frames are in part determined by massive objects, and they are moving relative to each other, means there's a lot to keep track of. There are lots of variables in play. Moving to the General Theory of Relativity does potentially make things easier for a defender of absolute simultaneity, however, as Gordon Belot (2013:196) points out:

> There are many physically interesting general relativistic spacetimes that can be thought of as falling in a natural way into instants of time – that is, there are general relativistic spacetimes in which simultaneity is absolute (rather than relative, as it is in Minkowski spacetime). There are a variety of ways this can happen.

He enumerates three examples of simpler scenarios in which it is possible to define an absolute relation of simultaneity, which we had no non-arbitrary way of doing in the Special Theory of Relativity. (We'll use the phrase 'foliation' to describe a way of splitting the Universe into times like this. Think about the Universe as a big book, with all the pages or leaves of the book representing different moments of time, running in order from the start of the Universe.) Belot's (2013:196) first example is that there are models of general relativistic spacetime which have some sets of inertial frames as privileged (i.e. relative to which the geometry of space is constant in time). These include cylindrical spacetimes, Schwarzchild spacetimes – spacetimes describing the region outside a non-rotating, neutrally charged, isolated massive body. There are sets of privileged frames on such models, where privileged observers end up agreeing with one another about the simultaneity relation. These very specific space-time geometries are not likely to be of use in foliating our own spacetime.

Belot's second example was based on the geometry of spacetime having symmetry, and 'the symmetries single out a privileged family of instants of time'. This is, effectively, a matter of finding a mathematical way of framing things such that a sequence of times falls out. Questions will naturally remain about how we are to interpret such mathematical solutions. If there are a number of different mathematically adequate descriptions of the theory, which differ in whether they have symmetry, and/or how they foliate the Universe, neither physicists nor mathematicians can choose between them without appeal to some expectations of which is more plausible as a description of the actual world. But neither do we have to accept that all mathematical models are *equally good* descriptions of the world, simply because they are all ways of describing the theory.[17]

Belot's third example is to consider General Relativistic models that have a start or an end:

> In spacetimes that have a beginning – as in the Big Bang Models that are the workhorses of modern cosmology – there is, for each point in the spacetime, a worldline that stretches from the initial singularity to that point and which has a maximum length among all such worldliness. So there is a privileged time-function in such spacetimes that assigns to each point the proper time measured by the longest curve that connects that point to the initial singularity. We often get a well behaved notion of singularity by considering two points to be simultaneous just in case they are assigned the same value by this cosmological time-function.

A few things to note here. These cases seem helpful to us because we have evidence that our Universe is one that has a beginning. Let's make sure we're clear on the ideas here. 'Proper time' is time defined solely in terms of 'ideal clocks moving along a worldline'.[18] Worldlines are simply the paths through spacetime that objects travel along. We might

17 That is, some mathematical models may be less lovely than others (cf. 4.2.). In §§6.2 and 7.6, we will be discussing 'Closed-Time-like Curves', where the geometry allows for loops to form such that some events are in the future light cone and past light cone of the same observer. We have no empirical evidence that this happens, but no empirical evidence that it is impossible either. Closed-timelike curves challenge a past-future distinction but present no special obstacle to McTchange in where in the loop one is.

18 Maudlin (2010:63) treats proper time as basic, with the result that 'instead of spatializing time, Relativity temporalizes space'.

think of this as a career of an object or the path of a light beam, but it will be some join-the-dots sequence of instantaneous events that are at least potentially causally related; from any point in a worldline, the rest of the worldline is either its past light-cone or its future light-cone. Ideal clocks are ones that (a) measure time and (b) do so in a way that give the laws of nature the simplest form given that way of measuring time.[19]

In the spacetimes in this example, then, we have a way of working out how much proper-time has passed since the start of the Universe,[20] and we can use the same technique for all causal paths from the beginning of the Universe to create a unit of time that can be used by all of them. This still leaves quite a lot undetermined; the geometry will be very complicated, even if we have a cosmological time function. But this *might* at least give us some way to make sense of spaces persisting through time, or times passing in the same space. Sort of. We won't recover the idea that space persists *without changing* through time or that time passes the same *no matter where you are*. Those are features of the Newtonian view that we're not getting back.

Getting a cosmological time-function seems like a very helpful thing. It would allow us to get a grip on what 'absolute simultaneity' might mean. But before we leap out of the armchair and spread the Good News, we should be mindful of the need for such a unit of proper time to give us sensible results. If we find ourselves in the distant past according to a measure of how much proper-time has passed, something has gone wrong. Indeed, it may be that this approach does not have all that much promise. Consider Belot's (2013:196) remark following his examples:

> Of course it is not entirely clear what philosophers interested in time should make of this situation. After all, it is fairly certain that our own world does not correspond to *any* of the possibilities countenanced by general relativity (since our world, in contrast to general relativistic worlds, appears to be fundamentally quantum in nature).

It is useful for discussing structural features of the laws of nature, he argues, but not so useful for considering specific accounts of how time may be foliated in our universe.

A brief engagement with this literature, and the mathematical complications I have glossed over, reveals starkly that the devil has firmly taken up residence in the details and won't budge for the benefit of our convenience. But, at least in this respect, the devil is egalitarian; the question of how we are to understand x being at the same time as y, when x and y are neither coincident nor contemporaneous, is up for grabs as the careful and detailed, philosophical work is done to make sense of these complex matters.

Earlier, we were presented with three options for dealing with this. We can be Quietists: there is nothing to say about x being at the same time as y, when x and y are neither coincident nor contemporaneous. We can claim that there is some objective, absolute standard, though we may be incapable of knowing it (perhaps in principle, perhaps merely pending further theoretical developments). Or we can claim that the

[19] Cf. Swinburne (1981:178). The thought is that an ideal clock needs to do something closely enough related to how we think about time for it to count as a clock, and it needs to be ideal in that it gives us the most regularity to the phenomena the laws attempt to describe.

[20] Or how much is left if the Universe is heading towards a single end.

Universe is simply incompatible with McTchange, given that McTchange *ex hypothesi* can't be merely relative to an inertial frame.

I'm not persuaded we must accept the incompatibility of McTchange and the results of relativistic physics. It's not clear that we have any reason to conclude that McTchange has been ruled out by either the Special or General Theories of Relativity. We have merely discovered the difficulties associated with scales at which contemporaneity is not a useful concept. But our conviction that there is McTchange is founded on our extremely firm grip on those cases where contemporaneity is useful; when I meet a contemporary for the second time, we both will have persisted through time in the interim. Insofar as proper time must be a measure of time to *count* as proper time, provided we have proper time, there can be McTchange over it. Insofar as observers can persist along worldlines, and act as clocks (ideal or otherwise), they must be able to persist through McTchange.

The biggest challenge we saw was this: we accept that we are inclined to extend our geometric intuitions about the medium scale further than they are useful, but I am content to claim that McTchange is not merely a perspectival phenomenon – relative to some inertial frame – but a change *simpliciter*. It is not that relativity has directly threatened McTchange, but it has perhaps undermined our confidence in extending our intuitions about our perspective-bound experience of the world to our best and most objective scientific account of the world. There are reasons to be cautious. But, as I expressed in 3.5.3, in such cases I'm prepared to stick with common sense until I discover it stops working for me, rather than to give it up over a mere possibility that it will fail me. Given the sense I can make of persistence, processes and of scientific inquiry by appealing to McTchange, discussed in Chapters 3 and 4, I have reason to hope it will serve me well yet.

5.4 Quantum gravity

Having shriven myself of any ambition to teach the physicist their business, I want to finish the chapter reflecting on the notable attempts by contemporary theoretical physicists to include McTchange in their views. Of particular note are Rafael Sorkin (1991; 2007) and Fay Dowker (2014). They use causal set theory, which assumes the passage of time (i.e. McTchange) as a heuristic, to make sense of time when quantum phenomena are combined with General Relativistic physics.

In this section, I will summarise some of the views that take the passage of time seriously and look at the philosophical critiques in the literature of them.

Causal set theory attempts to go from small basic causal sequences ('causets') to something like the smooth geometry that we have been discussing so far. We have some minimal events that stand in causal relations. These causal relations are reflexive, anti-symmetric and transitive. As a consequence they rule out causal loops and use causal relations to provide information about time ordering. But these causal sequences are *partially ordered*. Imagine some space-time atoms, the tiniest of regions of spacetime. Imagine that causal relations hold between these space-time regions.

We get causal relations between these regions that contain some ordering, but not a complete ordering. That is, A can cause C, and B can cause C, but there can be no fact of the matter about whether A is before B or B before A (see Figure 5.6). This is a consequence of the limit to the speed at which causal signals can travel, which we discussed above, and what makes it compatible with General Relativity.

In addition to these causal sets, with their partially ordered causal sequences, we have growth in the number of elements of the causal set. This growth involves adding more elements causally downstream of previously existing elements. This growth in the causal set, according to Rideout and Sorkin (2000) and Dowker (2014), *is* the passage of time. This gives a picture on which there is McTchange, insofar as the evolution of this causal set is an irreducible change in what there is. But this change is a very local one; it simply doesn't tell me about the activities of whatever is in absolute elsewhere.

On any particular way of imposing a metric on a causal set, we can get one ordering of events, and on another particular way of doing so, we can get another ordering. The causal set itself allows for 'gauge invariant' growth, such that, ignoring questions of which metric we are using, there is McTchange from one causal set to another larger one. It does so, however, without specifying which elements of the causal set exist at any time. It says that there is growth regardless of your perspective but does not specify what the elements of the causal set are in a perspective-independent way. This is simply not a feature of the physical theory.

Causal Set Theory has come under considerable criticism, including Wüthrich and Callender (2017:921) who describe the approach to the passage of time as 'radical and bizarre'. But Sorkin (2007:158) offers a defence of this feature:

> Our 'now' is (approximately) local and if we ask whether a distant event spacelike to us has or has not happened yet, this question lacks intuitive sense. But the 'opponents of becoming' seem not to content themselves with the experience of a 'situated observer'. They want to imagine themselves as a 'super observer', who would take in all of existence at a glance. The supposition of such an observer *would* lead to a distinguished 'slicing' of the causet, contradicting the principle that such a slicing lacks objective meaning ('covariance').

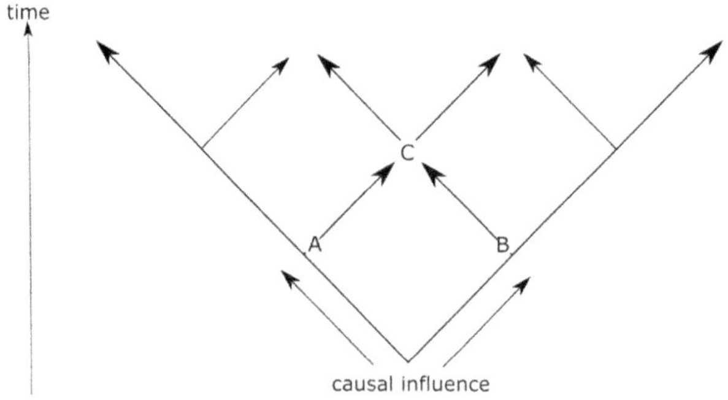

Figure 5.6 A causal set.

This is an issue we have discussed already.[21] If we had some way of giving sense to a distinguished foliation of the world, then we would have some sense to make about what was objectively at the same time as what (though, plausibly, we could not ascertain this in most cases). But we should be wary of the inference that in order to *make sense* of McTchange, we need to make sense of a superobserver, or God's eye view, from which we can describe McTchange. The crucial thing is that we can't make sense of a God's eye view of a world devoid of McTchange from which we can explain everything. If there is such a superobserver, it passeth all understanding.

When dealing with theoretical physicists, one needs to make sure that the claims they are making and the claims of philosophical interest line up properly. I argued above that the sense in which simultaneity is relative is divorced from the ordinary use I capture with contemporaneity. But when we consider causal set theorists, it is quite clear that it is McTchange that these physicists propose. Dowker (2014:22) says:

> What is missing within a [static] block universe view is the *occurrence* of the event: there is nothing in the block to correlate with 'the event happens'. The birth process of sequential growth models provides the missing concept, breathing life into the block's static view of the physical world.

Indeed, Dowker is careful to distinguish between the birth of a space-time atom and its existence. It is the coming into existence of space-time atoms that constitutes the passage of time on Dowker's view, not merely the existence of such atoms. It is clear that becoming has a processual aspect that would be lacking by the mere existence of spacetime.

Wüthrich and Callender (2017:920) raise a worry about the indeterminacy of events in a causet: 'The problem is that we are generally prohibited from saying exactly what elements exist at any stage of growth.' We can be sure about what's in the past light-cone, but not about the absolute elsewhere. But existence, they argue, cannot be vague. We have here two possible answers. One is that there is some fact about what exists and what doesn't, but it is one no empirical test can settle; there is a fact about causal relations that cannot be established because we have limited epistemic access to them. Alternatively, we can say that there is no matter of fact to be had, which is the way Wüthrich and Callender read Dowker. On this view, causal set theory makes the universe ontologically indeterminate, a consequence that Wüthrich and Callender find shocking. [22]

Note that asking the physical theory to settle the question of what exists at any time is unreasonable. Given the limitations we have discovered on what we can say in the absence of some objective standard for determining a strict temporal ordering on the universe, something has to give. And metaphysical indeterminacy is often appealed to in discussions of quantum mechanics, quantifying over *absolutely everything whatsoever*, and future contingents in the philosophy of time.[23] I don't see why prohibiting such

[21] §4.5.3.

[22] Commitment to ontic indeterminacy is not so uncommon in the philosophy of time, with Cameron (2015) defending a view with lashings of ontic indeterminacy.

[23] See Williams (2008) and Barnes for (2010) for a discussion.

questions is more radical and bizarre than denying that there is McTchange (and that there are processes, and moments ripe for intervention, and enduring subjects etc.).

There are a couple of concerns about causal set theory, sympathetic to it though I am, as it has been outlined here. One is that 'birth' is left rather under-explained. There are conceptual problems with identifying the present, as Dowker does, with the birth of some space-time atom, since the birth does not happen in stages. Causal set theory seems to run into Zeno's arrow paradox, which is normally stated roughly as follows:

> At any instant of time, the flying arrow 'occupies a space equal to itself'. That is, the arrow at an instant cannot be moving, for motion takes a period of time, and a temporal instant is conceived as a point, not itself having duration. It follows that the arrow is at rest at every instant, and so does not move. What goes for arrows goes for everything: nothing moves.[24]

So, an arrow's flight consists of a number of instants, during none of which the arrow flies. If the arrow doesn't fly during any of these instants, then it doesn't fly at all. Jonathan Lear (1981) suggests that an alternative translation of '*ta nun*' (usually interpreted as 'the instant') makes the paradox particularly pressing. If, rather than 'the instant', it is 'the present' we have the situation where the arrow can only fly when it is present, and it doesn't move when it is present because the present simply isn't long enough.[25] We have a series of present moments with no parts (i.e. they are atomic presents), *at which* a change is to take place, but changes cannot take place *at* atomic times. Even if we bring in some differentiation to define an instantaneous velocity, the instantaneous velocity seems to depend on the location of the arrow at other times, not vice versa.

In Dowker's case, we have space-time atoms that are born, but nothing happens during their birth since there isn't long enough for anything to happen (*ex hypothesi*; it is a space-time atom). But, the very idea of birth was brought in to contrast with the static existence they have after being born. And in this case differentiation will not save us. We can't come up with a mid-way gestational stage of a space-time atom; it's already as small as spacetime gets. Either the space-time atom has been born, or there is no space-time atom. There are no, so far as I'm aware, proto-space-time atoms in the theory. And even if there were, this would not help, for the question of their birth would arise. The identity between the present and the *birth* of a space-time atom makes the present a very hard thing to grasp. There is no need to make this move. One can make the present the limiting case of the past, one can view presentness as a property of processes (e.g. of successive space-time atoms being born), of activities, or of enduring things/stuffs rather than a property of space-time atoms.

A second problem is that Dowker seems to presuppose a perceptual access to the passage of time (i.e. the birth of space-time atoms) of the sort discredited in Chapter 2. Our awareness of our own persistence through time, as I argued in Chapter 3, seems

24 Sainsbury (1995:21), Cf. Hussey (1972), James (1911), Lee (1936), Owen (1958:199–222), Vlastos (1966:3–18).

25 I'll revisit this problem in §10.2.2.

more fertile ground than a claim that we can empirically detect the evolution of causal sets. We can certainly interpret the evolution of causal sets as requiring McTchange, but *evidence* for the dynamism will involve settling whether we interpret processes more generally as involving McTchange. This was the topic of Chapter 4. As Callender (2017:104) says: 'Here it seems we see the same old dance.'

5.4.1 Quickly rebuffing the quick argument for timelessness

The prospects for finding McTchange within quantum gravity are not as slim as sometimes depicted. But there are also prospects that things may be even worse than had been previously feared. Baron, Miller and Tallant (2022), following Wüthrich (2017), argue that many of the leading theories of quantum gravity have no role for McTchange, nor indeed time as we ordinarily think of it. They consider three versions of Quantum Gravity which, they claim, don't have anything that resembles time. Recall in §1.4 we discussed the A-series, the B-series and the C-Series. The versions of Quantum Gravity they discuss don't have anything, they claim, resembling an A-series, a B-series *or* a C-series. As they put it, 'a C-series is, at a minimum, an ordering of events in time that gives rise to a metric' (p.109). I disagree with them in taking a metric to be required to think we're dealing with a temporal dimension. The contrast between time and space doesn't seem to require this.

The three theories they discuss are Barbour's Machian approach to Canonical Quantum Gravity, Loop Quantum Gravity and Causal Set Theory (the very view that I just argued offers the best prospects for McTchange in Quantum Gravity!).

In the briefest and most superficial terms, there are a range of techniques to approach Quantum Gravity. Often, they involve starting with some established theory from either Quantum Field Theory or General Relativity, and then trying to adjust it to allow the other theory to be expressed within it.

Canonical Quantum Gravity uses techniques that have been previously successfully used in Quantum Field Theory to 'quantise' fields (i.e. give an understanding of them in terms of discrete units – 'quanta') and applies these techniques to the gravitational field. One of the key ways time normally gets into a physical theory of matter is in the 'dynamics' (i.e. the rules of how the system changes over time). But the canonical quantum gravity approach gives us a dynamics without McTchange, and indeed without any change whatsoever. As a substitute for change, it simply has variation of one 'degree of freedom' of a physical quantity with respect to another. A 'degree of freedom' is a parameter that defines the dimensions in which something can vary. In classical mechanics, these will include location and momentum in three physical dimensions. The classical picture has position and momentum evolving over time, such that position and momentum at one time give the position and momentum at all other times. But in the canonical quantum gravity approach, we don't get position or momentum changing over time, but any variation in one degree of freedom relative to another, where the difference between position in space and position in time has not been defined in absolute terms.

The Canonical Quantum Gravity approach gives us the Wheeler-DeWitt equation, which lacks a time parameter. As Lam and Esfeld (2013:287) explain:

> The Wheeler-DeWitt equation does not contain any explicit time parameter, unlike the Schrodinger equation for instance . . . this is one aspect of the 'problem of time', which finds its roots in the background independence and diffeomorphism invariance of classical [General Relativity] GR already. One way to deal with this 'problem of time' is to accept that it indicates time might not be a fundamental feature of the world.

Gryb and Thébault (2016:666) comment:

> On this view, it should be no great surprise that when the equations of a classically foliation-invariant theory are quantized, one arrives at a timeless quantum gravity formalism [the Wheeler-DeWitt equation], since, in essence, this facet is already implicit within the classical theory.

This is a concern we encountered even in the Newtonian picture, that assumptions built into the formulation of the theory come to have an impact on the 'findings' we get out of the theory. Gryb and Thébault (2016:667) continue:

> It is thus understandable that the radical morals with regard to change and time discussed above are often drawn. However, it is still true that within general relativity there exist fundamental temporal structures relating to ordering in time.

In their view, although a privileged foliation of spacetime into an evolution of three-dimensional spaces won't be found in Quantum Gravity, we still have a role for time:

> Things change and time, whatever it may be, certainly is at a minimum a means for describing this change. (p.684)

Loop Quantum Gravity has a similar approach to Canonical Quantum Gravity, in that the Wheeler-DeWitt equation is central to giving a timeless formulation of the theory. Whereas Canonical Quantum Gravity tries to 'quantise' gravity to get the Wheeler-DeWitt equation out, Loop Quantum Gravity starts with the Wheeler-DeWitt equation and tries to find an interpretation that works for General Relativity. Loop Quantum Gravity gives us a structure called a 'quantum spin foam', which is roughly a 'granular' four-dimensional spacetime structure. Given that it is an interpretation of the Wheeler-DeWitt equation, it is a 'timeless' dynamics. Nonetheless, as Le Bihan and Linneman (2019) point out in Loop Quantum Gravity there is a local split between 'quasi-space' and 'quasi-time'. Indeed, they conclude that

> all the candidates for a theory of QG include a difference between something quasi-temporal and something quasi-spatial. More precisely, they all feature some sort of split that maps onto – or could in principle map onto – the split between space and time present in [General Relativity]. (p.118)

Reading any of these discussions containing a defence of the role of time in some form within Quantum Gravity, they are supremely careful to stress that the space/time split we get isn't what we had in the Newtonian picture. And the physics doesn't contain anything corresponding to time's flow. Causal Set Theory gets treated as a threat to time by Baron, Miller and Tallant (2022), because, based as it is on causal structure, it doesn't give a metric for time, and we don't get an absolute ordering of events. Even Dowker, quoted above, admits that the process of space-time atoms coming into existence is missing from the equations. But on each of these theories, we do get a distinction between being in the past light-cone and the future light-cone, or absolute elsewhere. If you can persist through change, such that something is in your past light-cone that didn't used to be, we have some basis for thinking physics can allow McTchange.

In this chapter, the compatibility of McTchange and physics is all that is being defended. Although there are versions of Canonical Quantum Gravity that do away with change in any sense except variation along one dimension relative to another, this is not something that seems to be *forced* on us by the maths. And Gryb and Thébault argue that without a role for 'succession' our physics would 'simply fail to be descriptively adequate' (p.699–700). At the very worst, defenders of McTchange can insist we 'have very good reason to suppose that spacetime must be metaphysically emergent *even if we can't see how*' (Baron, Miller and Tallant 2022:153). Disputes in physics are not best settled by such insistences, however. It may be, as Baron, Miller and Tallant suggest, that we are on the cusp of a radical conceptual revision in light of new ways to understand Quantum Gravity. While McTchange isn't flatly ruled out by our best physics, how best to interpret Quantum Gravity is still very much up for grabs.

Where the likes of Callender search for the passage of time within physical theories, I do not. Given the history of taking the job of physical science to involve abstracting away from change, physicists can always exploit the tactic of 'running the process to infinity' (Dowker 2014:22), and treating their theory as describing a world of mere variation. When scientists presuppose that physics is devoid of McTchange, one should not expect them to discover it in their theories. But this chapter has shown that while relativistic physics might restrict our ambitions about what we can say (and what sort of God's eye view we can attain through physics), these theories do not show that the manifest image is mostly rubbish and do not undermine a commitment to McTchange.

Part II

The Arrow of Time

6

Time, entropy, memory and causation

The Moving Finger writes; and, having writ,
Moves on: nor all your Piety nor Wit
Shall lure it back to cancel half a Line,
Nor all your Tears wash out a Word of it.
—Omar Khayyam, trans. Edward Fitzgerald (1859:§71)

In Part I, I argued that we should accept that there is a difference between variation across space and change over time. Now, in Part II, I wish to move to a different difference – a difference within time – the difference between past and future.[1]

We might think of an investigation into the difference(s) between past and future as an investigation, by another name, into the differences between the earlier and later directions in time. After all, the past is that which is earlier than whatever we take to be now, and the future is later than whatever we take to be now. I have billed it as a difference between the past and future, partially out of a conviction that most differences between past and future will involve some story about why there is something special about that which is neither wholly past nor wholly future, i.e. the present. But our understanding of the present later in the book will come from understanding how it differs from the past and future, rather than starting with a primitive present and making the past and future simply distant from it in one direction or the other.

There are, of course, several differences between past and future. Horwich (1988:4–10) identifies ten,[2] and Callender (2017:12–14) appears to have identified even more, depending on how you individuate asymmetries.[3] In this chapter, we will be primarily concerned with explanations for the apparent differences between past and future as such. This will focus on the question of whether there *is* an objective difference between past and future as such. We will consider the alternatives that there is some other distinction (e.g. entropy or causation) that the distinction reduces to, and whether,

[1] Although, from this point on, I will help myself to McTchange in explaining phenomena, much of what I say will be open to those who think change is mere temporal variation. I will try to acknowledge when what I say is plausibly incompatible with such a reading.

[2] Viz. the movement of the now, the determinateness of truth, the entropic gradient, de facto irreversibility, knowledge, causation, explanation, counterfactual dependence, decision and value.

[3] E.g. asymmetries of mobility, and cosmological expansion. We might add to this list an asymmetry of reference: that we can refer to past things by singular reference in a way that we cannot refer to wholly future things. Cf. §7.4.1.

more radically, the distinction is an illusion produced by our language, our psychology or quirks of our local region of the Universe. In the following two chapters, we will be concerned with differences between past and future as they affect us. In Chapter 7, we will be concerned with the relation between intentional action and the openness of the future, and in Chapter 8, with the different emotional attitudes we have towards past and future.[4]

For all the asymmetries we observe between past and future, there are also some striking symmetries that have led some to question whether there really is a difference between past and future.[5] Many physical laws are standardly formulated in ways that are time-reversal invariant (i.e. work the same regardless of which temporal direction one is concerned with). This is often pointed out to be at odds with macroscopic phenomena we regularly observe, where we do not observe the temporal inverse of those phenomena. Soup left out on a cold day tends to gradually cool rather than gradually heat up; ripples tend to radiate out from a stone dropped in a pond, rather than radiating towards it and launching the stone from the pond; nail bombs tend to kill surrounding humans, but corpses tend not to shoot lumps of metal at a ball of fire and then start walking around. Given the laws of nature seem unconcerned with such matters, perhaps these macroscopic differences are to be explained away? Perhaps it is not some characteristic feature of the arrow of time that explains these striking observations, but some other, non-temporal, factor. Indeed, perhaps these phenomena are mere local anomalies.

We will proceed as follows. In §6.1 we will consider whether the differences between past and future are largely a result of arbitrary labelling convention. I will argue they are not; the asymmetries we encounter are more than mere products of our way of talking about them. In §6.2, we will consider whether the explanation for the differences between past and future is largely the result of a contingent epistemic asymmetry between past and future, such that we know much more about the past than we do the future. I will argue that the asymmetries between past and future are not simply about us; they are to be located in features of the world. In §6.3, we will consider whether the entropic gradient of the Universe offers a general explanation of the differences between past and future, and in §6.4, we will consider the prospects for causation – understood in terms of our capacity to intervene on the world – to offer such a general explanation for the differences between past and future. While the entropic gradient does explain the asymmetry between past and future, I will argue that the defence of this gradient itself presupposes a causal asymmetry. So while the entropic gradient is clearly part of an explanation of how the past differs to the future, it is part of a fuller explanation that appeals to effects being products of their causes. What's more, I don't think we can appeal to causal production without taking modality seriously. In §6.5, we will defend a particular, non-reductive way of understanding the relation between causation and McTchange.

[4] This way of dividing things up is somewhat arbitrary since these questions are related.

[5] E.g. Price (1996); Ismael (2016); Farr (2018); Kajimoto, Miller and Norton (2021).

6.1 A conventional labelling of axes

A radical sceptic about the difference between past and future could explain away the appearance of such a distinction as follows. We have a linguistic convention where we mark things earlier than now off from things later than now. There is no *real* difference between past and future, the sceptic would claim, and we are merely led to believe there is by contingent linguistic conventions.

In English, we read from left to right and from the top of the page to the bottom. But this is an arbitrary convention, and in other languages, the conventions can vary. Some languages, such as Arabic and Hebrew, are read from right to left. Some, such as Mandarin, are traditionally read in vertical columns from top to bottom, and the columns are read from left to right. The ancient method of *boustrophedon* involved lines alternating from left to right, and then right to left, so the text was continuous as it went down the page. So, we have an example of a way in which space seems to have a directedness that can seem natural and be widespread but does not reflect a genuine directedness in the world. That we treat the past as behind us and the future as ahead of us is just such a linguistic convention. Someone who is concerned with planning the future is described as 'forward-looking', but they are not literally looking forward.[6] There are some dead or dying metaphors about facing towards one's destination that seem to be at work here, but we could just as easily think of ourselves as being dragged backwards into the future in a perpetual bum's rush.

We, for example, distinguish between memory and premonition. A defender of human powers of premonition might object to a distinction between these. Premonition, they may argue, is just another form of memory that we have labelled differently so that we may better sneer at it. But as Roberto Casati and Giuliano Torrengo (2011:240) put it:

> Whatever the doubts one can have about one's memory, and whatever the confidence one may have one one's premonitions, it stands to reason that one's memory that P is prima facie evidence that P occurred, but one's premonition that P is not half as good evidence that P will occur.

When they say 'it stands to reason', I take it they are claiming that we are justified in treating premonition and memory differently. We have reason to treat memory as much more reliable evidence than we do premonition. The fact that, in general, we can remember the past and not the future needn't be simply that 'memory' refers to a way of knowing about things in the past rather than the future, but because memory is a widespread source of knowledge about the past and we have no analogous widespread source of knowledge about the future.

6 There is a convention in physics that later times are at the top of the page, in metaphysics that they are to the right, and in computer science that they are at the bottom of the page. Different cultures represent past and future using different spatial metaphors. Chinese speakers take the past to be 'up' and the future 'down'. Māori speakers take the past to be in front of us and the future behind. These differing conventions are highly historically contingent, as far as I can tell.

This is not to say that we lack knowledge about the future. We are often able to predict the future with considerable accuracy; precision engineering is based on this. But as David Albert (2000) points out, there is an important difference between memory and retrodiction. We can predict the future and retrodict the past.[7] Prediction and retrodiction work much the same way; we take how things are now, and some inference rules (e.g. our best approximations of the laws of nature) and extrapolate to some other state of the Universe either earlier or later than the present. Memory is interesting because it gives us access to information in a way that moves beyond retrodiction. Memory doesn't merely extrapolate from the present but relies on traces of past events.[8]

Memory is an instance of a broader phenomenon where traces of events enable us to work out what happened at a past time. We can use memories, or written records, or oral history, but we can also use records that are more reliable in the sense that we can more accurately infer from the state of the record to the conditions under which it was made. Memories are affected by the context in which we recall them, and oral histories often change with the retelling. Written records, even when written contemporaneously with the incidents they record, can be subject to biases from the author and be written for particular purposes to manipulate the narrative in which events are understood. Archaeology, palaeontology and geology provide us more directly with traces of the past, but even in these cases, they do not provide us with bits of the past. A flint spear-tip that I find in the ground is not a piece of the past; it is a present spear-tip that may turn out to be very old.[9] By understanding how present-spear-tips relate to past-spear-tips, we can come to know what the past was like. We either lack, or systematically fail to recognise, any analogue of traces of the future left on the present. This distinction between past and future, then, is not merely an arbitrary or contingent labelling convention.

6.2 An epistemic asymmetry

If the difference between past and future is not merely one of labelling, is it in fact simply a distinction of epistemic access? The fact that we have traces of the past and not of the future seems like it might be enough to set up a distinction between past and future, but does it explain the difference between past and future? I'm going to argue that the epistemic asymmetry is one of the ways we know that there is a difference between past and future, but it does not exhaust the distinction between past and future.

Though I don't know of anyone defending such a view, one can imagine a view on which the epistemic asymmetry, where we have epistemic access to traces of past

7 Here again we see a labelling distinction between past and future-directed activity.

8 Human memory, of course, does involve significant amounts of reconstruction and filling-in of gaps, but what separates memory from confabulation is that memory provides evidence beyond mere retrodiction.

9 I am assuming here that spearheads persist through being deposited and dug up. Cf. the discussion of persistence in Chapter 3.

events and not future events, is just a contingent feature of the world we inhabit. One needn't even claim that it is a feature of the Universe as a whole, merely of those parts of it we have encountered. It might be that other planets work differently than ours, or that there are creatures elsewhere on this planet with excellent premonitions but no memories. Alternatively, we could conceive of the Universe as having a fundamental symmetry, such that traces are nearer the edge than the centre. So, for us, memories are of events that take place earlier than our remembering them, and in the other half of the Universe, the subjects have premonitions of later events. But of course – those subjects being symmetrical to us – they think *they* have memories of the past and *we* have premonitions of the future.[10]

The view under consideration could be one where the epistemic asymmetry *constitutes* the past/future asymmetry, or it might be that it merely characterises and fixes reference to the past/future asymmetry. Either way, if we are in one of those situations, the past/future distinction is either local or subjective.

It doesn't seem to be the case, at least, that there are creatures on this planet known to science that rely on premonitions rather than memories. Perhaps science has missed them (because it presupposes that there aren't any), but having an epistemic asymmetry between past and future seems to be pretty universal for creatures with sophisticated cognitive abilities. What's more, the tendency for various disparate media to provide traces of the past suggests it's not merely a psychological quirk of humans that we experience such an epistemic asymmetry. Consider traces in cases from tree rings to wax cylinders, from hairs to rock formations, from levels of ^{14}C in organic material to Betamax, from footprints in the sand to (according to Sherlock Holmes) trifles.

Perhaps astronomical evidence, though scant compared to the evidence we have more locally, can suggest traces of conditions on other planets, in other solar systems or in other galaxies. We have some reason, then, to suggest that the epistemic asymmetry we notice between memory and premonition exists as a feature of our environment, but this is not enough to rule out the case considered above where the entire Universe is symmetrical in time.

What can we say about the view that traces will change from being primarily traces of the past to being traces of the future? One option is to rule this out as conceptually impossible; the past/future distinction just *is* the epistemic asymmetry created by traces. Something only *counts* as past if we know more about it than we do about the future. Linking the past/future asymmetry so tightly to the epistemic asymmetry in this way seems like a bad idea. It seems that, if you woke up in a strange place with no memory of how you got there, and no knowledge of what was going to happen (thus you were epistemically impoverished in both directions), other asymmetries, such as those involving intentional action, would still find you resolutely directed in time. Situations where we know much more about the future (because it is highly predictable) but not about the past (because it is all such a blur) are not unknown, and don't lead to confusion about which is the past and which is the future. Perhaps other sorts of creatures may not experience such an epistemic asymmetry, and we can

[10] This example is based on one given by Huw Price (1996), but in his example, the difference was entropy, not the presence of traces.

speculate about whether they would be creatures for whom there was any distinction between past and future to be drawn, but it is unclear, in the absence of such creatures to investigate, what conclusions we should draw from such speculations.

It seems reasonable there might be *some* link between the past/future distinction and the epistemic asymmetry, but there's no reason to think they are one and the same. As Baron and Miller (2015b) suggest, time may be whatever fills 'the time role' for us. Part of filling the time role is providing an epistemic asymmetry between past and future, but anything that explained sufficiently many other aspects of the time role would count as providing a past/future asymmetry even if, in some cases, there was no epistemic asymmetry.

In §1.3.1, I suggested that we might not be able to grasp what McTchange is without knowing what it feels like to McTchange. We might, equally, think that the past/future distinction is response-dependent. We can't grasp what the distinction is without some grasp of the role time plays for us. An epistemic asymmetry is an important part of the time role for us and may be very important to grasping the distinction between past and future. I'm going to argue that the asymmetry of control, where we can intervene on the future and not in the past, is at least as important as the epistemic asymmetry, though I don't know whether a creature that lacked agency could make sense of the past/future distinction on the epistemic asymmetry alone; such a creature is so alien to my experience that I don't know how to think about their experience in general.

The epistemic asymmetry seems important for explaining the past/future asymmetry – at least insofar as it provides one of our main ways to characterise it. But as we have shown, there are possibilities left open by all our evidence that the epistemic asymmetry reverses later in time. Given that most laws of nature, such as the laws of motion, are time-reversal invariant, there is no reason on the basis of our evidence to prefer an asymmetric Universe over a symmetric one, unless the epistemic asymmetry is based on some deeper feature of the universe.

6.3 Entropy

Let's go back to the problem with which we started; we see macroscopic phenomena that seem to give an arrow to time, and the laws of nature tend to be time-reversal invariant. One of the laws that isn't time-reversal invariant is the Second Law of Thermodynamics. Uhlenbeck and Ford (1963:16) characterise it as follows:

> in an irreversible or spontaneous change from one equilibrium state to another (as for example the equalization of temperature of two bodies A and B, when brought in contact) the entropy always increases.

Entropy is a measure of disorder. But that makes it sound rather mysterious, and perhaps morally pernicious.[11] So here's a more long-winded way of thinking about it.

[11] One can almost imagine an election campaign prophesying that voting for some candidate will lead to an irreversible increase in disorder. It will, but not necessarily in a politically relevant sense.

Imagine you have a large bucket of cutlery that you want to put in your cutlery drawer. If you randomly put each item from the bucket into one of the four partitions in the drawer, you might get a nicely ordered cutlery drawer, with the forks, the knives, the dessert spoons and the tea spoons in their own partitions. There are a limited number of ways in which four partitions get filled in this way.[12] But, with a 24-piece cutlery set, there are a huge number of ways of filling those four slots.[13] It is much more likely that you will have, in each of the partitions, a mixture of types of cutlery, simply because there are so many more ways of achieving this with a 24-piece cutlery set than having an unmixed set. What entropy measures is how likely specific arrangements are given all the possible ways of arranging things. A neatly sorted drawer has low entropy, because there are few ways of arranging things, and a drawer in which none of the partitions have just one kind of cutlery has high entropy because there are many many ways this can be achieved.[14]

This could give a natural link between McTchange and a past/future direction, for entropy increases in the direction in which things change. So, it could be that the increase in entropy over time – the 'entropic gradient' – is what plays the time role Baron and Miller described. It could be a discovery akin to the discovery that mean kinetic energy is what plays the temperature role; we were successfully using the concept of temperature before the discovery, but it turns out that the concept was getting at features that can be described in terms of kinetic energy. We have been making distinctions between past and future successfully, but it might be that the relevant features of the world can be explained by appeal to the entropic gradient.

The appeal to the entropic gradient comes in two flavours. Some, such as Albert (2000), Kutach (2011) and Loewer (2012), would describe themselves as accepting the reality of a temporal arrow and would base that arrow on the entropic gradient. Others think that the entropic gradient shows that there is no fundamentally *temporal* asymmetry between past and future; that the temporal symmetries we find are local products of a world that lacks a global temporal direction. Indeed this is the view of a number of philosophers, such as Price (1996), Kajimoto, Miller and Norton (2021) and Farr (2018), known as 'C-theory'.[15] The differences between the 'entropic B-theory' – those committed to an entropy-based arrow of time – and the C-theory can be difficult to gauge. Kutach (2011), for example, claims that the asymmetries we end up with are the most useful conceptualisations of phenomena with practical import.

The epistemic asymmetry, and other temporal asymmetries – explained by the entropic gradient – are of practical import for creatures like us, and so are a useful

12 4!=24.

13 24!= 620,448,401,733,239,439,360,000.

14 This example imagines that we're thinking about ways of arranging items of cutlery, which retain their identity through the sorting. But we could do this with quantities of stuff rather than things. I could have four flavours of squash which I'm randomly pouring into jugs. It's much less likely that I'm going to get four single-flavoured jugs at the end than it is that all four jugs will have a mixture of flavours.

15 Farr (2020) uses the term 'C-theory' for the view that there is no McTchange, and that there is no objective distinction between past and future, because the apparent distinction is a contingent (and perhaps local) consequence of the entropic gradient. This is based on McTaggart's (1908) 'C-series'; an ordered, but non-temporal sequence (cf. Ch.1.2). Price (1996), Kajimoto, Miller and Norton (2021) and Farr are all C-theorists in this sense.

way to think about the difference between past and future (though not the only one available). But there's not much ground between this view and the one in which there simply is no temporal asymmetry in physics. While there is a practical asymmetry at human scales in this part of the universe (which almost everyone agrees about), some scholars separate out the manifest image in which the practical asymmetry is relevant from the fundamental theory in which it is not.[16] With such a separation, the C-theorist can leave the practical asymmetry as being a product of the interests of creatures like us, rather than as a feature to be found in the world.

One worry about getting the entropic gradient out of the second law of thermodynamics, as Price (1996) and Albert (2000) point out, is that this law would predict that entropy increases in both directions from the present. Unless, that is, we already have a past-future distinction in place to give a direction to McTchange. For the law doesn't tell us which state to count as the initial one and which the one it will evolve to. One might think that the one we are currently observing should be the initial one, and one at another time as the one we change to. Unless we have some reason to pick out the start of the Universe as being a low-entropy state that evolves to a high-entropy state, the law wouldn't give us a past/future distinction to work with.

The second law of thermodynamics is compatible with a symmetric Universe that goes from a low-entropy state to a high-entropy peak, and then back to a low-entropy state (perhaps with an accompanying asymmetry of traces). This is because, as Price (1996) points out, there is no reason to demand that the direction in which we go from low to high entropy is constant over the duration of the Universe. Gold (1962) proposed a universe with low-entropy points both at the start and end, and the evidence that the universe *around here* has increasing entropy is not by itself evidence to rule out that we are at a 'Gold Universe'.

David Z. Albert (2000) takes the epistemic asymmetry we have discussed as data to be explained, and offers the entropic gradient, from a low-entropy starting point, to a high-entropy end of the Universe as an explanation for this phenomenon, and for the past/future distinction more generally. But to hypothesise a low-entropy starting point (and to rule out a Gold Universe) we need some inference that goes beyond the 'directly surveyable condition of the world', without appealing to the very records of the past that we are seeking to explain. As Albert (2000:119) says:

> The sorts of things that justify our *belief* in the past-hypothesis [that there was a low entropy start to the Universe] are (as before) precisely the sorts of things that typically justify our beliefs in *laws*.

We can appeal, that is, to it seeming like we can remember the past (that is part of the directly surveyable condition of the world), but such seemings hold both for genuine records of the past and for confabulations and other things that resemble genuine records but *aren't* records. We could explain these seemings by saying that they were put there by the evil forces of the mainstream media, by a deity wanting to convince

16 §4.5 was intended to put pressure on this separation of the manifest and the scientific; we'll come back to this in §6.5.

us the world was older than it is, or by appealing to some mechanism that, when in the appropriate 'ready' conditions, reliably turns processes into states from which the processes can be later inferred. Albert (2000:118) argues that the low-entropy start to the Universe gives us 'the mother (as it were) of all ready conditions'.

If we can allow that the Universe started with low entropy, we can then explain why the Universe was in such a state for the directly surveyable condition of the world to count as providing us, in addition to retrodiction and prediction, *records* of the past.[17] This gives us, based on the directly surveyable condition of the world, and some abductive reasoning, an explanation of why the entropic gradient, starting low and increasing forever, would generate an epistemic asymmetry between past and future, based on records of the past.

Records of the future that we haven't worked out how to use yet have not been ruled out by anything we've said. As Albert adds (2000:118–19):

> all this leaves quite open the possibility that there might turn out to *be* some as yet undiscovered future hypothesis . . . of which our experience may yet prove confirmatory, and whereby we might yet learn to record or even to remember the *future* as well.

On this view – and this view is shared by many B- and C-theorists – the distinction we take to divide the past from the future may be local, or contingent or both. If entropy is all that explains the epistemic asymmetry, we have no reason to rule out parts of the Universe where there is no past/future distinction to be had. In subsequent sections, I will argue that we have other explanations that are more plausible for the past/future distinction.

6.4 Intervention

There is very little dispute that there is an entropic gradient (at least around here), and that given such an entropic gradient, we should expect to see an epistemic asymmetry between past and future (at least most of the time). The challenge to the claim that the entropic gradient *is* the arrow of time, or explains why there appears to be one, is not a rival view on which there is *no* entropic gradient. Rather, it is a view on which the entropic gradient is *itself* explained by something else – causation. This, to avoid unnecessary suspense, is the view I will defend. But we need to build up to it; in this section, I will argue that one of the phenomena to be explained is an asymmetry of intervention. I will also present the case that this asymmetry is in turn to be explained in terms of the entropic gradient, so I can rebut it in §6.5. Then, in §6.6 I will respond to

17 It should be noted that records merely make inferences about the past *possible*. It may be very difficult to work out what those inferences should be, or whether this is a case where the record is reliable. The inference that ancient writers wrote about a character named 'Augustus Caesar' may be much safer than inferences that there was a man called Augustus Caesar about whom ancient writers wrote, for example.

objections to thinking it is causation that explains both the asymmetry of intervention and the entropic gradient.

The asymmetry of intervention is described by the claim that we can intervene on things in the future direction and not the past direction. It is an asymmetry of our capacity to control things.[18] Woodward (2003:11) describes causal explanation as 'an extension of our interest in manipulation and control from cases in which manipulation is possible to cases in which it is not'. It is not that our understanding of causation is *restricted* to those cases where we can manipulate things, but that our understanding of causation (and I would argue, modality more generally) requires a grasp of the capacity to manipulate and control in at least some cases. If we know that by wiggling something, we can make something else wiggle, we can understand, more broadly, how things that we cannot control can wiggle as a result of some previous wiggling. We can understand what it is for other people to cause a car to crash because we understand how to crash our first tricycle. But also we understand what it is for the natural formation of thick magma to cause a volcano to erupt from our experience of bad plumbing, *even though no agent was responsible for the formation of the thick magma*.[19] In the terms of §1.3.1, causation is a response-dependent property. Unless we had the experience of intervening in the world, we couldn't have the understanding of causation (or modality more generally) that we do.

If we have experience of intervening on things, we equally have experience of being unable to intervene on things. This can range from locking oneself out of one's house to realising that the email one has just sent went to the person it was about and not the intended recipient; from attempting to stop the incoming tide by means of royal decree to attempting to protect people from crime by means of sympathetic statements on social media. Just as we extend our interest in manipulation to cover cases where we cannot intervene in order to understand causes, we extend our grasp of impotence in those domains in which we are buffeted about by events in order to understand that there are constraints that apply more generally.

That we have a grasp of modality from our experience of agency doesn't yet show us whether there is an objective causal asymmetry in nature (indeed, one that gives rise to the past/future asymmetry) or whether we are merely projecting a modal dimension onto the world. Perhaps in reality, the world is, as Rudge claims in Alan Bennett's *The History Boys*, 'just one fucking thing after another'. The asymmetry of intervention is as important as the asymmetry of records to the time role. But perhaps they are *both* to be explained by the entropic gradient?

Albert and others who want to appeal to an entropic gradient are aware of the asymmetry of intervention. Albert says that the entropic gradient means that there are more routes to influence over the future than over the past. In an alternative phrasing of the same idea, he says that there are many more 'causal handles' on the future than

[18] I will use 'asymmetry of intervention' and 'asymmetry of control' interchangeably.

[19] Fernandes (2023) claims that accounts that understand causation in terms of agency have difficulty explaining causes in cases where we lack agency. This will be true of accounts that attempt to reduce causal claims to claims about agency, but Woodward (2003) for example makes a claim that we can understand cases not involving agency by analogy with cases involving agency, because we can grasp counterfactual claims about hypothetical interventions.

on the past.[20] Even ardent C-theorists acknowledge at least an apparent asymmetry of control between past and future and include it in their account. The fact that we can intervene on the future, and not, in general, on the past, is another asymmetry between past and future that has a central place in fixing the time role for us, alongside the asymmetry of records.

There is widespread acceptance, then, that both the entropic gradient and the asymmetry of intervention will typically line up with the direction of time. It should be noted, however, that C-theorists are open to the possibility that we have causal handles over the past. The asymmetry of intervention may be contingent and local.[21] Fernandes (2017) claims the asymmetry of intervention is to be explained in terms of another epistemic asymmetry: that deliberations come before decisions. The reason that deliberations come before decisions is that the knowledge we have of the past is only useful before we act upon it. That deliberation takes place before action is, according to Fernandes, in turn to be explained by the asymmetry of records, which itself is explained along the same lines as Albert (2000). It is based on our knowledge (including memories) that we form reasons for action and come to act. So, insofar as the asymmetry of records might be contingent and local, the asymmetry of deliberation may likewise be contingent and local. There may be cases where deliberations come after decisions.

There are some cases where we might even think that we have causal handles that allow us to intervene on the past, not merely theoretically, but in regular scientific practice. It can certainly be the case that there can be true counterfactuals, such that if this thing hadn't happened today, that thing wouldn't have happened yesterday. What makes such 'backtracking' counterfactuals distinct from *causal* counterfactuals? Huw Price (1996) argues that the difference is based on what we hold fixed and what we take to be an option.

Consider an example from Dummett (1964) of someone getting up three minutes before their alarm clock goes off every morning. After a sufficient number of mornings that this occurs, and a small number on which they sleep through and their alarm doesn't go off, they may begin to maintain that such a counterfactual dependence holds. We may appeal, of course, to what *would have happened* if they had not woken up – or some similar counterfactual conditional. In response, they might point out that we cannot observe non-actual cases. That we are allowed to appeal to counterfactuals in order to explain causes presupposes we can identify the genuine *causal* counterfactual conditionals. If we insist the genuine, causal counterfactuals are non-backtracking (i.e. don't allow the antecedent to be later in time than the consequent) we are simply begging the question against the person who insists that they cause their alarm to go off three minutes after waking up. We can take our having woken up to be held fixed, and then the alarm clock going off to be something that could, potentially, be intervened upon. The distinction between *fixtures* and *options* tends to go with both the asymmetry of records, and with the distinction between past and future. But if the

20 Fernandes (2023) describes these views as explaining the temporal asymmetry of causation in terms of statistical mechanics, and provides a helpful overview of these approaches.

21 Cf. §4.5.1.

asymmetry of records is, as C-theorists accept, potentially a contingent and local one, is the asymmetry between fixtures and options able to provide us with a less contingent and local distinction between past and future? Price (1996:169) argues that we have a choice to make here. The choice is not about what to hold fixed:

> So long as counterfactuals are to maintain their association with deliberation, in other words, our choice of 'hold the past fixed' convention is governed by the contingent but nevertheless unchangeable fact of our orientation as agents we have no choice in the matter.

We find ourselves with a temporal orientation, based on a (region of the) world in which the asymmetry of records, the fixtures/options asymmetry and the past/future asymmetry all line up. But Price says that *if* these three asymmetries failed to line up, we could take the fixtures/options asymmetry to vary with the asymmetry of records, rather than the past/future asymmetry:

> I think that if we found ourselves in a world in which the notions of the past and the epistemologically accessible came apart in this way – a world in which it was therefore important to resolve the ambiguity of current counterfactual usage – a resolution in favour of holding fixed merely what is accessible would be the more satisfactory. (p.179)

Price concludes that sometimes it might be reasonable to hold that there are back-tracking causal counterfactuals, where we hold fixed our intervention and find out how the past depends on it. This, he argues, can help us explain certain quantum phenomena.[22]

If Price, Fernandes and others who take roughly this line are right, then causal asymmetries are to be explained in terms of asymmetries of fixtures and options, which, in turn, are epistemic, and are to be explained by the entropic gradient in roughly the manner Albert (2000) defends. If the past contains only fixtures but no options, then we cannot intervene on the past; but if there are some fixtures in the future and some options in the past, we can intervene on the past. It is, for them, a contingent empirical matter based on facts about the entropic gradient from a low-entropy starting point with increasing entropy over time. The entropic gradient explains both why we can't (normally) remember the future and why we can't (normally) intervene on the past.

We have seen that *both* the asymmetry of records and the asymmetry of control have a role to play in explaining the (apparent) directedness of time. But we are left with the question of whether something causal, based around intervention, or something contingent (and perhaps local), based around the entropic gradient, is explanatorily prior in accounting for the asymmetries between past and future. In the next section, I will argue that it is something causal/modal.

[22] He has in mind the 'delayed two-slit experiment'. I don't propose to go into the detail of the quantum mechanics here. Suffice to say that Price's interpretation of these phenomena is by no means an established consensus amongst either physicists or philosophers of physics.

6.5 Realism about causation

We saw earlier that B-theorists like Albert (2000) and Kutach (2011) and C-theorists like Price (1996) and Fernandes (2017) think that creatures like us at human scales objectively find an asymmetry of intervention that is closely related to why time can be said to pass *from our point of view*. They will agree that causation is response-dependent, but conclude that the entropic gradient gives rise to our belief in an asymmetry of intervention. I, by contrast, think that the asymmetry of control is explanatorily prior to the entropic gradient. This is because an explanation of the entropic gradient *presupposes* a causal asymmetry in the world. This causal asymmetry is, as we all agree, a feature of our agential point of view on the world, but the claim I need to press is that our point of view is not to be contrasted with a more scientifically respectable account from which causation drops out.

6.5.1 Presupposing a causal asymmetry

The considerations marshalled in favour of a role for McTchange in science in §4.5 can be similarly marshalled in favour of causation and modality. Our ability to regulate inquiry involves a presupposition that we can intervene on the world and affect the outcomes of experiments as a result. An asymmetry of intervention seems to be built into the very practice of science.

Consider Susan Stebbing's (1937:195) objection to basing the arrow of time on the second law of thermodynamics:

> in order to discover the [second] law [of thermodynamics] scientists had to perform experiments. These experiments required that apparatus should be used by the experimenters, i.e. human beings; that observations should be made and recorded; that after these records had been interpreted the scientists had to formulate the law.

Stebbing was arguing that time is a prerequisite to scientific inquiry quite generally. But in arguing that, she appeals to the asymmetry of intervention; we set up experiments and then record the results because experiments produce results after they have been set up and not before. This is more or less the line I want to take; we rely on interventions where causes produce later effects to make sense of scientific claims. Stebbing is far from alone here. More recently, Weaver (2020) argues that our best interpretation of the General Theory of Relativity requires us to accept that the world fundamentally has a causal structure.

I want to focus, however, on a particular claim: the claim that the past hypothesis explains the asymmetry of records. The argument of the previous section was that the past hypothesis is 'the mother of all ready conditions'; hypothesising a low-entropy starting point to the universe would explain an asymmetry of records but does not presuppose anything beyond the directly surveyable condition of the world. I want to push back on this claim; surveying the world presupposes the passage of time and an asymmetry of control.

Think about it like this: we need some method of getting from a single moment of the world to the idea that there are other moments, and to the idea that information about how this moment is any kind of guide to how any of those other moments are. The past hypothesis already presupposes that there is some fact about the amount of entropy at some other moment in time, such that we can treat it as a ready-condition for this moment. Our experiences as of persisting through time, as of intervening on the world to create records (and as of remembering things) already need to be taken seriously to think that there is a world whose condition we can directly survey. Surveying the world takes time, and many of the things we find in the world – from mice to mountains, from meteors to mesons – only make sense as being things that last for some period of time. The very idea of the directly surveyable condition of the world requires some duration in which to observe it, and some temporal connection through which things persist.

In order to get the resources for Albert (2000)'s explanation of the past hypothesis, we need to think that the directly surveyable condition of the world points beyond itself to non-present times. J. L. Mackie (1980:218) argues that persistence can provide the modal element here. We can get to what's happening at other times by a presumption that changes need to be explained in terms of interventions:

> it may well seem to be expectable that it [a moving particle] should go on doing as nearly as possible the same thing . . . if nothing intervenes to make it do anything else.

As we saw in Chapter 4, advocates of processes want to reverse the explanatory ground: processes go on changing unless something intervenes to make them stable. But the point about persistence holds. We need some idea of what it is to be the same thing or process over time, and what it is to be an intervention, just to get from the directly surveyable condition of the world to some reasonable speculations about what it might be like. The idea that there is an entropic gradient is a particularly sophisticated explanation of what other times might be like. But epistemically speaking, we could not arrive at such a speculation starting from the directly surveyable condition of the world, unless we already had some abductive leap to a role for persistence and for intervention, both of which are modal concepts. Why conclude that the directly surveyable condition of the world arose as the result of a chain of events stemming from a low-entropy past rather than as a momentary blip? Because there's some modal/causal connexion between moments in time that makes chains of events much more plausible!

The appeal to causation is a particular variety of appeal to modality. There are debates in the philosophy of causation over what *the* causal relation is. I want to avoid that debate as much as possible.[23] What matters for me is whether there is any kind of *constraining* relationship between cause and effect.

[23] Causal *explanation*, it seems to me, is always relative to some explanatory project. In some broad sense, anything in the past light-cone of an event is causally related to the event, but explanations are more interesting if we can make some features of the past light-cone more salient than others.

The worry is that *built into* the phenomena-to-be-interpreted (i.e. an asymmetry of the directly surveyable condition of the world) are features that could explain the past/future distinction. That something counts as a trace *at all* suggests, at the very least, a counterfactual claim that had the past differed in some relevant respect, there would be a difference in present traces.

In the attempt to explain the asymmetry of the directly surveyable condition of the world through the entropic gradient, we are appealing to the second *law* of thermodynamics, which looks, again, to bring modality in. Not only that, but the phrase 'irreversible changes' (from the above statement of the Second Law of Thermodynamics) also appeals to modality. If something is 'irreversable' we have the modal claim that it *cannot* be reversed. We are appealing to something that doesn't merely describe how things are, but how they *must* be.[24] In setting up the entropic gradient as an explanation of the difference between past and future, we already seem to be appealing to a modal difference between past and future, where the future depends on the past, but the past does not (in the same respect) depend on the future.[25]

One might ask, as Beebee (2000) does, whether laws govern the world or merely describe its regularities. Perhaps we can account for the apparent causal connexion without any constraining of one time by another. Perhaps, as Beebee suggests, the laws merely summarise, but do not constrain how the universe evolves from one moment to the next. If we can deflate the modality away from causation and intervention, the appeal to our presupposing modality doesn't work as an argument for the asymmetry of control explaining the entropic gradient, rather than the other way around. In the next subsection, we'll examine the case for thinking that modality is not to be deflated in this way.

6.5.2 Detecting causation

I agree with Price about the analogy between causation and secondary qualities. I disagree about what we should conclude from this. Like Weaver (2020), I want to claim an irreducible role for causation in explaining the asymmetry arrow of time. Price (1996:169–70) thinks that once we realise how bound-up the difference between past and future is with our perspective on the world, and our agency within it, we should attempt to step outside this perspective for a more neutral description:

> I would like to compare this to the question as to whether it is an objective matter that lemons are sour and not sweet. Once we understand the rudiments of the biological basis of our sense of taste, it is easy to see that we might well have been constructed so that we found lemons sweet instead of sour. Moreover, it seems plausible to say that given differently equipped communities of this kind, there isn't an objective sense in which one is right and the other wrong. Does this mean

[24] As Beebee (2000) argues, one can have a purely descriptive conception of laws, rather than a 'governing' conception involving irreducible modality. I shall argue against this approach in §6.5.2.

[25] As an aside, the concept to which Albert often appeals, of 'phase space' where we take the positions, and instantaneous velocities of all the particles at a given moment, also appeals to modality because instantaneous velocity involves modality. There's a lot of it about.

> that it is not really true that lemons are sour? It seems wrong to say that lemons are *really* tasteless, rather than sour. The best option seems to be to say that in this form the question doesn't really make sense from the neutral perspective we adopt when we consider the possibility of differently equipped tasters.

This all seems true. That such questions don't make sense from a neutral perspective could equally count against the idea of such neutrality, rather than against the existence of sweet things or sour things, or of causes and effects. One could, as Price (1996) does, opt to find a neutral language which doesn't take up any particular perspective, from which to survey the perspectival claims, or one could reject the very idea of a view from no perspective. §4.5.3 argues that science is committed to taking up a changing perspective, and so science cannot be the source of a perspective-free description of the world; at least not without severely limiting the scope of science.

In more recent work, Price (2011) eschews the idea of such a neutral perspective, opting instead for *quietism*. Rather than contrasting a causal and temporal world with a more objective scientifically respectable one, Price (2011:249) says:

> quietism about a particular vocabulary amounts to a rejection of that vocabulary, for the purposes of philosophical theory. This may or may not involve a rejection of that vocabulary for other purposes.

Previously, Price viewed claims about causation and the passage of time as being fine for everyday use, but not the terms we use in our best theory. As a quietist, he will instead step away from the idea of a 'best theory'. 'Best for what?' he will ask. It is not, we will be reassured, that one theory is *true* and another *false*, but that one is useful for describing the manifest image, and another for describing the scientific image. The idea of a single 'best theory' that *gets things right* is precisely what the quietist must reject.

Price's view is not too far from my own. He and I will alike reject a God's-eye-view. The difference is that I claim that we don't need to think of the scientific image and the manifest image as being in tension with one another; they are useful for different purposes. I think the scientific image arises out of the manifest image as useful for doing, much more precisely, the kind of general sense-making we do with common sense. I agree that we must consider a best theory as best for some practical goal. I just think there's a coherent goal of *making sense of things*, of which metaphysics provides the most general attempt.[26] This sense-making goal is practical in that it is shaped by the kind of interests creatures like us have. But it is sufficiently general to count as metaphysics because change, causation, persistence and intervention are presuppositions of almost any more specific practical project we could have.[27]

If we lived in a world where there were different communities constructed in epistemically very different ways, it would be very difficult to talk generally about the sweetness of things or the sourness of things. If I say Christmas happens in winter, and

26 Cf. A. W. Moore (2012).
27 I elaborate this view of metaphysics in Forbes (2023b).

Price says it happens in summer, we can both be right because seasonal variation is local to the northern and southern hemispheres (and the celebration of Christmas is local to different cultures). When it comes to tastes, we can both refine our palates to become better at discriminating amongst flavours and experience variation, over time and across people, in the physiological impact that flavours and smells have on people. As it is, we can say things like 'that lemon is sour (because of the citric acid content) but will taste sweet (because of the miracle berry I fed you)'. We are happy to talk about the taste as a property of the lemon you may, for contingent reasons, be unable to detect. It is not merely an objective fact that the lemon tastes sweet to you, but also that this lemon is sour even if you can't taste that.

Tastes are quite widely accepted to vary amongst people, however. Sights and sounds are perhaps more obviously features of the world because we can be some distance away from the sources of the sight/sound to experience them (provided there is some medium through which the signals can travel without too much attenuation). We need not treat all response-dependent properties as equal. Colours may be treated differently to tastes, which in turn may be treated differently to comic properties, or moral properties if we accept that all such properties are response-dependent. We must distinguish between those cases where we are *projecting* properties onto the world and those where we are *detecting* properties in the world.[28]

Even that distinction between detection and projection is a little too apt to look like a sharp dichotomy, where on the one hand the world contributes nothing to our projections, and on the other hand, the detection just reveals how the world is, with the detective merely assembling the evidence and not adding anything to its content. But, of course one can accept that *everything* humans think, entertain or express is shaped in part by the cognitive and affective apparatus with which they do so, without thereby claiming that everything is a projection onto an unknowable world. Equally, one can accept that *any* way of narrating the world – no matter how distortedly – requires enough traction with features of the world for the narration to be coherent, without thereby denying that some accounts of the phenomena are works of fiction with barely a milligramme of truth to them.[29] The distinction between detection and projection can be useful without treating it as a sharp dichotomy.

I wish to argue our experience of agency provides reason to think we *detect* modality in the world and that a modal asymmetry explains the difference between past and future. Were we to discover a community that worked with a very different modal asymmetry from us, our grasp of modality would potentially be very different.[30] But given the robustness with which we can make use of modality in explaining the world, we should accept that modality is response-dependent but nonetheless detected, rather than projected.

[28] Note that this response-dependent sense of 'detecting', from Wright (1992), is very different from the sense of 'detecting' Prosser had in mind in §2.1.1.

[29] There are 64.79891 milligrammes per grain.

[30] In Chapters 7 and 8, we will consider our practical engagement with modality in more detail, considering such a community in particular in §§7.1.2 and 8.3.1.

6.5.3 Wright's marks of realism

Crispin Wright (1992) considers the problem of distinguishing detectivist from projectivist readings of response-dependent properties. He provides four marks of realism (and Deborah C. Smith (2011) offers a fifth) that we can use to distinguish between detection and projection of properties. So, even if we accept that questions of the reality of some properties simply don't make sense from a neutral perspective (as Price and I agree), we can still distinguish between the claims that these properties are projected onto the world and detected within it. I will explain Wright's four marks of realism (and Smith's fifth mark) and then show how each bears on the case of modal properties.

6.5.3.1 Evidence-transcendent truth

The first mark of realism is *evidence-transcendent truth.* The thought here is that if one is to be realist about a domain of discourse (as the detectivist is), one ought to accept that the reality of the phenomenon outruns our capacity to know of it. To take a simple case, we might think that we can know no more about a fictional character than there is presented or implied by the fiction. We know Hamlet wore a black shirt, but do not know the colour of his underwear. In the case of Hamlet, this is not a failure of knowledge, because there is nothing more to Hamlet than the fiction provides.[31]

Henrik, prince consort to Margrethe II of Denmark, by contrast to Hamlet, seems to outrun our capacity to know of him. If we take all the records of his life, there would still be lots of gaps in our knowledge. Indeed, if we added in records of him that have been lost (e.g. his own recollections of his life), there would *still* be lots of gaps. Perhaps if we were to include all *possible* observations of him throughout his life, then the evidence would no longer outrun truth.[32] But this only seems plausible if we read 'possible' to mean something broader than physically possible or humanly possible. For the act of observing from one angle might obscure observations from another; recording one event might overwrite the record of another. To use Wright's language, for the realist (and *a fortiori* the detectivist), there is a difference between truth and 'superassertability'. Superassertability is a property a statement has if it can be justified on the basis of some information, and would continue to be justified given all ways of improving on that information.

Superassertibility is a useful idea because it gives a way of characterising the aim of inquiry: to arrive at a position where further inquiry would not change what one is justified in asserting. If one is projecting properties onto the world, there is no more to those properties than one puts in, whereas if one detects them in the world, there may be more to them than one has yet discovered. But note that this very distinction seems to require a modal element, where the difference is understood in terms of what *would* happen or what *possible* evidence there is.

[31] Hamlet is, of course, a character in Shakespeare (1604/5). On this view, one can have the consequence that Hamlet wore some underwear and that it had a colour, but there is no colour such that Hamlet's underwear was of that colour.

[32] We might call the view that, even once all possible observations are included in our evidence, the truth outruns evidence 'hyperrealism'.

We at least often assume that there are modal facts that outrun our capacity to know about them. Appeals to counterfactual dependence, to laws of nature, to causes, to contingency and necessity have been cropping up throughout our attempts to explain the asymmetries of records and of agency. Huw Price (1996:159) is clear that appeals to such modal distinctions will be explained away later. He doesn't accept that he is committed to modality when the theory is put in the canonical notation. But we do have reasons to care, as Nolan (2013) argues, about counterfactuals for their role in explaining all sorts of facts about the past (and how it could have gone) that outrun our capacity, even with a considerable degree of idealisation, to know which counterfactual claims are the true ones. The role of modality may be explained away in the end (I will discuss this strategy in §6.5.4), but at least at a common-sense level, we have a motivation to think that there is evidence-transcendent truth for modal claims.

6.5.3.2 Explanatory ground

Wright's second mark of realism is explanatory ground in so-called 'Euthyphro' cases. In those cases where truth and superassertability are co-extensive, we might look at dependence relations between the superassertability of some claim and the fact that the claim is true. This references Plato's Euthyphro, in which Socrates considers whether the gods love pious acts *because* they are pious acts, or acts are pious *because* they are the sort of acts beloved by the gods. The gods, being omniscient and omnibenevolent, know what the pious acts are, and love all and only those acts. So the reality of the piety of acts can't outrun *their* evidence. But we can still distinguish between the claims that the acts provide the *explanatory ground* for the gods' love of them or the gods' favouring of some acts is *what makes them pious*. The detectivist holds that the world provides the explanatory ground for the response to it, and the projectivist holds that the response to the world is what makes the world have the properties it does.

We can vary what claims we put on the responders here. In Plato's case, we seem to be assuming infallibility, but we needn't. We have a few elements that can vary. The responders can be idealised to a greater or lesser extent; we can go from infallible knowers to merely competent users of the relevant concepts. The conditions in which the responder responds to the world can be varied; we can allow that any conditions count, to merely optimal conditions (again, this can be idealised more or less), to merely statistically typical conditions (though typical for whom?). The detectivist will tend to make the conditions under which superassertability is coextensive with truth more restricted the more fallible the responder is. So we might say that something is red iff a sighted, non-colorblind observer, who grasps the concept of redness, would sincerely judge that it looks red under typical daytime lighting conditions, where the thing to be judged is not occluded and is in line-of-sight of the observer. This allows us to explain a lack of consensus by failure of one or more of these conditions (e.g. it was dark, the observer was looking at something else, and had no eyes). To avoid making trivial claims, and to get an operationally serviceable distinction, we should be careful in how we set up the relevant conditions. Wright (1992:114) recommends a proposal to us:

> it is judgements of those who are *actually* statistically standard observers, in what are *actually* statistically standard conditions of observation, which count. Were we all colour blind, our judgements would, *ex hypothesi*, no longer be in line with those which, constituted as we actually are, we are disposed to make.

So, the fact that if there *were* other creatures, arranged differently to us, they wouldn't judge that effects come before their causes isn't relevant to whether effects in fact come before their causes. The contingency of our having the capacity to respond the way we do under the circumstances we do does not make our so judging merely one perspective amongst others. For what the terms 'cause', 'effect', 'past' and 'future' *mean* is based on the responses of *actual* observers in *actually* standard conditions.

If we were to make the case that the past/future distinction is merely a feature of our perspective, because the asymmetries of agency and of records are in turn merely features of our perspective, then a discovery that there is actual radical difference across communities would be compelling evidence for this. But that's not what we find. Callender (2017:18) makes the point emphatically:

> *The very evidence produced in favour of cultural variation turns out to be some of our best evidence that the core features* [of our common-sense grasp of time] *don't vary*. (emphasis original).

When we determine whose responses count, we need to be very cautious; it is easy to set things up so that we end up saying something preposterous or failing to say anything interesting at all. We presuppose competence with certain concepts to ask questions about the direction of dependence of explanatory ground. To possess these concepts doesn't necessarily require being anglophone (though I write about them in English), but it does require having the concept in whatever language. The judgements of those who lack the concept (in whatever language) don't fail to count because of any assumed inferiority of those who lack the concept. Indeed, there may well be concepts it is better not to grasp. The point is that those who are not competent users of a concept are simply not in a position to form competent judgements about whether that concept applies. It would be a mistake to conclude that, if a language community lacks a word for something, the community thereby lacks the concept, or – more extremely – does not experience relevant phenomena at all. For we often develop concepts or words in order to be able to express thoughts about some phenomena that we previously could not articulate.

As Callender observes, the evidence from the variation in how languages deal with matters of time, tense, aspect and the subjunctive mood in fact suggests that the asymmetry of records and of agency lines up with a past/future distinction very robustly across different cultures. This is, Callender (2017:18–19) says, 'to a large extent the natural by-product of solving certain problems common to our species'. Those who are actually statistically standard observers, then, need not be statistically standard well-educated observers from wealthy Western industrialised democratic countries. The 'we', here, when the asymmetry of agency and the asymmetry of records are discussed, need not even be specific to our species or to biological creatures. Any creature who is in a position to grasp the relevant concepts, and so to *count* as an interlocutor, is part of the relevant 'we'.[33][34]

6.5.3.3 *Cognitive command*

The third of Wright's marks of realism is *cognitive command*. The central thought here is that we explain the platitudinous thought that truth corresponds with the facts by appealing to the idea that the facts only permit certain ways of representing them. If two investigators disagree about some discourse, and that discourse exhibits cognitive command, then there are a limited number of explanations for that disagreement, viz.

a) different input;
b) unsuitable conditions;
c) malfunction/pathology/mistake;[35]
d) vagueness in the statement, in the application of standards, or variation in how confident one needs to be to form an opinion.

The projectivist can allow for faultless disagreement that doesn't have one of these explanations, since they allow that two different groups might project differently onto the world. Price, Callender, Fernandes and Ismael can all accept that, *relative to our perspective*, modal talk exhibits cognitive command. And they can accept that given the circumstances we are in, it is impossible to escape from living in this perspective. The point is that they don't accept that we can move from objective claims about what is true from our perspective to claims that our perspective corresponds to the facts. This is because they take our perspective to be *in competition* with a scientific perspective. But I have argued it is not. I argued in Chapter 4 against the image of science as abstracting from any perspective at all, and if science provides a view from anywhere, rather than a view from nowhere, then talk about past and future and about cause and effect both exhibit cognitive command. This is a mark of realism, and a point in favour of modality being detected in the world, rather than projected onto the world.

6.5.3.4 *Width of cosmological role*

Wright's fourth mark of realism is width of cosmological role. If the discourse appeals to some properties in lots of quite different explanations of the world, that is a mark in

33 We will consider a hypothetical case of creatures who don't experience the asymmetry of agency in the next chapter.

34 As an aside about the place of metaphysics in the humanities more broadly, it strikes me that metaphysicians could benefit from working with anthropologists and historians of other cultures much more than they do. One speculation as to why this doesn't happen is that anthropologists are often resistant to the use of Western epistemic concepts, such as justification (or indeed superassertibility), in engaging with other cultures. I broadly endorse Veli Mitova (2021) on this topic, when she says that we need some epistemic concepts that are used across cultures to allow for the possibility of disagreement between them. Another potential explanation is that metaphysicians are resistant to leaving their armchairs. The armchair method works best after a long day of talking to interesting and well-informed people.

35 'Malfunction' is a word to describe information processing systems, and 'pathology' seems better to describe organisms. 'Mistake' covers malfunctions that aren't dispositional defects of the organism.

favour of realism about those properties. If the only thing an appeal to some properties was to explain why sentences about properties occur in our discourse, that would be some reason to think that there were not *really* such properties.

But that doesn't seem right for modal properties, nor for causation. Modality/causation is *really useful* all over the place. Some might demur here. Russell (1912a:1) famously said that advanced sciences didn't talk of causes, and that philosophers only continue to talk of them because 'like the monarchy . . . it is erroneously supposed to do no harm'. This led to a movement, led by Price and Corry (2006) known as Russellian Republicanism, to get rid of discussions of causation, at least in our cleaned-up scientific description of the world.

I'm am sceptical of the idea that science, advanced or otherwise, when cleaned up, is going to get rid of causal talk. The work of Illari, Russo and Williamson (2011) suggests to me that the various sciences are committed to causal talk in many and various ways. But the issue is not primarily one of causal *talk*, but whether the commitments of the scientific theories appeal to irreducible modality. The appeal to laws, counterfactual dependence, agency, causal handles, possibility and necessity that seems to crop up in explanations, including Albert's (2000) explanation of the record asymmetry, is at least some reason to think that modal talk has a wide cosmological role and, crucially, has a wide cosmological role within scientific explanation.

Price offers Russellian Republicanism as a revisionary proposal, of course, so it is not much of an objection to point out that, as it stands science at least appears committed to modality. In §6.5.4, I will argue that we should not accept this revisionary proposal.

6.5.3.5 *Mind-independence*

Deborah C. Smith (2011:186) argues that, in addition to the four marks of realism Wright defends, we should add a fifth: mind-independence:

> the constraint that the facts to which true sentences of the discourse correspond must be robust in the sense of being constitutively independent of the existence and working of any and all minds.

Characteristic of the projectivist is that what is projected is mind-dependent in a way that what is detected is mind-independent. One might think that, insofar as some properties are response-dependent, they cannot also be mind-independent. But that is to confuse two different types of dependence. Properties that are response-dependent are those for which an appeal to responses must be made to *fix reference* to the properties in question. The properties thus identified can nevertheless be constitutively independent of any responses to them. The sky doesn't really get bluer, nor the flowers smell sweeter when you fall in love; you just respond to them differently.[36] And no one who accepts that there are any counterfactual dependencies will deny that there were some between volcanoes and lava before any sentient life was around. The modal relationships we are interested in are mind-independent ones, even if human interventions play a role

[36] Cf. Jollimore (2011:3): 'What looks to the lover like a transfigured world is really the result of a transformed eye.'

in our *grasp* of what it is to be a modal relationship. As I said in §6.4, I want to avoid talking about *the* causal relation. Causal explanations may well depend on the interests we have in explaining; what is a background condition and what is a variable might well depend on who wants to know and for what purpose. But the complex web of modal relations that we try to make more digestible in an explanation is independent of the purposes for which we are trying to explain. Indeed, it is notable that some explanations seem very compelling, and leave their audiences most satisfied, but are false. The explanations must be constrained by the mind-independent facts, and not the other way around. This is characteristic of detectivism, rather than projectivism.

6.5.4 Against Humean projection

The projectivist about causation, and, more broadly, the Humean about natural necessity, denies that the world contains irreducible modality. Causal or other modal relationships are projected onto the world by us. Nonetheless, they may sound awfully similar to the detectivist realist about causation. One option for them is to accept 'quasi-realism' about causation, as Simon Blackburn (2014) recommends. Quasi-realism is a view that tries to preserve the practices associated with a particular domain of discourse but avoids realist commitments. So they may be permitted to say realist-sounding things, provided in the canonical notation they are all explained away:

> A recognizably Humean picture of causation may go so far as to say 'it is the fact that A causes B that explains the convergence of opinion that [A causes B]'. Once the mind has 'spread itself on the world', it also regards itself as reading things off the world that it has projected. But a Humean also holds that there is a more illuminating and economical account of the convergence. (p.32)

For the quasi-realist, pointing out that we ordinarily talk like detectivists about causation is unpersuasive because talking like realists is part of their aim. This section, then, aims at denying that modality can be given a more illuminating and economical account than a realist one, where we grasp modality by our experience of acting in the world. This is obviously a challenge. It is very difficult to show a good explanation can't be given of something; failing to give such an explanation is hardly sufficient.

I will instead present two general reasons to reject the Humean picture, where causes are projected and laws merely describe exceptionless generalisations, rather than constraining the world to conform to them. The first reason is epistemic, and we have seen it before in §6.5.1; our explanation for the asymmetry of records itself seems to *presuppose* modality. The second reason is that Humeanism fails to offer explanations. We will spend a little time revisiting the argument that the asymmetry of records presupposes modality before criticising its lack of explanations.

The Humean will attempt to explain occurrences of modality away, but this will undermine their ability to offer the sort of explanation that Albert (2000) gives. For Albert's goal is to go from the directly surveyable condition of the world to some account of the past state of the world, such that it could have given rise to things that count as records. But how are we to go from the directly surveyable condition of the world to the conclusion that there *is* a past state of the world?

The Humean attempt to explain modal/causal connections away might be described as an attempt to climb a ladder (i.e. use modal talk) and then throw it away (by giving a non-modal account of the modal talk). But if we look at it from the point of view of one regulating inquiry, it looks like the Humean first commands that all ladders be committed to the flames and then insists that climbing the ladders you have just burnt is fine if you throw them away afterwards. There's nothing wrong with 'talking with the vulgar' when explaining the theory and then having a more sophisticated story for the ontology room. But one can't rule out abductively appealing to modality (because it is epistemically suspect) and then rely on just such abductive moves to explain why the local particular matters of fact in the past and future don't need to be connected by modal relations. Getting the entropic explanations for the asymmetries of records and of agency seems to require allowing ourselves to make use of the sort of abductive leaps that Humeans reject as unscientific.

The Humean can always appeal to the idea that we can progress an inquiry by undermining belief in things we previously accepted. We used to believe in phlogiston and aether and then realised we had a better theory that didn't involve them. But the current theory, in those cases, isn't *justified* by appeal to phlogiston and aether. Equally, the Humean can say that they are making an abductive leap towards the existence of a mosaic of local particular matters of fact, which *happen* to support generalisations that can be summarised by laws like the second law of thermodynamics, and *happen* to give rise to a macroscopic asymmetry of records and of agency. But compared to the abductive leap to accept some sort of modal relationship between the directly surveyable condition of the world and other times, we seem to be leaping towards something pretty miraculous.

To put the point another way, if one accepts a modal connection between this time and other times, one is doing no more than accepting that inductive inference provides some justificatory force; that there is no problem of induction. Once induction can be used, it can confirm, through holism of confirmation, the abductive moves made. The Humean position, however, replaces reliance on induction, which it cannot justify, with abductive leaps. This is because the Humean denies the very possibility of modal relations between distinct existences, except insofar as each distinct existence is compossible with any other. Russell (1912b:36) frames the problem of induction as follows:

> We have experience of past futures, but not of future futures, and the question is: Will future futures resemble past futures? This question is not to be answered by an argument which starts from past futures alone.

But the problem is *worse* for the Humean than Russell allows, because we must go from the directly surveyable condition of the world to the conclusion that we have experience of past futures! Abduction must be used somewhere to allow us knowledge of other times. It is better to allow the possibility of modal connection between distinct existences, which we are able to grasp through the exercise of our own agency, than to rely on our sheer luck in living in a world that can mostly be summarised by the second law of thermodynamics and the assumption of a low-entropy starting point. If we do this, we can at least allow that induction sometimes justifies, and we can learn when it does so through well-controlled inquiry. If we are to escape the directly surveyable condition of the world, we need to help ourselves to some modal concepts to do so.

The Humean alternative, of hypothesising miracle after miracle – that the directly surveyable condition of the world provides evidence of other times despite not being constrained by what happened at them, that the history of the world forms lawlike generalisations, and that it will continue to do so, and so on – seems to go against the very empiricist motivations for Humeanism. The story of explaining why our projections of modality are so successful must appeal to luck so frequently that a good Humean about epistemology should reject such miracles out of hand. Much better that they accept that there are modal connections which we can identify through careful yet fallible inquiry.

Perhaps it's unfair to link Humean Supervenience too closely to an epistemic claim. But without the epistemic claim that we don't experience connexions between distinct existences, the Humean eschewing of primitive modality seems under-motivated. As MacBride (2005:127) points out:

> it is a curious fact that the proponents of the contemporary Humean programme – Lewis included – having abandoned the empiricist theory of thought that underwrites Hume's rejection of necessary connections provide precious little by way of motivation for the view.

Jessica Wilson (2010) considers a range of motivations for the rejection of necessary connexions between distinct existences and argues that they are no more favourable to the Humean than the epistemic grounds considered above. Accepting the past hypothesis seems to put pressure on Humeanism as providing us with a possible epistemic route from our evidence to a theory on which it is true.

The second argument against Humeanism is that it fails to offer explanations since it is not ampliative. This is not to impugn the sort of explanation Albert (2000) offers for the asymmetry of records. I endorse that explanation. But when one attempts to remove all references to modality from the explanation, one must replace them by appeal to local particular matters of fact (matters of fact which, as I have pointed out, go far beyond the directly surveyable condition of the world). It was the striking uniformity of these matters of fact that was to be explained, but the response 'these are all the matters of fact' doesn't offer an explanation, it abjures one.

Aren't realist explanations equally poor? One may be reminded of Moliere's (1673) satirical appeal to the 'dormative virtue'[37] of opium as the explanation of the onset of sleepiness when taking opium. An appeal to dormative virtue is a poor explanation, but is nonetheless superior to no explanation at all.[38] For it does make some commitments:

[37] The original phrase was in Latin: *virtus dormitiva*.

[38] Cf. Molière (1666), where the character Sganarelle gives a 'diagnosis' in the form of a series of gibberish words followed by 'and so that is why your daughter is mute' ('*Et voilà pourquoi votre fille est muette*').

that there is something about the opium (a mechanism?) that causes sleep. The opium is no placebo, and though no mechanistic explanation is given, the suggestion is that one is to be found. So too the modal realist. There is some connection between this time and other times, and though this commitment doesn't say what that connection is, it adds the suggestion that a more precise explanation of it is to be found. The contrast with the opium case is that everyone in that case accepted that there was something about the opium that caused sleepiness, whereas the possibility of such a modal connection is what the Humean seeks to deny. The appeal to a modal connection is a better start than an appeal to unconnected patterns in the Humean mosaic of local and particular matters of fact, which preclude the possibility of anything more than a summary of the mosaic.

The entropic gradient is undeniably a feature of the world that we live in. But it can be explained by a modal asymmetry; that effects tend to have higher entropy than their causes, as I will argue in §6.6.1. Attempts to do away with modality from causation are epistemically suspect and explanatorily poor. In the next section, I will consider objections to the view that causation can explain both the asymmetry of records and the asymmetry of control.

6.6 Productive causation

Having argued that causation (and by extension modality) should be treated as response-dependent, and that the Humean doesn't have a more illuminating and economical account to offer than the detectivist, I shall now give a positive account of the role of modality in explaining the distinction between past and future.

There are a number of specific realist accounts of causation, laws, dispositions and counterfactual dependencies available. The project of settling on some particular story of the modal aspects of reality is beyond the scope of this chapter and will likely depend on considerations elsewhere in one's theory, such as their potential width of cosmological role. In the next subsection, I will say something at least a little more specific about what I take modality to be. My account assumes that once we have some modality, there is a method of getting to the other modal features to which science might wish to appeal. Briggs and Forbes (2017; 2019) talk of 'the nomological package' as being the union of categories of existence 'conspicuously absent from the Humean mosaic: laws of nature, counterfactuals, dispositions, and chances' (2017:930). As Briggs and Forbes (2019:509) argue, there may be a number of different ways to go from different elements of the nomological package to causation, but I'm assuming here that a realist view of causation is, *as such*, committed to realism about some elements of the nomological package.

6.6.1 Modality as the basis for the asymmetries of records and agency

Modality can explain the asymmetries of records and of agency. Or, more accurately, a modal connection between earlier and later events can be part of an explanation of the asymmetry of records and of agency. Let's start with the asymmetry of agency: we can affect later times and not earlier times because causal influence is propagated through time, from earlier to later. The asymmetry of agency is a result of this combined with the fact that action always takes place in the present. So the past cannot be causally influenced because it is earlier than the present, and the future at least sometimes can be causally influenced. The appearance of simultaneous causation is always, in fact, contemporaneous causation, where there are multiple things exerting causal influence on each other over some duration.[39] Causation by omission is the result of the action of persisting processes, where there was a background assumption that a process of mitigation of that action would also persist.

Many people before me have seen a link between causation and the past/future distinction. Hans Reichenbach (1956), for example, offers a causal theory of time, in which the past/future distinction is reduced to that of the causal ordering of events. Time is no more than the 'proper time' of Chapter 5, and allows for time to be compatible with relativistic physics. The asymmetry of records is then explained, roughly as Albert does, but without the need to cash out the appeals to modality in non-modal terms. We start with the directly surveyable condition of the world and abductively accept that we have persistence through change and that at least some of the directly surveyable condition of the world is modally connected to other times at which we have existed and can act as a record of those times. We then infer from the records we have that there are other times at which we weren't present, and that entropy has been increasing over time. Once we posit a low-entropy initial condition for the Universe, the explanations of why causal interactions (in the form of irreversible processes) generate an increase in entropy for the system can be the standard ones from statistical mechanics.

Causal theories of time have been criticised in roughly three ways.[40] These objections go roughly as follows.

> *Circularity*: The reduction of the temporal to the causal fails because the causal asymmetry itself presupposes a distinction between before and after.
>
> *Regress*: The reduction of the temporal to the causal depends on the explanation of the causal asymmetry in terms of another modal asymmetry, which in turn stands in need of justification.[41]
>
> *Dogmatism*: One just assumes by definition that there is some metaphysical, empirically undetectable, modal asymmetry between past and future.

Astute readers will notice that these objections are mainstays of the Pyrrhonian scepticism associated with Agrippa and Sextus Empiricus: refusing to accept any claim because it is dogmatically asserted without justification, the justification fails to end or is circular.[42] Phyrrhonians argue that these objections can be levelled against any claim to knowledge, but let's consider their merits in this case. We should nevertheless bear in mind that we should not set the bar so high that we commit ourselves to

[39] 'Contemporaneous' in the sense used in Chapter 5.

[40] E.g. Sklar (1977), Price (1996).

[41] Price (1996) calls this the 'third arrow strategy'.

[42] See Hookway (1990).

Pyrrhonian scepticism across the board, nor so low that we can justify anything from our armchairs, regardless of the evidence.

The argument that our grasp of causation does seem to presuppose a grasp of the distinction between before and after seems a good objection to any attempted *reduction* of the temporal to the modal. We should not attempt such a reduction. But that the temporal presupposes an understanding of the modal, and that the modal presupposes an understanding of the temporal seems to push us into the objection from circularity. But if we accept that there is an unanalysable distinction between before and after, in what sense have we *explained* the asymmetries of agency and of records? It looks like we are in the grip of Agrippa.

We should be cautious about *a priori* reasoning dictating what phenomena there are in the world; but this cuts both ways. The modes of Agrippa are *a priori* reasoning, in this instance against using causation to explain the asymmetry of records, but an appeal to experience shows that explanations offered do appeal to modality to explain time, and to time in order to explain causes, counterfactual dependences and the operation of laws.

In this case, we have a story that explains why our grasp of both might depend on each other. It is our grasp of being agents who intervene on the world, and being patients who persist through change, that gives us an understanding *both* of the distinction between past and future, and the distinction between cause and effect (and other modal notions such as counterfactual dependence, necessity, possibility, propensity and tendency). Our understanding of the temporal is informed by our understanding of the modal and vice versa.

This isn't quite a causal theory of time.[43] It does not prioritise modality over time but presents a package in which they both fit. Both are core features of our grasp of the world we live in and are presupposed by our inquiries into it and interventions on it. And because of neutrality over the priority of different aspects of the nomological package, I can't appeal to some most fundamental modal concept, because I am neutral as to what that could be.

This looks like an acceptance of dogmatism: we explain the asymmetry of agency and of records by *fiat*; there just is an asymmetry between cause and effect which is coextensive with the asymmetry between earlier and later, because causes only effect things in the future, and never affect the past. There are two problems with such dogmatism. First, it looks unmotivated, because it is arbitrary, and second it looks unscientific, because it is not subject to empirical confirmation or refutation and rules out as impossible some situations apparently permitted by the laws of nature, such as closed timelike curves. I will respond to both of these objections to show why my view isn't simply dogmatic.

Justification must end somewhere, and an appeal, based on our ability, as agents and patients, to intervene on the world and endure its interventions on us, makes a commitment to a distinction between past and future and between cause and effect a presupposition of the practice of inquiry and of intervention. Both of these practices

43 Though I fear its critics may deride it as a *casual* theory of time, it is because of my desire to avoid what I consider to be needless formalism.

are themselves presupposed by the scientific enterprise, in the cause of which we must reject the arbitrary. Indeed, the very practice of seeking empirical confirmation and refutation seems to presuppose such practices of inquiry and agency from which we get these distinctions. Secondly, because of holism of confirmation, these presuppositions can be confirmed or disconfirmed by the success of the science they allow for.

6.6.2 Backwards causation

Recall the discussion in §1.5, where we discussed Callender's arguments that different views of time provided merely verbal disagreement. We might worry whether there are differences between the views on offer here. Linking time to modality in the way I propose – the past/future distinction is the distinction given by proper time – rules out the possibility of closed timelike curves. If we find ourselves dealing with phenomena where we could make better sense of them by appeal to closed timelike curves, that will be a reason to change the presuppositions of our theories. Although my view is not subject to *empirical* refutation because, for example, the unreliability of our instruments can always be offered as an explanation for irksome results, the usefulness of the presuppositions of our theories in allowing us to explain the phenomena gives a check to the abductions that we rely on. It is noteworthy that the models of spacetime that allow for closed timelike curves don't tend to resemble the spacetime we find at this world.

Indeed, Causal Set Theory, which we discussed in §5.4, relies on equating time and proper time, and the success or otherwise of this theory as a scientific explanation of the world will reflect heavily on the presuppositions of a modally charged distinction between past and future that it relies on. If we accept that the past/future distinction is that given by proper time, which is a fundamentally modal distinction, we have a basis in physics for accepting a distinction between past and future and for accepting that this distinction is modal.[44]

This chapter has relied on an appeal to agency which, whether we are detectivists or projectivists, has a good claim to be the source of our grasp of the distinction between past and future. In the next chapter, we will explore what agency reveals about what makes the past different from the future, the future different from the present and the present different from the past.

[44] Though, as noted in §5.4, Causal Set Theory won't generate a strict ordering of facts.

7

Agency, fate and the open future

Even God lacks this one power alone:
to make any deed undone once it's been done.

—Aristotle Nicomachean Ethics 1139b10

In the previous chapter, I argued that the central difference between the past and future was an objective modal one. I considered the resources in physics, and the entropic gradient specifically, to give us such a distinction. In this chapter, I want to consider the difference between past and future back at a more macroscopic level at which our everyday experience might be more useful.

It is common amongst those who think there is an objective modal difference between past and future to describe the difference as between the settled past and the open future. In particular, there is meant to be a relation between agency and the open future, such that a future which is not open, in the relevant sense, undermines 'freedom' or 'agency' in some sense. One of the big pragmatic consequences of denying an open future, advocates of an open future may maintain, is *fatalism*. Fatalism, whatever it turns out to mean, is *bad* and should be avoided. In the fields of public health, health education and risk prevention, there is concern that people around the world with fatalistic beliefs are dying preventably because they fail to perceive risks properly, engage in risk-taking behaviour or neglect to use safety measures.[1] Curiously, what the openness of the future amounts to is highly disputed, with virtually all philosophers claiming that, on their own view, the future is open *in the sense of 'openness' that matters* and fatalistic worries are thereby avoided.[2]

I'm going to outline a number of worries that go by the name of fatalism, and try to get a sense of what is at stake when fatalism is invoked. I will then discuss various proposals for capturing the difference between the openness of the future, showing how a causal/modal difference does well here. I will consider a dilemma that Linda Zagzebski (2015) articulates: how do you get a causal asymmetry that doesn't allow the past to be open, but does allow the future to be if the laws of nature are deterministic.

[1] E.g. Kayani King and Fleiter (2012); Norenzayan and Lee (2010); Franklin, Schlundt, McClellan, Kinebrew, Sheats, Belue, Brown, Smikes, Patel and Hargreaves (2007); Dixey (1999); Kouabenan (1998).

[2] This is similar to the phenomenon noted in Chapter 1, where virtually all philosophers deny an error theory about time and claim that time passes on their view, in the sense of 'passes' that matters.

I will respond by revisiting the modal change that is involved in the occurrence of processes; it is this change that allows us to resolve Zagzebski's dilemma. I conclude the chapter by considering two issues: how the modal change from potential to actual fits with our ability to carry out intentions and how this rules out the possibility of time-travel to the past.

7.1 The fatalistic worry

'Fatalism' names a number of attitudes where something about our agency is undermined by something about the future. So it could be that the Yoruba creator of life, *Oludamare,* has chosen a future for you, and therefore you won't die in a car accident even if you don't wear a seatbelt. It could be that your genetic makeup means that you are going to die of cancer, so giving up smoking won't make a difference. Or it could be that the initial conditions of the universe and the laws of nature mean that you never exercise any choice about what to have for dinner. I'll try and tease apart some different versions of fatalism. I'll be fairly dismissive of a few, before focussing my attention on a version of fatalism on which the actuality of the future means that we can't attempt to intervene on it. But I should say now that my being dismissive of bad arguments is not to be confused with a dismissiveness of the real consequences of bad fatalistic reasoning.

The link between bad fatalistic reasoning and perception of risk, risk-taking and neglect of safety measures has received some support. It is hard to generalise, however, because fatalistic beliefs are often tied to local religious practices, which make cross-cultural comparisons difficult. Fatalistic beliefs, in some cases, may be *responses* to a situation of perceived impotence rather than simply a cause of risky behaviour, which makes generalising about the causal relationship between behaviour and beliefs difficult. Indeed, as Kayani, King and Fleiter (2012) argue, fatalism may often be a coping mechanism to a traumatic event, which in turn shapes how future possibilities are anticipated. Nonetheless, fatalistic beliefs, in their various forms, tend to make a link between the attribution of responsibility, credit or blame, and some modal connection between past or present and future.

Fatalistic beliefs are at least correlated with tens or hundreds of thousands of deaths a year from apparently preventable causes. There are genuine and complex policy challenges for reducing the number of casualties in contexts where the correlation between fatalistic belief and preventable death is high. There are also genuine and complex sociological/anthropological challenges in working out how the beliefs relate to behaviours and fit within the local cultural contexts. These challenges are not ones I will attempt to deal with here. But it is a presupposition of dealing with many of those challenges that belief in a divine plan, or auguries, does not provide a good reason to believe that wearing a seatbelt is irrelevant to the likelihood that you will be fatally injured in a car accident. It seems that getting the metaphysics right matters.

7.1.1 Bad arguments for fatalism

In Shakespeare's (1599) *Henry V* (IV.3:20–2), the eponymous Hal criticises Westmoreland for wishing for more warriors:

> If we are marked to die, we are enow
> To do our country loss: and if to live,
> The fewer men, the greater share of honour.

This has elements of a classic argument for fatalism. We shouldn't wish to intervene on our situation, because either it is futile or it is redundant. This is because we are either marked for death or marked to live, and being so marked would be unaffected by having reinforcements. This argument is *really bad*.[3] It relies on believing that one's fate is independent of what one does, and there is no reason to believe this.

Consider a slightly better bad argument for fatalism, this time from Somerset Maugham's (1933:100) play *Sheppey*. A personification of Death recalls an anecdote to the protagonist, Sheppey:

> There was a merchant in Bagdad who sent his servant to market to buy provisions and in a little while the servant came back, white and trembling, and said, Master, just now when I was in the market-place I was jostled by a woman in the crowd and when I turned I saw it was death that jostled me. She looked at me and made a threatening gesture; now, lend me your horse, and I will ride away from this city and avoid my fate. I will go to Samarra and there death will not find me. The merchant lent him his horse, and the servant mounted it, and he dug his spurs in its flanks and as fast as the horse could gallop he went. Then the merchant went down to the market-place and he saw me standing in the crowd and he came to me and said, Why did you make a threatening gesture to my servant when you saw him this morning? That was not a threatening gesture, I said, it was only a start of surprise. I was astonished to see him in Bagdad for I had an appointment with him tonight in Samarra.

In this case, there is still an appointment, but the appointment already takes into account any interventions that take place. So it does not assume that what will happen is independent of what one does, but that what one does contributes to the inevitability of what will happen. Seeking to avoid an outcome is actually just part of bringing it about. In this case, too, there is a suggestion that there is some plan involved; that some (divine?) agent has been marking or appointing the outcomes in advance. This might be a team effort, as in the ancient Greek pantheon; there are three Fates:[4] *Clotho* who spins the thread of life, *Lachesis* who measures the thread and *Atropos* who cuts

3 Noting that it is a bad argument doesn't settle how we should interpret the use of this argument within the play. The fiction could be interpreted such that the king knows it is a bad argument and doesn't care, for example.

4 *Moirai*.

the thread. Alternatively, on a monotheistic conception, any marking or appointing of fates might be the responsibility of a single deity.

Fatalism needn't have any agency involved in it. There is a significant tradition concerned merely with the logic of future truth, or with the role of foreknowledge (divine or otherwise) of future truth, causing tension with people's responsibility for their lives. Logical fatalism starts by assuming the necessity of the past – at least insofar as there is nothing to be done about the past now – and concludes that there is nothing to be done about *any* time.[5] This is because statements about the past entail statements about the future.

So the argument for logical fatalism goes something like this:

(1) If it was the case that *p* then it is necessary that *p*.
(2) It will be the case tomorrow that *p* if and only if it was the case yesterday that it will be the case in two days that *p*.
(3) If it will be the case tomorrow that *p* then it is necessary yesterday that it will be the case in two days that *p*.

Any truths about what will happen a day hence generate truths about what, yesterday, was going to happen two days hence. If truths about the past are necessary, then truths about what, yesterday, was going to happen two days hence are necessary. So any truths about the future are necessary truths in just the same sense that past truths are.

This argument can easily be avoided by denying (1). Those who claim that the past is necessary tend only to accept (1) if it is further specified that *p* contains no embedded claims about what will be the case or was the case. The conclusion (3) may be true, if certain conditions obtain, but we do not establish its truth on purely logical grounds, but because of some worldly connection between the past and the future, for example laws constraining how the universe evolves. As we shall see in §7.3, the sense in which the past is necessary but the future is not is hard to capture. But if we end up with the claim that the future is also necessary in just that sense, just as a matter of logic, that's a sign we've failed to establish the relevant apparent contrast between past and future that talk of 'necessity' was in service of.

7.1.2 Reasons, agency and modality

We can move on now from *bad* arguments for fatalism to some with a little more promise. We'll begin with a classic argument, which Cicero attributes to Democritus, Heraclitus, Empedocles and (inaccurately) to Aristotle:

> If all things take place by fate, all things take place with an antecedent cause; and if desire is caused, those things which follow desire are also caused; therefore assent is also caused. But if the cause of desire is not situated within us, even desire itself

[5] Diodorus Chronus' 'Master Argument' appears to have been a form of logical fatalism along these lines, though there is considerable interpretative difficulty in establishing exactly what the argument was. See Long and Sedley (1987:230–6).

> is also not in our power; and if this is so, those things which are caused by desire also do not rest with us. It follows therefore that neither assent nor action is in our power. From this it results that there is no justice in either praise or blame, either honours or punishments. (Cicero *de Fato* 40)[6]

Determinism is a thesis about the causal relationships between times. Cicero's argument suggests that from the fact that things are causally determined, there is some threat to our ability to be responsible for neither assent nor action and to deserve neither praise nor blame. That is, there is a link from causal determinism to a loss of moral responsibility.[7] This argument is less bad than those discussed in the previous section, though it is less clear whether it should be understood as an argument for *fatalism*. If not strictly an argument for fatalism, it is fatalism-adjacent and the link between time and modality is in any case instructive.

Causal determinism is susceptible to a number of formulations, depending on who how it is intended to fit in one's theory. One might frame it as a relationship of entailment between propositions about the Universe, as Mackie (2003:674-675) does:

> Where H is a true proposition specifying the intrinsic state of the world at some time in the remote past, and L is a true proposition specifying what the laws of nature are, the conjunction of H and L entails that I do A today.[8]

Alternatively, we could think of determinism as a relationship of necessitation between times themselves. Given our concerns in Chapter 5 about simultaneity, we can refine this to say that a particular 'phase-space', given the laws of nature, necessitates some other phase-spaces, where phase-spaces give the position, velocity, charge, mass and so on of everything in a foliation of the Universe.

Cicero's argument can easily be turned into an argument from foreknowledge, provided the foreknowers have epistemic access to the future they foresee.[9] We can do this in roughly three ways:

- by having a deterministic picture, like the one Cicero describes, and some plausible story about how the foreknowers are well-informed enough to predict the things they foresee;

6 Van Inwagen (1983) offers a contemporary version of roughly this argument: If you never had any choice in the laws of nature, and you never had any choice in the state of the past a million years ago, and the two entail the truth of some proposition *p* about the future, then you never had any choice about *p*.

7 I have deliberately avoided talking about 'freedom' and its relation to moral responsibility here. There is a significant literature on freedom, including contributions by Dennett (1984) and Pereboom (2001), disputing what exactly the term means and how it relates to moral responsibility. I have nothing to add to the dispute about what 'freedom' means or ought to mean, so I have only discussed more directly what I take to be the connection between moral responsibility and time/modality about which I have something to say.

8 In a footnote, Mackie acknowledges similarities with formulations by Lewis (1981:291–2); van Inwagen (1975,§1; 1983:70).

9 Etymologically speaking, 'fate' relates to that which is foretold.

- we can have creatures who have records of the future (as we have records of the past);[10]
- we can have creatures for whom the directly surveyable condition of the world simply includes future times.[11]

This third option, which relies on a knower who sees the world *atemporally*, presents a different argument for fatalism and doesn't rely on a worldly connection between different times. I shall take this up later in §8.3.1. For now, let's focus on the versions that rely on causal determination. In these cases, unlike the atemporal knower, *knowledge* isn't doing any work; that someone knows truths about our future doesn't add to the challenge for our assent nor action. We shall therefore just stick with the connection between past times and future ones for now, as in the case where causal determinism is supposed to lead to a loss of moral responsibility.

The Stoics resisted the fatalist conclusion, placing a great emphasis on the distinction between moral responsibility and determinism. Roughly, the Stoics distinguished between *internal* and *external* causes, where internal causes were required for action and assent. Moral responsibility requires that we are responding to these internal causes rather than merely external causes. The distinction between internal and external causes is murky.[12] But the position-that-may-have-been-Stoic I want to consider is the one where internal causes are nonetheless necessitated. The thought is that there is something about internal causes that ties them to moral responsibility that is distinctive from causes elsewhere. I will offer two thoughts about how this might work.

First, we have a thought from Aristotle about what is distinctive of skills generally (including moral virtues) that makes skills the appropriate locus of praise and blame; they are 'in our power' or 'up to us'.[13] Something is up to us if we have a 'two-way capacity'[14] – an ability to do something or refrain from doing it. So the medical skill can be used by a doctor to make someone better, or to kill them; the same skill is used in both cases, and it is up to the doctor what they do with their skill. The idea that the skill can be exercised in different ways needn't conflict with causal determinism, because it

10 This suggestion seems to be made by Augustine in *City of God 4*.

11 This seems to be the view of Boethius in *The Consolation of Philosophy* V.

12 See Sorabji (1983) ch.4 for a discussion of eight different interpretations of the Stoic response. Bobzein (2001; 2021) discusses some of the changes that happen within Stoicism, for example, a change of focus from backward-looking assessments of moral responsibility to forward-looking acts of planning (2001:333). As Bobzein shows, the concerns that motivated the Stoics to think about these issues were against the background of a very different conception of nature than we have today. She also notes that our contemporary use of 'freedom' is often connected to the internal/external causes distinction, and the Stoic notion of (political) freedom, '*eleutheria*', which were linked by Epictetus and used almost synonymously in later antiquity, but apparently not linked before him (2001:330). Freedom of the will, in the sense of *eleutheria*, is a kind of self-mastery that only sages, if anyone, can achieve, and is not what is at stake in debates about fatalism.

13 Greek: '*eph hauto*'. Cf Aristotle *Nicomachean Ethics* III.5, and Metaphysics Θ.2, though the purported conflict between causes and responsibility arrives after Aristotle. Makin (2012) §V has a helpful brief discussion.

14 Capacity is a translation of the Greek *dunamis*. Steward (2020) uses the term 'power' rather than 'capacity' in her recent defence of this concept in agency.

may be that, amongst the things that are determined, are desires to cure, rather than kill. Indeed, as Makin (2012:407) observes:

> it is necessary that if an agent has a two-way capacity, and is in the right relation to a patient with the appropriate passive capacity, and the agent chooses to exercise the capacity in *this* way, then the capacities are exercised, and *this* change occurs.

It would be an odd complaint about a doctor that they cannot be praiseworthy for their excellent treatment because there was simply no chance that they would have decided to deliberately kill their patients. When choosing a physician, we might actively seek someone who we think is, by nature, motivated to give the best care they can to their patients. We can do this without thinking that the physician is thereby exempted from responsibility for the treatment they provide. A pharmaceutical vending machine, by contrast, lacks responsibility even if it provides the same medicine. This is because physicians are not stimulus-response systems, like vending machines, but skilled practitioners.

There is, then, some sense of things being 'up to us' which isn't about the *likelihood* of our doing something, but whether we have the ability to be the origin of a range of actions.[15] That is, when we come to explain the action, agents form an appropriate stopping place in an explanation, rather than being a mere instrument of some other origin. That we are predictable, and consistent, in using our skills doesn't undermine our having the *ability* to use them for other reasons. There must be some sense in which we 'could have done otherwise', but this might be *counterfeasible*.[16] That is, it may be that we are permitted to consider situations in which the history of the world had been different and we do otherwise to make sense of the claim that we have the ability to do otherwise.

The second thought is that reasons are not a specific variety of causes.[17] This idea that the skilled can do or not do something might give rise to the view that there need to be two actual possibilities, where it is still unpredictable what they will do. But trust in doctors seems to rely on their being, at least to some extent, predictable. It is not the *unpredictability* of the doctor that makes them responsible, but their *reasons responsiveness*; they are not merely *causally* responding to the situation, but they are responding to reasons they encounter and acting to achieve certain aims.[18] This cannot be said of the vending machine. Reasons are importantly distinct from causes, even if there can't be a distinction in reasons without a distinction in the causal history of the Universe and/or the laws of nature. Reasons are not supernatural features of the world

15 Greek: *kurios*.

16 Cf. Briggs and Forbes (MS). Von Wright (1963:50–1) distinguishes between the 'can do' of success and the 'can do' of ability. This latter sense, the 'can do' of ability, is the one I claim can be counterfeasible.

17 Cf. Anscombe (1963:§16) When the Stoics talk of 'causes' this will cover different types of explanation, including efficient causation (which is something like the contemporary notion of causation) and teleological causation (i.e. explanation in terms of ends, aims or purposes). My proposal is roughly that these are different sorts of explanation.

18 Cf. the quote attributed to Martin Luther: 'Here I stand; I can do no other.' This is usually regarded as a paradigm of *taking* responsibility for one's actions and assents, not of denying that one has any.

somehow in conflict with the causal order, nor are they a label for a special type of cause. Reasons are simply concerned with different sorts of explanatory projects than causes are. Reasons are involved in justification, practical rationality and motivation, and often are in the service of achieving some aim or purpose, whereas causes are aimless and don't offer justification.

People are parts of the causal world, and the actions of people at one time are, on the determinist view, necessitated by features of another time plus the laws of nature. So it may be that *causal* explanations of people's actions go back to the origin of the Universe, but agents are the origins of their actions in the sense that the action is justified or not based on the way the agent responded to reasons. There need be no conflict between moral responsibility and causal necessity; reasons can necessitate a particular course of action (in which case that is the only *morally responsible* course of action) or they can underdetermine courses of action. Provided any causal necessitation of people's actions and assent involves responsiveness to reasons that necessitate (or at least permit) that action and assent, then moral responsibility has not been threatened. If we look to contemporary accounts of moral responsibility, such as David Brink and Dana Nelkin (2013:292) we can see the role of such a capacity being invoked:

> [R]esponsible agents must be *normatively competent*. They must not simply act on their strongest desires, but be capable of stepping back from their desires, evaluating them, and acting for good reasons. This requires responsible agents to be able to recognize and respond to reasons for action. If so, normative competence involves *reasons-responsiveness*, which itself involves both *cognitive* capacities to distinguish right from wrong and *volitional* capacities to conform one's conduct to that normative knowledge.

Brink and Nelkin are explicitly interested in the possession of capacities, rather than the use of such capacities, because they do not think that normative competence depends on determinism being false.

It is clear that the issues around the relation between causal determinism and moral responsibility are of more sustained philosophical interest than the arguments for fatalism previously discussed. I have tried, here, to show that there are plausible responses to the concern that there is something about causal determinism that undermines moral responsibility. I still think there is a fatalistic worry to be addressed, however. This worry is distinct from that of causal determinism. I think the future must be 'open' in a sense in which the past is settled, even if causal determinism is true. I shall therefore now explore what is meant by the future being 'open' and argue for a way of understanding this that avoids fatalism of a sort, even if causal determinism is true.

7.2 Characterising the open future

What's going on when we talk about the open future in contrast to the past, which is 'settled' in some sense? The next section will pose a dilemma about whether any of these versions of openness can give us the asymmetry between past and future that we want (i.e. a modal/causal asymmetry that avoids a problem that we might think of as a fatalistic worry). For this section, let's just browse some options that are on the market.

I have been referring to the settled past. I should like to note, in passing, that there are two ways in which the past might be settled: it might be 'fixed' where something is immutably the case, and it might be 'closed', where something is immutably ruled out. I shall say times are settled if they are fixed or closed, and they are open if they are not settled.[19] So the 1989 where the Berlin wall came down is fixed, but the 1989 where it was instead reinforced with titanium is closed. That the wall came down is immutably part of the past of the world (whether historians forget about it or not), and that the wall was reinforced with titanium is immutably ruled out from the past of the world. Whether a new wall will be built in 2089 is, perhaps, not settled.

The distinction between an open future and a settled past has been subject to a variety of analyses and characterisations. In §7.2.3 I will begin to sketch my own view, which will be the subject of discussion in the remaining sections, but before that we will consider semantic and branching approaches to the open future.

7.2.1 Semantically open futures

One way for the future to be open is semantic; for propositions about it to be non-bivalent. That is, bivalence could fail for certain propositions about the future because of a gap in truth-values (i.e. some proposition is neither true nor false).[20] In particular, we might be interested in propositions that are about the actual future and are neither necessarily true nor necessarily false. We will call these 'future contingents'. If determinism is true, and a complete description of the Universe is entailed by some past description of the Universe and the laws of nature, then there are no future contingents. So concerns about the semantics of the open future are often concerns about the semantics for future contingents and the logic that governs them.

Alternatively, one could claim, along with C. D. Broad (1923) and Patrick Todd (2016;2021) that all future contingents are false. This preserves the sense that if you say something will happen, you are committed to it happening, but has the (undesirable) consequence that it is impossible to know that it's safe to cross the road. Yet another option is to accept that future contingents are bivalent, but it be indeterminate which of the two truth-values applies. Barnes and Cameron (2009; 2011) defend such a view. This takes the indeterminacy away from truth-value, but puts it into metaphysically indeterminate future events. In theory, one could hold that future contingents are both

[19] I am thinking of times, here, as complete instantaneous descriptions of the world, following Bourne (2006) and Briggs and Forbes (2012). As such the same time cannot be both immutably the case and immutably ruled out.

[20] See Besson and Hattiangadi (2014); and Todd (2021) for discussions of the logical issues around bivalence. In this sense, the present or the past might be open, if propositions about present things are non-bivalent, for example due to vagueness. Even if there are some such truths in the past and present, we can still make a useful contrast between the extent to which the future, past and present are open.

true and false – a view of truth-value gluts, rather than truth-value gaps – or that all future contingents are true, though I don't know of these views having been defended.[21]

I'm confident that there are tractable ways of semantically representing the open future, and I don't think my argument here hangs on working out which we adopt. Following Stephan Torre (2011), I think that semantics for the open future isn't *enough* to capture the difference between the open future and the settled past. As Torre (2011:361) puts it:

> If openness of the future just amounted to failure of bivalence for future contingents, we would need to come up with some explanation for what intuition these theorists [who think there is an asymmetry in openness between the past and the future] claim to accommodate. . . . Perhaps the openness of the future is best understood as the claim that the future is indeterministic. . . . Again, this way of characterising the open future intuition seems problematic. First, the open future intuition seems largely independent of views about whether the world is deterministic or indeterministic. . . . Secondly, if our world is indeterministic, then the current state of the world and laws are compatible with multiple futures. However, given indeterminism, the current state of the world and the laws are also compatible with multiple pasts. So indeterminacy alone fails to capture the *asymmetry* in openness between the past and future.

I think questions of the truth-values of future contingents are tied to whether the world is a deterministic one. I think the asymmetry between past and future is distinct from this, however. One reason for believing this is that even if one accepts that determinism is false for the Universe as a whole, there may still be local deterministic contexts such that in those contexts determinism holds. We can imagine a carefully controlled laboratory setup where once the initial conditions are set and the apparatus is switched on, only one possible outcome will follow until the apparatus is switched off. We shouldn't accept that in such a context there is no asymmetry between past and future.

More importantly, I don't think determinism is the important issue. As J. L. Mackie (1980:179) once remarked, '[t]he full freedom of choice requires a metaphysical non-fixity in its object that goes beyond the mere absence of causal determination.' We are not, here, primarily concerned with freedom of choice; I prefer to talk of responsiveness to reasons.[22] But the openness of the future, I claim, requires a metaphysical non-fixity that could obtain even in situations where determinism holds, and descriptions of the

[21] In work I've done with R. A. Briggs (2012), we give three semantics for future truth on the Growing-Block view – a view on which no future events exist. Each semantics (the supervaluationist semantics, the intuitionist semantics, and the Łukasiewicz semantics) deals with cases of non-trivial truths about the future and propositions about the future such that neither they nor their negations are true (i.e. truth-value gaps). We (unlike Broad 1923) are perfectly happy for statements about how the future will turn out to be true. If the laws of nature and the state of the universe leave only one feasible outcome for the Universe, then there are truths about what will happen. Nonetheless, for any of the semantics on Briggs and Forbes' view, future reality could fail to speak either way to how the future will go, but it could nonetheless be inevitable that it goes some particular way.

[22] We must not conflate what's up to us with sage-like self-mastery. Cf. fn12.

future aren't notably all true, all false, all indeterminate or entirely lacking in truth-value.

7.2.2 Branching time

The approaches I have discussed so far have tended to assume that the future *lacks* something compared to the past, whether this is determinacy, existence or something else. But there are approaches that think the asymmetry is captured through a *superabundance* of branches of spacetime. Storrs McCall (1994) has a view on which there exist numerous future possibilities branching off from each other. These possibilities drop out of existence as they cease to be possible. The present, on McCall's 'shrinking-tree' view, is the earliest point at which branching occurs. I will discuss McCall's view more in Chapter 9.2.4, since it is a view on which there is both McTchange and a past/future asymmetry. Here we can briefly observe that McCall makes the openness of the future tied to lack of causal determinism in the way Torre (2011) has suggested can't capture the asymmetry we're after. It is the fact that the future *branches*, rather than indeterminism, that is of interest about McCall's view. Branching futures are compatible with a number of different semantics of future contingents, since there being multiple different volumes of space-time doesn't yet tell us how to describe what happens when the volumes diverge from each other.

Branches needn't be associated with indeterminism. On the Everett interpretation of quantum mechanics, as defended by Simon Saunders and David Wallace (2008), there are multiple branching futures, all of which are collectively determined by previous states of the universe. Each branch is an inevitable consequence of what's gone before (we either don't know which branch we are on, or who we are). Thus branching creates an asymmetry between past and future not based on a lack of necessitation from the past and the laws of nature to the future. The 'static shrub'[23] view Saunders and Wallace defend doesn't, strictly speaking, require indeterminism but does rely on there being multiple possible futures, however. If we're looking for an asymmetry between past and future that doesn't depend on there being multiple future possibilities, branching views may prove tricky.

Some defenders of the open future believe in what is known as 'the Thin Red line' (e.g. Belnap and Green 1994; Belnap, Perloff and Xu 2001). This does not refer to the 93rd Sutherland Highlanders at the Battle of Balaklava,[24] but rather to a branching future with real and incompatible alternatives where one of the paths through this branching future is privileged, since it alone will be what *will* happen. This view tries to present the future both as open and as settled. The future is open in the sense of being causally underdetermined but settled in the sense that bivalence holds for propositions about the future, the future exists, and it is determinate what the future will be. I follow Jiri Benovsky (2013) in thinking that the privileged actual future, in such a case, is

[23] So called because it is like McCall's Shrinking Tree, except it doesn't shrink, and it branches all the way down. Not to be confused with the 'glowing shrub' view I discuss in §9.2.6.

[24] Cf. https://www.military-history.org/regiment-profiles/the-93rd-highlanders-and-the-thin-red-line.htm

not open in a particularly interesting sense. Whatever asymmetry we are trying to capture between past and future, such a view will not be any more or less open than a view, like Mellor's (1998) view, on which causation is not deterministic but the future (determinately) exists. The branches in the future that won't happen, on the thin red line view, have more to do with counterfactual possibility than they have to do with how the future will unfold.

7.2.3 Causal fizz

In the previous chapter, I suggested that the asymmetry between past and future was modal/causal. I think it is this modal/causal relationship we should focus on to understand the fatalistic worry around the past/future asymmetry, rather than the semantics of future contingents or issues around whether space-time branches. The modal/causal asymmetry is better placed to give us the kind of asymmetry we need.

Just as we saw in the previous chapter that philosophers of physics, such as Hans Reichenbach, have defended a connection between causation and the past/future asymmetry, metaphysicians have also done so. Jeffrey (1979), is one example. Dorothy Emmett (1984) defends a view of 'immanent' causation as 'effecting' temporal becoming. J. L. Mackie (1980:225) defends the view that the future is open, and causation '*extrudes*' the future from the present and the immediate past. Peter Forrest (2004) describes the present as having a 'causal *frisson*' that the 'dead' past lacks (p.359). Though they may differ on some details, it is something much like what these philosophers seem to have in mind that I wish to defend. I shall use the term 'causal fizz' (alternatively 'causal effervescence') as a label for a particular aspect of the nomological package that explains the asymmetry between past and future.[25]

Labelling something 'causal fizz' is likely to spark a host of questions: Is it an empirical phenomenon, or a theoretical posit? Is it fundamental, or derivative? Is it a stuff, or a process, or an object, or a property, or a relation or what? For now, let's introduce it as a placeholder concept for whatever the modal/causal element of the nomological package is that we appeal to in explaining the past/future asymmetry, if there is such a modal/causal element. I intentionally leave these questions unanswered here, so we can further specify as the argument requires, rather than confecting a very specific account that doesn't answer to reality.[26]

In some ways, invoking causal fizz provides quite an intuitive picture. It's very easy to talk of the current state of the Universe 'producing', 'bringing about' or 'generating' successive states of the Universe. But an appeal to a modal difference between past and future is not straightforward, as we shall see in the next section. It's very easy to talk of the past 'producing', 'bringing about' or 'generating'. As Mackie (1980:86) observes:

> But it is not enough to reiterate such words as 'necessity' and 'production' and 'generation'; we need some clear account of what this necessity is, or of what producing or generating can be.

[25] Thanks to David Braddon-Mitchell for suggesting the term 'causal fizz'.
[26] We will return to causal fizz in §§7.4.3 and 10.4.

It's not merely that what we mean by those words is mysterious, but that the very logic of them poses challenges.

7.3 Zagzebski's inconsistent triad

Linda Zagzebski agrees that the difference between past and future isn't to be explained by indeterminism. As has been discussed above, this would not generate the relevant kind of asymmetry. What we need is something that generates an objective modal asymmetry between past and future, and that may be hard to find, as Zagzebski (2015:195) explains:

The root of the problem, then, is that it is impossible for there to be a type of modality that has the following features:

(a) The past and future are asymmetrical in that the past is necessary with respect to this type of modality, whereas the future is contingent with respect to this type of modality.
(b) There are propositions about the past that entail (or are logically equivalent to) propositions about the future.
(c) [Transfer of Necessity Principle] (or equivalently, [Transfer of Possibility Principle]) obtains.

Let's start with two of the standard modal notions: *possibility* and *necessity*. Consider the sentential operators '$\Diamond$' ('it is possible that') and '$\Box$' ('it is necessary that'). Zagzebski's appeal to the Transfer of Necessity Principle is one about the logic of these sentential operators. She uses subscript '$_t$' to indicate that we are dealing with the temporal kind of necessity. If p is a proposition about the past, and p is true, then $\Box_t p$. If some proposition is necessarily true, and that proposition entails some other proposition, the necessity will also be transferred. So, inferences of the following form are valid:

$\Box_t p$
$\Box\underline{(p \rightarrow q)}$
$\Box_t q$

This, as Zagzebski (2015:191) notes, is logically equivalent to the claim that inferences of the following form are valid:

$\Diamond_t p$
$\Box\underline{(p \rightarrow q)}$
$\Diamond_t q$

Let's accept, for the sake of argument, that these inferences are valid. Zagzebski's worry is that if there are truths about the future that are entailed by the past (i.e. determinism is true for at least some contexts), then any necessity the past has will, as a matter of logic, be transferred to the future. Given we are accepting that determinism is a separate issue from the issue of the modal asymmetry between past and future, if there is such a modal asymmetry, it ought to be compatible with Zagzebski's premise b. So we either need modal notions that don't transfer in this way, or we need to abandon the asymmetry. We can still, of course, accept Price's distinction between fixtures and options (introduced in §6.4), where the distinction between open and closed is a contingent and perhaps local feature of our embeddedness in time, and get rid of the link with the past/future distinction.

I will pursue the horn where we reject the transfer of necessity principle. The sense in which the past is necessary must not be one that transfers in the way $\Box_t p$ does.

7.4 Modal change: Potentiality to actuality

My proposal is that we think of the past/future asymmetry as being an asymmetry between *potentiality* and *actuality*. We might say of a talented young student that they have the potential to be a great composer. Here we contrast a claim about what they actually are – a talented young student – and a claim about what they could be – a great composer. The sense of 'could' here is one, as we might put it, of *actual possibility*; it's not merely that there could be a great composer who resembles this talented young student, but *that this very student* could become a great composer. Equally, it's not just that this person *could have been* a great composer, but that they actually possess the potential to do so.

Though we talk quite naturally in these terms, these are quite different modalities from possibility and necessity. This is what allows us to avoid Zagzebski's difficulty. Rather than thinking the past is necessary, we should think of the past as actual. Rather than thinking of the future as possible, we should think of the future as potential. The passage of time, then, is a change from the potential to the actual. Actuality isn't a specific kind of necessity, and though the actual is always possible, the actual cannot be potential.[27] Actuality and potentiality simply don't stand in the same relationship to each other that possibility and necessity stand in. That's why we avoid the transfer of necessity principle; actuality doesn't transfer in the way necessity does.

When we talk about 'the past' or 'the future', this is a convenient shorthand but is ambiguous. It disguises a shift from the possibility or necessity of *propositions* to the potentiality or actuality of *things*. One of the ways the distinction between potential and actual is different from the distinction between possibility and necessity is that potentiality and actuality are *de re* modalities. That is, actuality and potentiality are modalities of worldly stuff, rather than of truth-bearers such as propositions. So, it can

[27] This differs from Vetter (2015:162). Vetter claims potentiality is implied by actuality. This means that I will not be able to follow her in defining possibility and necessity in terms of actuality and potentiality.

be necessarily true that it will be the case that there is a sea battle, without there being anything such that it is *actually* a sea battle that will (of necessity) happen tomorrow.[28]

7.4.1 The actuality/potentiality distinction

Actuality[29] and potentiality[30] are associated with Aristotle and function somewhat differently to some other modalities. As Makin (2012:401) observes:

> Some (im)possibilities involve no reference to a *dunamis* [potentiality] possessed by some bearer or other. It is possible that no-one attends my party, it is possible that there be a sea battle tomorrow, it is impossible now that I died in childhood. We are not drawn in this sort of case to seek some capacities or abilities possessed by various objects in virtue of which the (im)possibilities obtain.

Indeed, this difference between the actuality/potentiality distinction and the possibility/necessity distinction has been a worry for contemporary defenders of *de re* modalities. Barbara Vetter (2015) worries, for example, about cross-temporal potentialities, such as the potential for Vetter to have been taller than Socrates. Jessica Leech (2017) has argued against Vetter's (2015) defence of potentialities, because of the difficulty of accounting for possibilities, especially *de re* possibilities concerning what could, but hasn't yet happened.

The difficulty in saying *of* anything in the future that potentially won't happen will be exactly what I will use to get the past/future asymmetry I'm after. I will disagree with Vetter's account of potentiality, then, to make some of the 'problems' for her account *features* of mine. I wish to claim that it can be necessary that it will be the case tomorrow that there is a sea battle, without there being some sea battle that possesses the potential to happen tomorrow. (There might be some actual geopolitical tension that has the potential to escalate into armed conflict, however.) Likewise, I wish to claim, of past sea battles, that they lack the potential not to have happened. This is true, I wish to claim, even if it was that case last week that nothing settled whether there would be a sea battle or not.

Actuality and potentiality are clearly related, but not as straightforwardly as our interdefinable sentential operators $\Box$ and $\Diamond$. Makin (2012:402) notes that Aristotle argues at length that actuality is primary relative to potentiality.

We seem to end up with a number of features of actuality:[31]

[28] When we model the semantics for future truth, we may well appeal to descriptions of future times that involve names, so in such semantics we may well quantify over future sea battles. But as R. A. Briggs and I (2012:294) argue, 'Quantifying over everything that is or was or will be is like quantifying over all detectives, real and fictional.'

[29] Greek: *Energeia* cf. §4.1.1. where it was translated as 'activity', but its wider range was acknowledged.

[30] Greek: *dunamis* cf. §§4.1.2 and 7.1.2, where the alternative translation of 'capacity' was also given.

[31] The scholarship on Aristotle is highly debated here, as Makin notes, so think of this as introducing my own term of art, which may or may not be what Aristotle had in mind.

- Actuality has explanatory priority: we explain potentiality in terms of the actuality that is potentially manifested.
- Actuality has ontological priority: potentialities must be potentialities *of* something actual,[32] but actualities needn't be actualities of anything potential.
- Nothing can have a potentiality earlier than a time at which there is an actuality for the potentiality to be grounded in.
- Actualities often arise because of an attempt to realise some potential. So I form a plan (perhaps by contemplating what I have the potentiality to create) and the result of that plan is an actuality which is no longer dependent on my plans (once I have realised my plan, I don't need to keep planning it).

Vetter (2015) explains potentialities at some length, explaining some useful features we can add to the previous list:

- Collections of things can have joint potentialities. So two people can have the joint potentiality to play a game of pool against each other. One person can have the potentiality to play a game of pool with someone who themselves has the potentiality to play a game of pool, but no potentiality to play a game of pool *simpliciter*.[33]
- Something can have iterated potentialities. So Vetter has the potentiality to be a grandmother in virtue of the potentiality to have a daughter, which would have the potentiality to have a daughter.[34]

This can sound a bit mysterious, so here's an Aristotelian-style example to illustrate the concepts: building a house. If I wish to build a house, I come up with a plan, and in planning and executing that plan I am constrained by the things I can potentially do. In such a situation, there is no house such that I can say *of it* that I'm planning to build it. We might occasionally appear to say of some drawing or model that we're going to build it, but we usually mean that we are going to build the thing that the thing we pointed to represents. One can imagine a facetious remark, where you point to a scale model of a house and say, 'I'm going to build this house', and I peer down at the tiny house and reply, 'I don't think you'll fit in it'. The scale model is not the house that you are expressing the potential to build; merely a representation of it.

So, I've got this plan to build a house, but no house. I then embark on a process of house-building. I still have no house such that I'm building it, but I do have a process of which I can say that it has the potential to result in a house. One can imagine my construction technique being dismissed with the comment that my work doesn't have the potential to result in a house. But this seems to amount to the same as saying either that the process will not reach any conclusion at all, or that the result of the process won't be a house (but e.g. a shed, or a fortress, or a deathtrap or whatever).

32 Or potentialities of something actual, etc.

33 One could solitarily pot pool balls with a pool cue on a pool table. But this isn't the same as playing pool.

34 Vetter need not have a daughter to have such an iterated potentiality, but Vetter could not come to have a granddaughter without first having had such an iterated potentiality.

Now, this housebuilding process not only has the potential to result in a house but also has some actuality. If the process is taking place, there will be some changes taking place – processes are changes which change, after all. Indeed, thinking back to our discussion of processes in §4.1.1, housebuilding is a characteristically *telic* process, in that it aims towards some end (a house) but involves lots of other processes as parts. The process of getting planning permission, of digging the foundations, of pouring concrete, of building walls, of making the structure weathertight, of snagging, etc. Some of these processes are telic and others are atelic, but the process of building a house will be made up of lots of smaller changes. These changes will be changes from potentiality to actuality. So, I can say of a hole in the ground that it has the potential to hold the foundations of a house, and I can say of some wet concrete that it has the potential to be level, and I can say of a structure that it has the potential to be weathertight. The housebuilding involves an accumulation of actualities which themselves realise the potential of the housebuilding process to result in a house. Before pouring the concrete, I can say that the foundations have the potential to be level, but there are no foundations *of which* I can say that they are potentially level; I'm really saying it's possible that I will have a house with level foundations. One could view this as an iterated potentiality: the housebuilding process has the potential to result in some foundations, and those foundations, when such there are, will have the potential to be level.

One key feature of this example is that there is a change from potentiality to actuality, but also, once we have new actualities, there are new potentialities, for those new actualities can have potentialities of their own. But these new potentialities will not be inconsistent with what was possible before; Vetter cannot have previously lacked the iterated potentiality to have a granddaughter *and* have a daughter who has the potential to have a daughter.[35] What's new is that there is a daughter *of whom* having a daughter is a potentiality.

7.4.2 Asymmetry of reference

In Chapter 6, when discussing asymmetries between past and future, one of the asymmetries mentioned was one of reference. We can see how that might come in here. J. J. Valberg (2012) claims that being, once it is, cannot 'unbe'. From this, he gets to the claim that there must be a difference between past and future, in that not even God could make the past unbe, but God could, miraculously, prevent something that is causally determined to happen in the future. Once something has actuality, it is available for singular reference. That is, we can refer demonstratively to *that* thing rather than merely referring via particularised descriptions that happen to be satisfied.[36] Valberg's

[35] There's a wrinkle here. Given that I don't accept that potentiality implies actuality, if Vetter has an actual daughter, the potential for that daughter to have a daughter will not result in an *iterated* potentiality for Vetter, in the sense that it will no longer be the potential to have a potential, but will be a joint potentiality depending also on the actual daughter. However, Vetter must previously have had the potential to have a daughter before it was actualised.

[36] Cf. Kaplan (1989), who introduces the term 'dthat' for demonstrative reference. The distinction between singular thought and particularised thought may be helpful here. Fitch and Nelson (2018) characterise this difference as follows:

thought seems to be that once we have something available for singular reference, it cannot cease to be a possible object of singular reference (though it is possible that no one is able to refer to it). The impossibility of God's making something 'unbe', according to Valberg, seems to be the impossibility of making it the case that that thing is no longer a possible object of singular reference. He offers this asymmetry of reference as an explication of the modal asymmetry between past and future.[37] That all the actual things are in the past would generate just such an asymmetry of reference – in order to singularly refer to something we must have some actual thing to refer to.

This asymmetry of reference could be achieved in two ways. One is that, because of a causal asymmetry, we only stand in the right sort of causal contact with past things to singularly refer to them. Another is that there is an asymmetry between past and future in terms of what is there to be referred to, as Valberg suggests. Valberg pushes for an asymmetry of being to go with the asymmetry of reference, as a way of explicating the modal asymmetry between past and future. This is because he considers the possibility of an omniscient God referring, in order to avoid the claim that the asymmetry of reference is merely a product of our embeddedness in a particular epistemic position. Valberg's argument isn't going to satisfy anyone who thinks that the apparent asymmetries between past and future are mere products of our embeddedness in a world with an entropic gradient, but it does give us some link between a modal asymmetry and an ontological asymmetry of some kind.

If the modal distinction between past and future is that between actuality and potentiality, and we accept the claim that actuality has ontological priority over potentiality, then it follows that there are no merely potential things to which we can singularly refer because there can be no merely potential things. We are in danger of getting ahead of ourselves, however. Although I am going to endorse the claim that actuality has ontological priority over potentiality, discussion of an asymmetry of being can wait until Chapter 9. For now, let's explicitly allow that the actual/potential distinction is compatible with either actuality or potentiality being more fundamental, or of the two concepts mutually depending on each other.

> 'A singular proposition is *directly* about an individual whereas a particularized proposition is *indirectly* about an individual in virtue of that object satisfying a condition that is a constituent of the proposition.'

The asymmetry of reference *could* be framed in terms of an asymmetry in singular propositions; there are singular propositions about past/present entities, but only particularized propositions about merely potential future entities. Singular thought requires actual entities. Framing it in this way however, would commit me to a particular view of the metaphysics of propositions. The asymmetry of *availability* of entities for singular (rather than particularised) thought is what matters here, rather than the particular account of propositional content that makes the distinction between singular and particularised thought. We'll return to some of these issues in §9.3.

[37] Note, however, that he doesn't frame the modal asymmetry in terms of the actual/potential distinction.

7.4.3 Causal fizz revisited

Let's consider Zagzebski's dilemma again. If the past has actuality and the future is the potentiality of the Universe to evolve in certain ways, then the distinction between past and future is broadly modal. What's more, if the past has actuality, and the potential to evolve in only one way (i.e. any alternatives are ruled out), then there will be truths about how the Universe will evolve, but it does not follow from that that the future evolution of the Universe has actuality. We have a modal difference between past and future that doesn't transfer in the way that $\Box$ and $\Diamond$ do.

One particularly satisfying feature of using the actuality/potentiality distinction to get the past/future asymmetry is that we get an account of presentness out almost for free. I suggested in my (2016:705):[38]

> Ongoing (i.e. present) events seem naturally placed between the potential of the future, and the fixed actuality of the past; ongoing events are precisely those that have some fixed actuality (i.e. they have begun) and some potentiality (i.e. they are unfinished).

Perović (2021) has since persuaded me that it is *processes* (as I used the term in §4.2), rather than events, that I want to describe as having a mixture of potentiality and actuality. Processes, as irreducibly dynamic goers-on in time, have some actuality and some potentiality. We might think of events (as I use the term in §4.2) as being temporally extended actualities that are excreted by processes.

The present, then, is where actuality and potentiality meet and is the locus of change from potentiality to actuality. But we needn't think of the present as a time, but rather as a property of processes, things or stuffs that can be the proper subject of change from potentiality to actuality. Thus, if you buy the argument of Chapters 3 and 4 that the change in question must be McTchange, the passage of time itself is a McTchange from potential to actual.

In Chapter 6, I argued that the asymmetry between past and future was causal/modal. I appear to have given a modal distinction here, concerned with metaphysical modality. What has it to do with causation? In §7.2.3 I introduced causal fizz as whatever the modal/causal element of the nomological package is, if such there be, that we appeal to in explaining the past/future asymmetry. So, at least in my nomenclature, I have suggested that something causal is going on. I will now try to flesh out the concept of causal fizz a little, so the relationship between the causal notion of proper time, from Chapter 5, and actuality and potentiality can be seen.

Causal fizz: the causal process whereby the potential becomes actual, in accordance with the laws of nature.

Causal fizz is part of the nomological package. It is crucial for understanding causal fizz to acknowledge that causal fizz is intimately related to other parts of the nomological package. The proposal is not to add an independent feature of the world

[38] Kroll (2020:255) defends the view that 'time passes and it does so in virtue of there being events in progress', which I take to be very similar to the view I defend here.

in addition to the rest of the nomological package. Rather, the proposal is that laws, chances, counterfactual dependencies, and dispositions are all to be understood as manifestations of the same modal reality, and causal fizz is a further manifestation of that modal reality. As I said in §6.6, there might be a variety of stories of how the different elements of the nomological package fit together, but the facts about the laws could hardly contradict the facts about chances, counterfactuals or dispositions. Whatever the fundamental modality is, the various forms of nomic necessity we are committed to must play nicely with each other.

Although actuality and potentiality are metaphysical modalities, the necessity of change from potentiality to actuality is a nomological modality. So while it might be necessarily true that some potential *will be* actualised – made true with the necessity of laws of nature – it can still be the case that some actuality possesses multiple potentialities that will not be actualised. It can be metaphysically possible that my ice-cream melts, for it has the potential to do so, even if, as a matter of nomic necessity, it will not, because my freezer will keep it frozen.[39] If my ice-cream has melted – *actually* melted – then it lacks the potential to never have melted. Although it is metaphysically possible to have *some* ice-cream that never melts, it is no longer metaphysically possible for *this* ice-cream to never have melted. We can say it *could have* stayed frozen, but not that it could stay frozen; that's impossible now. And this isn't a matter of impossibility-given-the-laws-of-nature, but something stronger. Even God, who can miraculously override the laws, cannot make what's been done never have been. This explains why there can be nomic necessities for the future, and the future still be open. The future is metaphysically open, since there is no actuality to the future, but (if determinism is true) the future is nomically settled.

To return to Zagzebski's difficulty, we can have necessary truths about the future. The actual past and the actual laws of nature can constrain things so that there is only one nomically possible future. In this sense, necessity can transfer. But because it does so through the laws of nature, it only does so with the modal strength they provide: nomic necessity. The actuality of the past doesn't transfer into the actuality of the future.

The change from potentiality to actuality has to be a change in the potentiality *of* something. An obvious candidate is the causal process of change itself. So, just as an actual house is a product of a process of housebuilding, an actual past is the product of a causal process of producing, bringing about or generating effects. The emphasis on the process of bringing about effects, rather than the actual past, as the thing that has the relevant potentiality is, perhaps, rather artificial. As discussed above, there are joint potentialities, and the actual past and the process of bringing about effects are jointly going to have the potentiality to generate new actuality. In §1.3, we discussed Fine's (2005) objection to a frozen world that has a solitary dynamic fact. The irreducibly dynamic causal process of bringing effects about is important because this process is meant to provide the difference between a static world and one that, as a matter of nomological necessity, McTchanges changes.

[39] Mackie (2006) points out that it is important to distinguish what it is metaphysically possible for an actual thing, for example this ice-cream, to *become* from what it could have been *instead*. The claim here is restricted to how it is possible the actual ice-cream could change.

I described causal fizz as 'producing' effects. Ned Hall (2004) makes a distinction between two aspects of causation: dependence and production. Hall explicates the notion of dependence as follows: 'event *c* is a cause of (distinct) event *e* just in case, had *c* not occurred, *e* would not have occurred' (p.225). Dependence is the familiar sort of counterfactual dependence that forms the basis of counterfactual theories of causation (e.g. Lewis 1973; 2004). Hall (2004:225) has more difficulty in characterising production. He says, 'we evoke it when we say of an event *c* that it helps to *generate* or *bring about* or *produce* another event *e*'. Causal fizz therefore evokes the productive aspect of causation. Causal fizz is a productive process. It is a process of bringing about. But what does it bring about? Actuality. It is a process of change from potentiality to actuality.

Wesley Salmon (1998), and Phil Dowe (2000) defend 'process' views of causation. They think that there are certain interactions that involve a productive element – the causal interactions – and, on Dowe's view, 'causes and effects are linked by a causal connection by virtue of the fact that a quantity exchanged in the "cause" is possessed by a causal process that links the cause to the effect, where it is also exchanged' (p.171). On their views, we are to think of two 'worldlines' – paths through space-time – that interact in such a way that there are conserved quantities. These paths through space-time don't McTchange, however. The causal interactions are plausibly productive in the relevant sense. Salmon and Dowe's views are likely to count as having a productive element by Hall's lights; although Hall doesn't reference them, he is clearly interested in similar cases. For Hall, as for Salmon and Dowe, production is a feature of *individual* causal interactions and is used to explain which of a number of specific phenomena are causally responsible for bringing about an effect.

Production, for Hall, Salmon and Dowe, is a question of identifying a causal relation in amongst a complex web of causal dependencies and identifying some of the causal antecedents as producers of the effects. We might call these individual productive interactions 'biffs' following Lewis (2004). To take the case of one billiard ball striking a second and causing the second to gain momentum, we might say that the first billiard ball *biffs* the second, transferring momentum to it (such that momentum is a conserved quantity across the interaction). We might expand from cases of billiard balls to cases of agential responsibility. When Suzie bombs an enemy target, she has *biffed* it through a causal chain of her biffing the release button, and that leading to the biffing open of the bomb-bay doors, and the bombs biffing the enemy target. But although the bombing was causally dependent on Billy's shooting down enemy fighters (without which Susie's bomber would not have got through), Billy did not biff the enemy target. He did not bring about the bombing; Susie did (cf. Hall 2004:229).

There is no incompatibility between causal biff and a productive aspect of causation as changing potentiality into actuality. We may, for example, identify Suzie as making the bombing actual by biffing the button, rather than Billy. But appeals to causal biff don't always go with McTchange; Salmon's and Dowe's views seem to allow for biffs without any role in their views for McTchange. What we need, then, is a slightly different kind of causal production that seems more fitting to play the role of changing potentiality into actuality. Biff is usually used to talk about something that is present in individual causal interactions, in contrast to some others. Not all cases of causation have biff in this sense (e.g. causation by omission). The difference between causal biff and causal fizz is as follows: biff is for individual elements within a causal system, whereas fizz is a process of the system as a whole.[40] Fizz is that minimal productive aspect of causation required for there to be any effects at all.

Recall our billiard ball example. One billiard ball biffs a second billiard ball. The explanation for the second billiard ball gaining momentum is that it was biffed by the first, but the explanation for anything happening at all is that causation brought it about that there continued to be a table, that gravity continued to obey an inverse square law, that the second ball didn't do something miraculous, like turn into a pigeon.

Biff is usually brought in to explain a certain sort of necessity, the necessity provided by one billiard ball to another, or by one agent to the bombing of an enemy position. These necessities are interesting and worth explaining. But necessities in need of explanation abound. We know, as a matter of nomic necessity, that the second billiard ball will not turn into a pigeon. When we watch a magic trick in which this apparently happens, we conclude that there has been sleight of hand, not that we need to revise our views about whether billiard balls sometimes become pigeons. And we are right!

Our knowledge seems in need of explanation. As Dorothy Emmet (1984:84) says:

> Causation is invoked to explain changes, and persistence might be the simplest kind of change, from one phase of, say, a system, to another. In another sense, this could of course be a non-change, if the properties of the system remain the same. But even that is a change from being the system as it was to being the system as it is.

Causal fizz is meant to help us explain how there can be a non-transferable causal/modal asymmetry between past and future that is compatible with causal determinism settling truths about how the future will go. Not everyone will accept change from one phase of a system to another as being properly called *causal*. Cartwright (2020:297) says, 'On the process/interaction account, *causation* – one thing *making* another happen – is something over and above things happening as laws describe.'

Cartwright, unlike Emmet, does not think of the evolution from one phase of a system to another as being a causal relationship. As Cartwright (2020:297) explains:

> So earlier events in a process do not cause later ones. . . . For more complicated – more realistic – cases, features essential for producing the effect in the later interaction *evolve* from those produced at the earlier one. . . . For this to make sense, of course, we need to accept that there are natural processes of evolution for some features, described perhaps by natural laws, and that these are not processes in which earlier states cause the ones that evolve later. This is just what this feature does over time, much as we suppose that a body in motion tends to remain in

40 Cf. Emmet's (1984:86) distinction between *transeunt* causation (the action of one thing on another) and *immanent* causation (activities within a system as a whole, where the product is not separable from the activity producing it). I don't know whether Emmet means what I mean, but the distinction seems to be related. There is a question of what we mean by 'the causal system'. I will leave this open to contextual determination, but presumably, this will come down to causal sets (cf. §5.4), or sets of such sets, when we're dealing with causal fizz in fundamental physics.

> motion – that is its natural evolution with no causal interactions necessary to get from earlier to later states.
>
> Although causality is missing, there is no suggestion that it is arbitrary what features are where when.

This non-arbitrary evolution from one state of the universe to another is exactly what causal fizz is intended to explain. But Cartwright argues that no causal explanation is needed for such temporal evolution. I suspect the disagreement here may be largely terminological. No explanation in terms of causal biffs is needed to explain the evolution of a process according to the laws of nature. And Cartwright, Emmet and I agree that the evolution of the universe from one state to another is non-arbitrary.

I think causal fizz does do explanatory work for us. Let's begin with my (2015) suggestion that we need explanations for why time continues to pass, why time passes without skipping times, and why there couldn't be past events that were never present. Causal fizz looks well set to help us with each of these. Why does time not suddenly grind to a halt? Time continues to pass because there is some process that, as a matter of nomological necessity, keeps it passing. Why does time pass in an orderly manner, rather than skipping ahead to the distant future, or adding events to the distant past? Because causal fizz is a law-governed process that changes potentiality into actuality by building on what's actual (or 'extruding from it' as J. L. Mackie phrased it). It is characteristic of causal relationships that proximate causes screen off more distal causes, so – being a causal process – the effects that causal fizz makes actual are explained by proximal actual causes.[41] If one accepts this view of causal fizz, it is natural to say that *what it is* for events to be past is to have been brought about by causal fizz. So the question of there being past events that were never in the process of being brought about doesn't arise.[42] As Simons (2018:54) succinctly puts it: 'There is time because stuff happens, and time is directional (at least above a certain granularity) because the causation that makes stuff happen and moves processes along is irreversible.'

Is this appropriately called *causal* explanation? I think it is at least plausible that it is, but if Cartwright wants to reserve the word 'causal' for biff rather than fizz, I won't object too heavily. What's important for my purposes is that the change that is constitutive of time passing happens systematically with nomic necessity, McTchanges changes.

7.4.4 An evolving universe

What *is* causal fizz, though? In §7.2.3 we were left with a number of questions. I'll spend a moment trying to answer some of them. Causal fizz is a process, rather

41 Cf. Hausman and Woodward (1999), who discuss this screening off in the guise of a condition on representations of causal relations – the Causal Markov condition. See Forbes and Parkkinen (MS) for a discussion of apparent exceptions to this.

42 This is true, at least, for events that have multiple temporal parts. Events without temporal parts will be the limiting cases of events, and will never be present; when they come into existence, they will immediately be limiting cases of the past. Cf. Zeno's Stadium paradox, found in Aristotle's *Physics* 239b33, and discussed in Davey (2007).

than a stuff or a relation, and therefore is irreducibly dynamic. It is a process of nomically constrained McTchange. Given that it is a process of McTchange, it will be irreducible to anything that is a mere variation of states and properties. It's a manifestation of the nomological package. Some element of the nomological package will have to be fundamental, though I don't want to commit to causal fizz being that element. Equally, there needs to be a fundamental enduring entity or process that persists through McTchange in what time it is, such that we can make sense of temporally extended agency and activity discussed in Chapters 3 and 4. But it might not be that causal fizz is that fundamental persisting process, but that the Universe as a whole is.

Causal fizz is not straightforwardly an empirical phenomenon; I don't anticipate physicists setting up an experiment to detect it. In Chapter 2, we considered concerns about detecting McTchange generally, and being part of the nomological package, we would also have to detect hypothetical properties around what could have been actualised but wasn't. Such hypothetical properties are going to be compatible with a deflationary treatment if enough accommodations are made elsewhere in the theory. But this is not to say that causal fizz is theoretically inert from a scientific point of view. We have phenomena – apparent causal constraint in the evolution and persistence of systems – that stand in need of an explanation. Causal fizz is offered as the start of an account of those phenomena. As we come to investigate the propagation of causal constraint through causal sets, for example, we may have empirical grounds on which to refine our theory of causal fizz.

We have, by now, quite a grand picture of the Universe developing. From Part I we have the claim that there is McTchange, and from Chapter 6, we have the claim that there is some modal/causal asymmetry between past and future.[43] In this chapter, I have put forward a candidate asymmetry – an asymmetry between potentiality and actuality – that might satisfy the thought that there is something settled about the past that is open about the future, even if the future is causally determined. I have then invoked a causal process of changing potentiality into actuality to explain how this asymmetry is tied to the change in what's past and what's future.

Causal fizz ties the asymmetry between past and future and the passage of time together quite intimately, then. The passage of time is part of a unified nomological package that can determine what will happen, what would have happened and what must happen. This picture also gives us a distinction in metaphysical necessity between past and future, since once something is actual, it cannot cease to be so. If we think back to the discussions of causal set theory in §5.4, the growth of the causal set is limited by the propagation of causal signals through it, such that the growth of the causal set is a process of sequential causal growth. But, the growth of the causal set is irreversible, because the change from potentiality to actuality is.

[43] I should emphasise that the arguments in Part II assume we have accepted the arguments from Part I; I offer no reasons here to accept causal fizz if you don't think there is McTchange. Likewise, the arguments I offer for causal fizz depend on one's accepting the argument from Chapter 6 that the distinction between past and future needs some kind of modal explanation. If you disagreed with those claims, I have no further defence of them besides the elegance of the package of views provided as a whole.

7.5 Intending the inevitable

What of fatalism? In §7.1, we motivated an open future in terms of a fatalistic worry, which we spent some time trying to characterise. Now we have a proposed asymmetry between past and future; we must examine if it answers our initial motivation. How is causal fizz, a process of turning potentiality into actuality, supposed to connect with the concerns about agency with which we began?

We previously discussed the connection between causing and intervening.[44] The asymmetry between past and future in terms of our capacity to intervene on future events, but not on past events is tied to the direction of causation as given by the change from potentiality to actuality. Having found a way out of Zagzebski's dilemma, we have a directedness to causation that gives a robust difference between past and future, such that we can intervene in one direction but not another. This is because potentiality can become actual, but actuality cannot become mere potential. The change from potential to actual is, *ex hypothesi*, irreversible. We have a distinction between fixtures and options that isn't merely relative to some particular agent but sets the conditions for agency. Options must be potential, and actuality should, for most practical concerns, be a fixture.

In practical reasoning, we may include unknown features of the past amongst the options. Dummett (1964) uses the example of hearing on the radio that a ship has sunk, and one's praying for one's son who was on the ship to have been saved. In such a case, the prayer relies on our having some doubt about whether the reports are true. Dummett insists that this is not a prayer to change the past, merely a prayer that the past turn out not to have been one in which the son drowned. To put it in the language of actuality, the prayer, to make sense, must be interpreted as a prayer to the effect that the son being saved *is* actual, not that there is a change from his being actually drowned to actually saved. The retrospective prayer, as Dummett observes, is therefore something of a special case. In our terms, it is an action that seeks no change in what is actual, rather than seeking to effect a change from potentiality to actuality. Indeed, it may seem that retrospective prayer is equally as practical as crying over spilt milk, and for the same reasons. If something is actual, it is *a fortiori* not optional. Not knowing that it is actual may excuse expending energy on it, but it does not make expending such energy a good use of one's time.

Is it not also true that *future* fixtures are not options for deliberation? Price (1996), Ismael (2012), Fernandes (2017) and so on may argue that the difference between fixtures and options relies on us having multiple epistemic possibilities, and so deliberation doesn't involve an asymmetry between past and future, except contingently. To put it another way, the actuality/potentiality distinction can only be the right distinction for avoiding Zagzebski's dilemma if this gets us a practical difference between past and future. If the relevant practical distinction is a merely epistemic fixtures/options distinction, the problem of fatalism won't relate to a distinction that is between the past and future *as such*.

[44] §§4.5.1 and 6.4.

We saw in the discussion of two-way capacities of §7.1.2 that potentiality (the Greek for which is cognate with the Greek word for capacity) need not be confined to nomological possibility but applies also to our potential to respond to reasons. There are often situations where there is only one practically rational option, but reasons responsiveness requires that we have the potential to do the practically irrational options. Given that whether we are practically rational can still be a fixture, we can still be practically rational when we have only one option. For instance, compare the case where we intend to do something, and that thing is the only option, versus the case where we intend to have done something, and that thing is what we actually did. In each case, the outcome is, in some sense, settled in advance of our action. Nevertheless, one seems a paradigm case of practical rationality, and the other case a paradigm of futility. The difference is that one involves a change from potentiality to actuality that is reasons responsive, and the other case, even if we are very sensitive to the relevant reasons, does not change potentiality into actuality.

I will explore the connection between our emotional responses and the change from potentiality to actuality more closely in the next chapter. For now, we need to observe that the difference between actuality and potentiality, between past and future, provides an asymmetry of intervention, such that there is something futile about attempts to intervene where there is no potential to make something actual, and intervening is reasonable in cases where it is inevitable that there will be an intervention, provided that intervention will change the potential into the actual.

7.6 Time-travel to the past

Considerable ink has been spilt on whether a time traveller called Tim can kill his own grandfather.[45] This case is of interest to us because it is illustrative of the actuality/potentiality distinction in action. The thought is that, given that Tim's grandfather is causally upstream of him, there is no way that Tim can succeed in changing something causally upstream of him, given that we know Tim exists (and does so as a result of things his grandfather hasn't yet done by the time of the assassination attempt). The 'grandfather paradox', as it is known, trades on the thought that if determinism is no barrier to 'free will', or actions being 'up to us', in the case of future events, it can be no barrier to our being the author of our actions in cases where the future and past coincide, as in the case of causal loops.[46] If we want to say that there is something that means that Tim *cannot* kill his grandfather, rather than that he merely *does* not, we need some account of why this case is different to the future-directed case. The modalities of actuality and potentiality give us a link between a sense of cannot (lacking potential) and the fact that he does not (his not doing so is actual).

[45] See e.g. Lewis (1976); Vihvelin (1996); Smith (1997); Rennick (2015); Garrett (2016); Effingham (2020).

[46] In §§5.3 and 6.6.2 where we discussed closed time-like curves. In fn. 11 we discussed Bobzein (2001) on the tendency to run freedom and what's up to us together. I suspect this debate is affected by that conflation.

The actuality/potentiality distinction gives us an explanation since it gives us the asymmetry between future inevitabilities and past actualities that we need to see why Tim is *unable* to kill his grandfather despite having the two-way capacity to do similar things (like murder his children) that he certainly will not, in fact, do. That Tim's grandfather does not *actually* die before siring at least one of Tim's parents means that Tim lacks the potential to kill his grandfather, because there is no potentiality to make actual. Tim does have the potential to kill his children, and, if he has children, they have the potential to be killed by him, *even if, as a matter of nomological necessity, Tim will not do so*.

There are alternative accounts of why Tim can't kill his grandfather. Fernandes (2020), for example, offers the fixtures/options distinction as doing the explanatory work. Tim cannot, as a matter of practical rationality, deliberate about killing his grandfather, because that he exists is not an option for him; it is a fixture. Note that this is merely a constraint on Tim under certain descriptions: Tim cannot, while being practically rational, deliberate about killing his grandfather *under that description*, but he can deliberate about killing Ortcutt the milkman, or the man in the hat, without knowing that either are his grandfather. But in these cases, Tim cannot succeed in killing his grandfather. According to Fernandes, this is because how we evaluate claims about what Tim can and can't do depends on, in assessing the counterfactual, what we hold fixed and what we don't. In this case, we hold Tim's causal relationship to his grandfather fixed (because we need to ask whether he can kill his grandfather). The context of the question does much of this work for us. So Tim cannot kill his grandfather, because there isn't a relevant possibility in which Tim kills someone, and that person was going to be (but was not yet) Tim's grandfather.

The distinction between fixtures and options is going to be enough to rule out the grandfather paradox, without requiring the actuality/potentiality distinction. The actuality/potentiality distinction allows us to say something stronger, viz. that we are stuck with a past in a more robust way than we are stuck with the future options; there are, in fact, no past options, only future options.

The actuality/potentiality distinction I have outlined rules out, *a priori*, intervention on the past. It does so, in part, by claiming that what it *is* to be past is to be intrinsically *metaphysically* immune from intervention in a way that the future is, at best, merely nomologically immune from intervention. Since Fernandes and I both agree that Tim cannot actually kill his grandfather, we cannot tell between our views by seeking empirical confirmation of grandpatricidal time-travellers. Indeed, the difference between nomological and metaphysical necessity is hard to observe, given that we only observe the actual, not the merely possible.

Some time-travel examples may help us, however. Imagine I am trying to write a book.[47] One technique that might appeal is to travel into the future, acquire a copy of the book I wish to write, and then copy it onto the computer and email it to the publisher. If we are considering such a time-travel scenario, we know writing the book

[47] If you are reading this, I am not trying to write *this* book anymore; this book is by now an actual book, not a potential book being actualised by my writing process. But I am probably trying to write another book.

is an option for me. Fernandes can therefore countenance such a scenario. But in this scenario, we seem to be left with the possibility of a causal loop, where the book is never written *for the first time*, but there is a cycle that always involves a completed book being turned from hard copy to soft copy to hard copy, without ever adding or removing spelling errors.

I think there is something wrong with this scenario that isn't troubling about writing a book under conditions of causal determinism. The troubling thing is that the causal-loop scenario doesn't involve me writing a book! I self-plagiarise, but the book has no *author*. My explanation for why this example seems troubling is that authoring a book is a causal process of turning potential into actuality. Because this is a process that starts with an actual book and ends up with the same actual book, it cannot be a process of *authoring* at all.

Given that one intuitive way of expressing the fatalistic worry is that fatalism means I am not the author of my own actions, we can see that this worry about authoring applies to actions more generally. It is not merely counterfactual dependence that we need to account for and be constrained by. The causal loop of self-plagiarism involves a consistent circle of counterfactual dependence. The problem is that we need a change that constitutes the actions taking place or, as Dowker (2014:22) puts it, 'there is nothing to correlate with "the event happens"'. Such causal loops are ruled out because there is no room for anything to *happen* in them. Causal fizz is the process of *happening* that we're missing.

An event happening cannot merely be some time having a property of presentness. We learnt this from Fine's (2005) frozen present argument. And writing a book, or acting more generally, cannot merely be occupying some node in a network of counterfactual dependence, with a set of fixtures and options that shape the practically rational constraints on deliberation. Doing things (and enduring things that are done to one) involves a change. In the previous chapter, I argued that a modal/causal change gives direction to time. In this chapter, I have argued that that change is a change from potentiality to actuality. But this means that the very idea of time travel to the past is ruled out because if you can travel there, there must be some potential you turn to actuality by doing so, and so it will always be present when you get there.

8

Thank goodness that's over!

What is complete does not need to be spoken about. What has run its course does not need admonishment. What is gone forever does not need censure.

—Kongzi, *Analects 3.21*

We have already discussed a number of asymmetries between past and future. In addition to the ones that philosophers of physics and metaphysicians may be interested in, which we have discussed in the preceding two chapters,[1] there are asymmetries in our attitudes towards the past and future. It is to these that we now turn.

Even if we think time is to be understood primarily as a feature of the natural world – a world that humans exist within – it is uncontroversial to observe that time is a significant part of our human lives. Indeed, the philosophical interest in time would be much less if it didn't seem to impact our attitudes, feelings and the meaning, such as it is, of our actions and lives. Human, or subjective, time does not, as I argued in §2.1, allow us to immediately infer a metaphysics. But it does give us information about what time is and how it works; in turn, it is constrained by the objective nature of the phenomena of time.

In this chapter, I will explore a further asymmetry between past and future – a value asymmetry where we prefer pleasures to be future and pains to be past – and consider arguments that such an asymmetry of preferences is irrational. We can accept a view that says we cannot prefer outcomes just because of how far away they are or what their location in time is, provided that these considerations don't involve changes in uncertainty, the narrative features of our lives, actual (rather than potential) acquaintance with things we value or disvalue, or transformative experiences. In practice, however, time causes us to change our minds in all sorts of ways. This will show up in why we treat the time after our death differently to the time before our birth, how we treat loving relationships, and why suicide is often bad for the victim despite the fact that they won't regret it.

In addition to considering whether a value asymmetry is *irrational*, I will consider whether we can make sense of the appropriateness of different emotional responses. In particular, I'll discuss Prior's (1959;1962b) 'Thank Goodness That's Over!' argument that feelings of relief only make sense if something is past, rather than merely earlier in

[1] Or the succeeding two chapters if you are reading the book in reverse.

time. I will argue that the picture is more complex than Prior's argument and responses by Mellor (1981) and MacBeath (1983) make out, revealing a picture of emotional attitudes that relate a range of factors from our epistemic state to our control, to merely imagined pasts, no longer feasible futures and lost future pasts.

Finally, I will return to our fatalist worries from the previous chapter by considering the attitudes of creatures for whom there is no past/future asymmetry of actuality and potentiality. The creatures I will discuss are the alien Tralfamadorians from Kurt Vonnegut's (1969) novel *Slaughterhouse 5*, whom I briefly discussed in §5.2. They have an attitude of equanimity to all events – 'So it goes.' is their only comment on disasters and triumphs alike. The Tralfamadorians represent, I will argue, a fatalist resignation that is a response to having no room in their metaphysics for a change from potentiality to actuality.

I will conclude the chapter by reflecting more generally on the relationship between action and passion, and actuality and potentiality, arguing our emotional states are not merely to be understood as passions we endure, but as active responses to a changing, value-laden world.

8.1 The temporal value asymmetry

Decision theorists have tried to account for a widespread tendency to treat outcomes differently depending on whether they are past or future, and how distant from us they are. In this section, I will set up the framework in which decision theorists operate (§8.1.1), and outline the argument that changing our minds opens us up to exploitation in ways that it is irrational to accept (§8.1.2). I will then consider ways in which this argument either idealises things or fails, such that we still, in practice have many permissible ways in which the passage of time explains our reasonably changing our minds. These will include the fact that our lives are narratives rather than simply sequences of intrinsically valuable moments (§8.1.3), that we relate differently to actual people than merely potential people (§8.1.4), and that we undergo transformative experiences that change who we are and what we value (§8.1.5).

Consider, as an example to get us going, a thought experiment by Derek Parfit (1984:165–6), which I shall quote in full:

> I am in some hospital, to have some kind of surgery. Since this is completely safe, and always successful, I have no fears about the effects. The surgery may be brief, or it may instead take a long time. Because I have to cooperate with the surgeon, I cannot have anaesthetics. I have had this surgery once before, and I can remember how painful it is. Under a new policy, because the operation is so painful, patients are now afterwards made to forget it. Some drug removes their memories of the last few hours.
>
> I have just woken up. I cannot remember going to sleep. I ask my nurse if it has been decided when my operation is to be, and how long it must take. She says she knows the facts about me and another patient, but that she cannot remember

> which facts apply to whom. She can tell me only that the following is true. I may be the patient who had his operation yesterday. In that case, my operation was the longest ever performed, lasting ten hours. I may instead be the patient who is to have a short operation today. It is either true that I did suffer for ten hours, or true that I shall suffer for one hour.
>
> I ask the nurse to find out which is true. While she is away, it is clear to me which I prefer to be true. If I learn that the first is true I shall be greatly relieved.

This is quite an involved thought experiment, so let's unpack it a little. We are comparing pain with pain, and the pains are meant to be qualitatively similar enough that we can see that, all things being equal, more pain is worse than less pain. The situation is meant to be one in which uncertainty and future painful memories are not relevant (though our ability to imagine these away may be less than perfect). Our *relief* at discovering pain is in the past is meant to reveal a *preference* for pain to be in the past rather than the future. That is, we have a preference asymmetry between past and future, such that we are biased in favour of the future.

I'm going to come back to different attitudes in §8.2. For now, let us consider this comparatively straightforward case of bias towards the future set up in terms of pains at different times. Greene and Sullivan (2015) and Sullivan (2018) have argued that such time biases are irrational. In the next subsection, I will set up a fairly basic framework for decision theory, so we can understand why avoiding time bias is purportedly A Good Thing.

8.1.1 Decision theory

In addition to this future bias, we also have a bias towards the proximate in time. We would rather pain was in the distant future than in the immediate future (or in the distant past, rather than the immediate past). This seems widespread. Caruso et al. (2008) and Burns et al. (2019), for example, have found that people will assign different monetary values to the same hypothetical scenarios (e.g. compensating someone for work undertaken or for wrongful injury, thanking a friend, extending a holiday) depending on whether the scenarios described things as happening in the future or the past. Clearly, such biases exist. But are they rational?

To frame the question a bit more precisely, we can delve into Decision Theory. If you are already familiar with Decision Theory, you should expect greater utility if you skip to the next subsection; I'm just explaining some key concepts like *expected utility*, *logical dominance* and *money pumps*. Decision Theory uses information about preferences and expectations to arrive at a view about what the best decision is. I'm going to start slowly, partially because I suspect some readers will be resistant to trying to put numbers onto things in the way Decision Theory assumes we can.

Let's assume that you prefer having more money to having less money. Imagine two options, A and B. Option A has a 50 per cent chance of getting you £10 (and 50 per cent chance of losing £2), and option B has a 20 per cent chance of getting you £5 (and 80 per cent chance of getting nothing) (see Table 8.1). You have to pick one.

Table 8.1 A Simple Decision Problem

	A wins (50% chance)	**A loses (50% chance)**	**Average money**
Choose A	£10	−£2	(10 × 0.5) + (−£2 × 0.5) = £4
	B wins (20% chance)	**B loses (80% chance)**	
Choose B	£5	£0	(£5 × 0.2) + (£0 × 0.8) = £1

You can see that you are more likely to have more money on option A than you are on option B.

You should expect, on average, to get £4 on option A, and on average to get £1 on option B. However, the worst outcome on option A is worse than the worst outcome on option B. It looks like you would be rational to choose A over B; you are more likely to get more money. But if you were particularly risk-averse – that is, you are concerned to avoid the worst outcome – you should choose B over A, since you are not possibly going to lose any money.

But we don't want to assume that each bit of money matters to me as much as any other. For some, £1 is the difference between eating or not, and for others, it is an insignificant rounding error. So, let's stop thinking about money and start thinking about some quantity that represents how much value an outcome has for me. We'll call this value the *utility* of a particular outcome. If we want to go from cash values to utilities, or feelings of pleasure and pain, or from any other real-world outcome to utilities, we will need some mapping of outcomes to utilities called a *utility function*. We might think that this mapping is revealed by your preferences (i.e. how much you would value the outcome) in different situations, but we don't have to assume that. It might be that there are some objective values that you should adjust your preferences to match. It might be that the utility function is mathematically quite neat, such that you value the next unit of currency half as much as you did the one before it, or it might be that it is much messier. There might be a sum of money below which you barely value money at all, and a sum above which you actively disvalue it. The reason for using utilities is that you can insert whatever utility function you want into the examples, and the results will still work for those utilities.

Similarly, what you should rationally do might be based on your epistemic situation, rather than what is in fact the case. Although you might have reason to do whatever happens to be good for you, in working out what you should do, you can only appeal to information that you have. So rather than the *chances* of outcomes, we might be interested in *expectations* of them, that is how likely we should think they are, given our evidence.

So, we can combine utilities and the expectations to get an expected utility, as in Table 8.2. We can then compare the expected utilities of the various outcomes to work out which option has the highest expected utility.

It is always rational to do what maximises expected utility. So, option C here has higher expected utility than option D, and it will always be rational to choose C over D. In fact, C is *logically dominant* over D; no outcome on C is worse than any outcome on D, and C has an outcome that is better. So in that case, you would, according to decision theory, be rationally compelled to choose C over D and irrational to choose D over C.

Table 8.2 A Decision Problem with Utilities

	C succeeds (expected 50% of the time)	**C fails (expected 50% of the time)**	**Expected utility**
Choose C	utility of 12	utility of −2	(12 × 0.5) + (−2 × 0.5) = 5
	D succeeds (expected 50% of the time)	**D fails (expected 50% of the time)**	
Choose D	utility of 8	utility of −2	(8 × 0.5) + (−2 × 0.5) = 3

What about choosing B over A (to recall the previous example)? You might not be worse off; you could choose A and lose £2, and you could choose B and get nothing (but lose nothing). Indeed, you might be unlucky; 10 per cent of the time you'll choose A and lose £2 when choosing B would have resulted in you winning. You won't have been wrong to have chosen A (given the information you had), but you will be worse off 10 per cent of the time.

There are some special cases where you are guaranteed to be worse off as a result of your choices. In particular, there are cases where you can be offered a series of bets that will exploit your beliefs so that, if you consistently choose to maximise expected utility at each step, you would be led to sure ruin. These cases include ones where your expectations are incoherent because your levels of confidence (known as your 'credences' or 'degrees of belief') in your expectations that the different outcomes will happen don't obey the axioms of probability. You might think each of three incompatible options has a 35 per cent chance of happening, for example. This openness to exploitation is known, rather offensively to turf-accountants of the Netherlands, as a 'Dutch book'.[2]

It's not just degrees of belief that can be incoherent. If you have preferences of a certain kind, you can be open to exploitation. In particular, if you have *cyclic* preferences, such that you prefer *x* to *y*, *y* to *z* and *z* to *x*, you can be exploited such that you are always paying a small cost to go for a preferred option. If the expected utility of changing to a preferred outcome is higher than the cost of changing (and there is no risk involved), you are rationally compelled to change. Since someone with cyclic preferences will always be in a situation where there is some expected utility to changing to a preferred outcome, there will always be some cost they are prepared to pay to do so. So, someone could set up a *money pump* that extracts that cost simply by rotating which outcomes are on offer. Note that *money* is not of great importance here. 'Money pump' has

[2] See Vineberg (2016) for a discussion of Dutch book arguments. The axioms of probability theory are as follows:
1) No proposition can have a negative probability.
2) If a proposition *p* is a necessary truth, then it has a probability of 1.
3) If proposition *p* and proposition *q* are incompatible, then the probability of (*p* or *q*) equals the probability of *p* plus the probability of *q*.

become a generic name for a situation in which cyclic preferences can be infinitely exploited, but we could have 'utility pumps' that extract utility for any utility functions that involve cyclic preferences. Philosophers often permit eccentric utility functions – it takes all sorts to make a world, after all – but inconsistency, whether in beliefs or in preferences, is something that they tend to worry about.

Preference reversals suggest a kind of time inconsistency of particular concern to Decision Theory. Let's introduce a technical term: '*Consistency*'. Consistency is avoiding preference reversals over time.[3] It is clear that some examples of inconsistency are bad. Indeed, many real-life examples of inconsistency are associated with worse economic, social or health outcomes.[4] Nonetheless, real-life examples of inconsistency abound.[5] There's a problem of explaining when, if ever, such preference reversals – failures of our technical sense of Consistency – are permissible. We'll examine this question in the next subsection.

8.1.2 Diachronic rationality

Imagine you are planning a trip to visit the libraries of the Netherlands. Let's suppose you value it less the further away it is, so if it's happening tomorrow you value it a certain amount, and if it's a month away you value it 67 per cent as much. That provides a 'discounting function' on your future preferences based on how far in the future they are. Some discounting functions will have points on which my priorities change, whereas others will be consistent. In the example I gave, let's say things get discounted by 20 per cent in a week, but only by another 10 per cent after another week. I could sell you a trip scheduled for two weeks' time, and buy it back from you a week later, and sell it back to you on the day of departure, making a profit overall. At each step, you would think you're getting a good deal, but I would have extracted a profit from your getting the same outcome you would have had if you had done nothing for two weeks. Unless there is a good reason for exposing yourself to such exploitation, let's assume it is irrational to do so.[6]

There are discounting functions that avoid these static preference reversals, and they have a property called *Stationarity*.[7] Ranking two outcomes, I'm stationary with

[3] For those interested in formal treatments, I borrow this formulation from Callender (2021: 12):

> Consider outcomes $x, y \in X$ whose values are real numbers, and $t, t' \in T$, the set of dates, such that $0 \leq t, t'$, and delays $\Delta_2, \Delta_1 \geq 0$. A set of time preferences is consistent if

Consistency $(x, t + \Delta_1) \sim_t (y, t + \Delta_2) \Longleftrightarrow (x, t + \Delta_1) \sim_{t'} (y, t + \Delta_2)$.

[4] As with the discussion of the problems of fatalistic reasoning in Chapter 7, the problems of preference reversals may be a response to a bad situation, rather than simply a cause of bad outcomes. See Story, Vlaev, Seymour, Darzi and Dolan (2014) for a review of some of these complications.

[5] See Urminsky and Zauberman (2016) for a review.

[6] Vineberg (2016) surveys the debate about whether we should take such exploitation as a hallmark of irrationality.

[7] Again, for those interested in formalism, this formulation is also from Callender (2021: 8):

> Consider outcomes $x, y \in X$ whose values are real numbers, and $t, t' \in T$, the set of dates, such that $0 \leq t, t'$, and delays $\Delta_2, \Delta_1 \geq 0$. A set of time preferences is stationary if

Stationarity $(x, t + \Delta_1) \sim_t (y, t + \Delta_2) \Longleftrightarrow (x, t' + \Delta_1) \sim_t (y, t' + \Delta_2)$.

respect to my preference ranking if my ranking depends only on the values (expected utilities or whatever) of the outcomes, and the delay between them. If a delay of a week is discounted the same, regardless of whether it is from today to a week from now, or from a week to two weeks from now, that discounting function is constant ('stationary') through time.

Before moving on to dynamic preference reversals, let's just discuss discounting functions briefly. Greene and Sullivan (2015) and Sullivan (2018) have argued that (non-zero) discounting functions are irrational. Even stationary discounting functions should be avoided. Uncertainty is already built into the expected utilities, so there is no need to add a discounting function just to accommodate distance from the present. Callender (2021) defends some circumstances in which we reason well when we adopt discounting functions that violate Stationarity. My strategy, in what follows, is to grant Greene and Sullivan (2015) and Sullivan (2018) that mere delay, once we've controlled for uncertainty and changes in the utilities, should not get discounted. I side with Callender (2021), however, in accepting that in practice, there will be all sorts of reasons that we get situations that look superficially like static preference reversals, because it's hard to separate out what we value from the costs of delay. That is, we are very rarely in the case where we are discounting on the basis of time without concern for other reasons to change our utilities.

In addition to static preference reversals (which, as Callender points out, are not genuine reversals at all), we can have dynamic preference reversals. We avoid these when our set of preferences is not a function of what time it is; that is when our preferences are 'invariant under time translation'.[8] The idea is that position in time doesn't make a difference to how we value things. Whereas Stationarity dealt with distance from me, Invariance deals with *when* something happens. The difference between past and future that we encountered in Parfit's hospital example is a case where we take the pastness or futurity of the operation to be relevant. As such, it is an example where our preferences are not Invariant.

In addition to claiming that we should reject any discounting function based on *distance* in time, Greene and Sullivan (2015) and Sullivan (2018) claim that we should not discount things because they are *past*. That is, we should accept the limiting case of Stationarity, where we discount at a constant rate of zero, and accept Invariance, where the *pastness* of something does not affect its value. A central argument for this is the Arbitrariness argument for Past Neutrality (Sullivan (2018:108):

(1) At any given time, a prudentially rational agent's preferences are insensitive to arbitrary differences.
(2) Being past rather than future is an arbitrary difference between events.

Note that this differs from Consistency in the placements of t and t' on the right hand side.

[8] Callender (2021: 13). I again borrow his formulation, continuing with the same terminology as fn.7:

A set of time preferences is invariant if

Invariance $(x, t + \Delta_1) \sim_t (y, t + \Delta_2) \Longleftrightarrow (x, t' + \Delta_1) \sim_{t'} (y, t' + \Delta_2)$

Once again, this differs from Consistency and Stationarity in the placements of t and t' on the right hand side.

(3) If you are future-biased, your preferences are sensitive to when an event is scheduled relative to the present.

(4C) So, at any given time, future-biased preferences are not rationally permissible.

Let's grant premise 1, for the sake of argument. Premise 3 is just a claim about what it is to be future-biased (i.e. to discount the past). The action is in premise 2. As Sullivan (2018:110) says:

> The defender of future bias owes us a theory for why being in the past is normatively significant when applied to pains and pleasures but not when applied to other sorts of events. The temporal property by itself does not automatically provide a reason to discount.

Future bias is not permissible because it involves dynamic preference reversals (i.e. failures of Invariance). These open us up to exploitation. We considered that it might be *near bias* (i.e. a failure of Stationarity) that is doing some of the work in the trip to the Netherlands example. The *nearness* of the trip rather than the pastness of the paying for it makes you pay for the upgrade. Perhaps I valued the extra convenience in the near future, over the distant hours down the philosophy mines I will spend/have spent in order to pay for it. Greene and Sullivan (2015:959) discuss a case that arises if one has future bias and no near bias:

> *Fine Dining*: Jack wins a free meal at a fancy French restaurant on Monday morning, and he must schedule the meal for a night some-time next week. Given his flexible schedule, every night is as good as any other. Therefore, Jack schedules the meal for Monday night. As expected, it is an incredibly delicious meal. On Tuesday morning, Jack strongly prefers that his restaurant experience were in the future, rather than the past. And so he regrets scheduling the meal for the previous night.

The thought is that by scheduling the meal at the start of the week, he has limited the amount of time it is in the future, and future pleasures are *ex hypothesi* preferable to past ones. In fact, if there were no reason to prefer any day over any other (and we allow ourselves to ignore factors like ageing, loss of facility to enjoy food, uncertainty about whether the chef will still be there, etc.) the result comes out that you should infinitely postpone the meal to keep the pleasures in the future, not the past.

This is a dynamic preference reversal that is based on an arbitrary preference for the futurity of pleasures. If it were non-arbitrary, that would be another matter. The normative significance might be worth it. But, the argument goes, if we ignore the other factors like ageing, death, uncertainty and so on, there are no non-arbitrary reasons to reject Invariance.[9]

[9] In fact, as Greene (2021) points out, the non-arbitrariness argument is specifically for the limiting case of Invariance, where the discount function is zero.

I'm going to discuss three broad sorts of considerations that push against accepting Consistency across the board. Even if we accept temporal neutrality, we should accept that there are cases where we have preference reversals. Given the narrative structure of our lives, delays at some times will matter more than others. Given the change in what's potential and actual, we'll change what we value as we come to value things for being actual rather than valuing them as potential. Given changes in who we are and what we value, these narrative features of our lives and transformative changes in what we value will come together to make preference reversals a common and permissible feature of our lives.

In §8.1.3, we'll discuss why the narrative of our lives leads to failures in Stationarity. In §8.1.4 we'll discuss how changes in what's actual lead to failures of Invariance, and in §8.1.5 we will discuss on how transformative experiences lead to failures of both. Most, if not all, of what I'm saying is that Greene and Sullivan (2015), Sullivan (2018) and Greene (2021) can accept, provided that we don't view *time* as being what's responsible, but rather the role of uncertainty, ageing, change of circumstances or transformative experience as doing the work. Callender (2021) argues that we simply cannot separate these things off from time.[10] Nonetheless, these considerations, regardless of how they show up in the technical formulations, are going to produce preference reversals for reasons which, I shall argue, are closely bound up with the account of the passage of time as change from potentiality to actuality.

8.1.3 The shape of a life

Stationarity was the view that we need to treat the temporal distance between things the same, no matter how distant from us. But so far we have been dealing with very simple cases: pains and financial costs. I want to show, in this subsection, how things get more complicated. Our lives often don't treat utilities as things that hold the same value outside of the context in which they take place.

If all we care about is pleasure and pain (i.e. we're *hedonists*), it's tricky to know what to say about future pains and past pleasures to begin with.[11] We can feel pleasure in a pain being over, or bitterness that we had to endure it at all. Or, we can not think about it at all. We can anticipate future pleasures with delight, or become anxious that they will pass so fleetingly. They are not in themselves either pleasurable or painful, and it

10 Greene (2021) replies to Callender (2021), rejecting the claim that the argument for temporal neutrality only works if we accept 'pure' time preferences. But Greene seems to read these as being solely motivated by considerations of time, whereas Callender seems to read these as being considerations where it is possible to treat time as a separate variable from considerations like ageing, the entropic gradient and so on that are manifestations of human temporality. So, I don't think Greene actually responds to the point at issue about the difficulty of disentangling the temporal variables from the non-temporal contributors to the utility function.

11 It, of course, matters whether it is just our own pleasure and pain we care about, or pleasure and pain generally. Misak (2020:318) cites Frank Ramsey as distinguishing between individual temporal discounting, where we might accept that people simply don't care as much about future generations as they do about people now, and temporal discounting in public policy, which Ramsey found unacceptable. From a public policy point of view, no pains hurt because public policy is a third-personal endeavour, and pains hurt first-personally.

is possible to have attitudes towards them which are pleasurable or painful to feel.[12] Indeed, the ancient school most closely associated with hedonism, the Epicureans, proposed training our emotional responses so that we don't let these attitudes bother us: 'what's pleasurable is easy to obtain, what's painful is easy to endure.'[13] Ideally, we would like to extend present pleasures and diminish present pains. But non-present pleasures and pains matter for the hedonist only in their tendency to produce present pleasures and pains in us. Now, future pleasures and pains have a tendency to produce present pleasures and pains, in that the future becomes present. This is not true of past pleasures and pains. Our complicated emotional relationship with non-present pleasures and pains is not to be explained by treating them as having fixed values in a utility function, which we then discount by distance, but by understanding that their value depends in part on the context in which they occur. In order to assess the value of pains, we need to place them in a diachronic context.

We care about past and future pains insofar as we care about what will happen to us and how we might control our future by learning from experience. Thinking of our future bias as a difference in the *degree* to which we prefer pains depending on their location elides important differences in kind. Our attitudes towards experiences and anticipated eventualities are of a different kind to each other, and to present experiences.

Even when it comes to regrets, things are complicated. For one thing, we may regret the actual things we did and the lost potential of the things we didn't do. Indeed, we may regret the things we did *and* value them as learning experiences that are, in some sense, constitutive of who we now are. We may confront past trauma, though it is unpleasant, because it has a continuing malign influence on our future, and because we foresee that it will be more damaging not to deal with it.[14]

The non-traumatic past pains of Parfit's example seem to be very different from the past pains that revisit us in the middle of the night, giving us nightmares and leaving us sleep-deprived and so less able to prevent future unpleasantness. For a Decision Theorist, the job of the utility function is to take all these complexities into account; it is a mapping of utilities to outcomes, and it can assign different utilities to superficially similar outcomes. But the ineffability of our emotional responses might be captured by an incomparability of utilities. It might be, for example, that the utility function gives a series of partial orderings to outcomes such that some outcomes are strictly worse than others, but it can nonetheless be the case that it is indeterminate, for a different pair of outcomes, whether one outcome is worse than another.

In fact, we can have very complex emotional responses to an event. Imagine you have listened to a musical masterpiece for the first time. You may both cherish the experience that you had listening to it and wish that you had the opportunity to listen

[12] Or both. Cf. Juliet's line in Shakespeare's *Romeo and Juliet* (2.II): 'Parting is such sweet sorrow.'

[13] Philodemus of Gadara, in Inwood and Gerson (1994:vi): Greek: *kaì tagathòn mèn eúktēton/tò dè deinòn eukartérēton.*

[14] Cf. Baldwin (1966:174): 'History, as nearly no one seems to know, is not merely something to be read. And it does not refer merely, or even principally, to the past. On the contrary, the great force of history comes from the fact that we carry it within us, are unconsciously controlled by it in many ways, and history is literally present in all that we do. It could scarcely be otherwise, since it is to history that we owe our frames of reference, our identities, and our aspirations.'

to it for the first time again. We can think that our youth was the happiest time of our life but have no desire to repeat it. And in these cases, we don't take it to be a failing that we have these complex emotional responses. Rather, it is a sign that we have grasped something that exceeds our power to express in words or even a utility function that compresses them into a single positive or negative figure.

These considerations might make us prefer certain events to happen at certain times, but all of this can be done in a way compatible with temporal neutrality: the claim that we shouldn't discount based on time. For though these will be violations of Stationarity, it's not because we are discounting time in a non-stationary way, but because what we value depends on its temporal context. Stationarity assumes that we can trade-off events in a time period without affecting overall goodness, but many events depend on their temporal context for having the character they do. If we took every event in my twenties and moved it back a decade, there's a question about whether those events would be as valuable in a different historical context, and a question about what I was doing in my gap decade.[15] These considerations of context push us away from treating sequences of pleasures and pains as the components of our utility function to thinking about what kind of story our lives tell.[16]

Marya Schectman (1996) points out that our self-interested concern for our own future is related to the possibilities of compensation and moral responsibility. It makes sense to trade utilities across times because we think of them as being part of a life that we narratively constitute. As Schechtman (1996:157) claims, this concern for our future is the basis on which it makes sense to think a 'happy outcome can reasonably be taken to compensate for a rough patch'. But it matters how such compensations and sanctions relate to the narrative of a life. Not just any distribution of pleasant and unpleasant things can be understood as counting as compensation.

How well someone's life goes isn't simply a matter of adding up the pains and pleasures within it, as Velleman (1991:72) argues:

> How the value of one's life is affected by a period of failure combined with a period of success, for example, cannot be computed merely from the timing of these periods and the amounts of well-being they contain. Their impact on the value of one's life depends as well on the narrative relations among the successes and failures involved.

15 The fact that some of the places have been bulldozed since, and the people aged and scattered across the world, also seems to be a reason to think we couldn't delay those events just for me. The scenario is going to get tangled up in all the other lives that have actually been affected by the decisions I took, and all the lives affected by them and so on. Lenman (2000) points out that this is a reason to think that calculations of expected utility cannot provide the sole basis for working out what actions are good.

16 Readers may notice that I move from hedonism to accounts of meaningful, flourishing lives without discussion of desire satisfaction views. I'm not, here, really interested in the question of what we should maximise (I'm no consequentialist), but in the sorts of considerations relevant to time that affect what combinations of preferences we have. The move here is to think about narrative features as making purely temporal delays of a set of preferences less obviously applicable to the decisions we make on how to live our lives.

Global 'narrative' features of a person's life matter. Indeed, these features might provide the basis for whether regret and relief are appropriate or not.[17] As these global features change (as the person lives their life), the assessment of how well their life is going can change dramatically in a short interval. This will produce situations that look very much like preference reversals. They might not count as violations of Stationarity because once we hold a narrative fixed, and remove uncertainty, it wouldn't matter if we delayed it in time. If we added another ten years to the Cretaceous period, it may not affect our preferences, for example. Nonetheless, in our actual lives, a realisation that the heroic epic we had planned is turning into a cautionary tale may involve our rationally agreeing to bear some costs to change the prospective cautionary tale into a story of redemption that would not have been borne if one had set out to craft a story of redemption from the start.[18]

What kind of features count as the relevant sort of narrative features? Wolf (2015) suggests a number of features. A good life, she argues, will tend to include activity, loving engagement and the pursuit of objective value, rather than passivity, alienation and a concern only for the subjective. These features will perhaps come together in the pursuit of projects, which pursue and partially achieve things of objective value, rather than disvalue.[19] Antti Kauppinen (2012:368) builds on Wolf's view of meaningfulness to emphasise the role of projects coming together to create a *point* to one's life, though Lloyd (1993) and Setiya (2022) think it a mistake to think our lives should have a single narrative point. We needn't endorse the details of Wolf's, Velleman's or Kauppinen's views here. What matters for our purposes is that we are interested in the sorts of relationships, projects and activities that a life contains, how those relationships, projects and activities relate to one another, and to the life as a whole, and the fact that these can change, for better or for worse, over time. Crucially, the value of some long-past episode can plausibly change now, and change again in the future – even after we are dead, depending on how these narrative features evolve.[20]

On the assumption we are dealing with such a non-hedonic conception of what makes a life well-lived, we still might be interested in some criteria about what makes decisions rational. Even if we accept that the change in value of episodes fluctuates over time, in a way compatible with Stationarity (because the utilities are what fluctuate over time), this merely complicates the decision theorists' project, rather than undermining it. We should still, insofar as we can, achieve Consistency in the face of such flux in utilities, where we lack non-arbitrary reasons to reverse our preferences. But how does it apply to the richer, more complex cases of lives viewed as unfolding narratives?

Sullivan (2018) considers three forecasting principles that might apply to a non-hedonic conception of a good life: Strong Forecasting, Weak Forecasting and Balanced Forecasting. Strong Forecasting says that you ought to prefer any option you know you

[17] Though, of course, we do get some say in shaping how to narrate our own lives to ourselves!

[18] Setting out to act badly and repent later may be rational, but it is a failure of practical rationality, however. It's not clear that you intend to repent if you also intend to do the things for which you will have to repent. See Kavka (1983) for a similar puzzle about practical rationality.

[19] I will note here that 'activity' and 'passivity' are cognate with 'actuality' and 'potentiality'. I'll say more about the relation between these later in the chapter.

[20] See, e.g., O'Neill (1993) for discussion of how we may be harmed *post-mortem*.

will never regret in favour of one you know you will eventually regret (provided you have and always will have all the relevant information about the options available). This, as she remarks, is perhaps too strong, since anything that you might fleetingly regret is now rationally impermissible. Weak Forecasting says it is merely permissible to prefer any option you know you will never regret over one that you know you will regret. Balanced Forecasting extends Weak Forecasting to cases where regret is inevitable. On Balanced Forecasting, it is permissible to prefer options you will mostly never regret over those you will mostly regret (with the usual provisos about always knowing what the relevant options are – provisos that often won't hold in practice).

Sullivan considers an objection to these forecasting principles. [*Warning*: the remainder of the subsection, and the following subsection, involves discussions of suicide.] The objection is that those forecasting principles make suicide rationally permissible. (Indeed, Strong Forecasting appears to *mandate* it in a number of circumstances.) Sullivan is quick to acknowledge that the epistemic assumptions that people know what options are available are very often not met amongst actual suicidal people. And there may be moral reasons not to commit suicide, for example, that act as further constraints.

Suicide is, it must be admitted, not a course of action the person who commits it will regret if successful. One solution Sullivan suggests is that we have a forecasting principle based on preferring options where we *affirm* our choice to continue living, over ones where we regret so doing. And I think she is right that such affirmations (considered to be a positive endorsement counterpart to the negative endorsement of regret) should play a role in our rational forecasting. Even taking such affirmations into account, she concedes that:

> if it turns out you will always prefer to have died or be indifferent to living, it may very well be rationally permissible to die (modulo all of the important moral and psychological questions raised above). (2018:100)

Sullivan accepts these principles while asserting temporal neutrality. Nonetheless, it seems like I have a rather obvious asymmetry of valuing over time, associated with some obvious preference reversals. We can sometimes choose between futures on which we will value things, and futures on which we are not around to do the valuing. We can't make such choices with respect to the past, however. If there's one way to secure invariance, it's to remove oneself from susceptibility to change. Given that I think there is a metaphysical asymmetry between the actuality of the past and the potentiality of the future, I seem to owe some explanation of this rather profound, if morbid, asymmetry of valuing.

8.1.4 The actual and the potential

The Epicureans had an argument that one should not fear death because one had no more reason to fear being dead than one had to fear not being born yet.[21] This

[21] Cf. Lucretius 3.830–911, in Long and Sedley (1987:151).

creates a challenge: why is the time after our death different in value from the time before our birth? Wherein lies the badness of death? Not because of an arbitrary difference in your temporal location, surely. Yet this difference seems to involve something really important about what it is to value the future rather than the past.

One popular answer is that death deprives us of some good things in a way that not having been born earlier does not.[22] To make sense of deprivation, we must make sense of how things would have been if I had not died, or had been born earlier. These comparisons are somewhat tricky, because it's hard to know how much to hold fixed.[23]

One possible answer to this asymmetry problem is that death deprives an actual person of the potential for a valuable life, whereas being born earlier compares an actual person to a merely possible (and not even potential) person. It may not be obvious why this should count as an *answer* to the puzzle. Why should the difference between the actual and potential make a difference?

An answer lies in the ways we relate to the actual that we do not relate to the merely possible. Actual people are not *fungible* in the way that mere possibilia are. When comparing possibilities, it would be odd to prefer a less than ideal one over an alternative that was better in all the ways that matter. But when it comes to actual people to whom I am attached, it would be odd to think of 'trading-them up' for a better alternative. As Kauppinen (2012:364) puts it:

> In contrast to other worthwhile goals, there are goods that only you can confer to people with whom you are in a loving relationship, whether it is your partner, daughter, or grandfather. Someone else can make a wooden assault rifle[!] for your son or daughter, but it will not have the same significance as a handmade gift from a parent. No one else can take your place in such activities. This is important even when the activities in question are valuable as such. . . . A person who stands only in impersonal relationships to others – as a customer, sales assistant, taxpayer or the like – can easily be substituted for, as long as there are others who share the capabilities involved.

Death involves the loss of potential of an actual person, whereas you never had the potential to have been born earlier, and the merely possible, qualitatively similar person who was may have had a better life, but they do not stand in the actual loving relationships that we actual people do.

To return to the puzzles of forecasting of the previous subsection, we can now make sense of a puzzle due to Elizabeth Harman (2009:185):

> Consider the woman who became a mother at 14 years old. She may well recognise that things would have been better if she had waited to become a mother later and if she had had a different child. . . . She may recognize that if she had waited, she would have come to have a wonderful child that she would have loved and that that

[22] E.g. Brueckner and Fischer (1986). Deng (2015) explicitly explores the connection between this strategy and the passage of time.

[23] Briggs and Forbes (2019; MS) discuss some of these issues.

> relationship would have been *as good* as her current relationship with her child. She may even grant that it would have been *better*: there would have been far fewer strains, because she would have been mature enough to be a responsible parent. She may conclude that things would have been, overall, *much better*, if she had waited. Nevertheless, she loves the child she actually has, and does not wish that she had not conceived – she prefers things as they are. That preference I claim, is utterly reasonable. (emphasis original)

Why would such a preference be reasonable, given that one prefers one situation over one that one acknowledges would be much better? Because when one compares two *possible* situations, it is clear which possibility is preferable, but the fact that one of those situations is actual makes all the difference. Because, in Harman's case, the mother loves her *actual* daughter, and the actual daughter is not fungible; another daughter wouldn't do just as well.

The fact that we have attitudes both towards possibility and actuality explains some of the complexity that surrounds them. For we may regret the actual trauma that we or others have been exposed to, but value the actual life we have, the narrative properties of which depend on the trauma that we regret. We may regret the pain of heartbreak when a romance ends, but value the friendship that it leads to. Heartbreak is particularly interesting, because it often involves a painful loss of hope for a strongly desired future, without which the current relationship no longer makes sense. The meaning of our most valued relationships, even now, may depend on narrative properties that are unstable and in flux. But if it is love, rather than mere limerence, the desires must relate to the actual beloved.

If we were creating a fictional narrative, perhaps by playing a computer game, we might easily give up on a particular narrative because it hadn't been going well, or simply out of boredom. This makes sense as a response to fiction: it needs to be of a certain standard and to maintain a certain level of interest. But to commit suicide because our actual life had merely failed to engage us recently would be widely, and correctly, condemned. This needn't be because of any views about the sanctity of life or the intrinsic value of human beings, but simply because we should relate to our own lives, and the lives of our loved ones, in a different way to the way we relate to the lives of mere possibilia or to abstract artefacts.

This leaves a puzzle about, for example, bad things that happened before our conception, on which our conception causally depends. We might think that it's bad that they happened, but since we are glad that some relationships with loved ones resulted from it, must we be glad of the causal antecedents of those relationships? I'm not sure what to say about such cases. I certainly feel conflicted and confused when I think about all the atrocities that are causally antecedent to everyone I've ever loved even existing. I think, however, that the actual loving relationship we have is rather insensitive to the causes of the relata. We can love another and feel nothing, or many different things, about the circumstances that made such love possible, without any kind of rational failure.

Defenders of temporal neutrality, such as Sullivan, think they can accept that in cases such as Harman's example, it's reasonable to prefer worse alternatives because of actual attachments. Presumably, as we form these attachments, our preferences change, and we lose Consistency. But what this shows is that if we accept temporal neutrality, we still have a temporal asymmetry of valuing based on the metaphysical asymmetry between actuality and potentiality.

Invariance has failed here, but this isn't an objection to temporal neutrality by Sullivan's lights, for it isn't a *mere difference* in time that has changed. But given I claim that what the passage of time *is* is a change from potentiality to actuality, it looks like a case where the passage of time is going to be pretty central to the explanation of why Invariance doesn't hold. Callender (2021: 21) quotes Galberti and Strulovivi (2014), who write:

> we adopt the view that anticipated consumption is radically different from actual consumption, a view inspired by the fact that that the former consumption is purely imagined, while the latter has specific physiological and sensorial components.

Callender comments:

> Put like this it seems very demanding to insist on Consistency between these types of consumption.

Separating out the change that time contributes from the change in how we value things as contexts, knowledge and actuality change seems increasingly difficult to justify. In any case, treating anticipated consumption and actual consumption seems to be importantly different, and it seems a task of the philosophy of time to acknowledge this. If this is a case of Invariance failing, and so of a lack of Consistency, then we can hardly claim this to be *arbitrary*. Thus much I agree with Callender. But note that for me it is not merely a practical difficulty in separating pure considerations of temporal discounting from these other features. What time *is*, on the account I have been defending, is a change from potentiality to actuality. Separating time from a change of potentiality to actuality isn't merely practically difficult, by my lights, but in principle impossible.

8.1.5 Choosing for changing preferences

We've discussed cases where the narrative features of our lives mean that we get preference reversals due to changes in the value of events in context. We've also looked at changes that take place as potentiality becomes actuality, and we have the possibility of standing in new relationships to things; these, in turn might lead to non-arbitrary preference reversals. I wish to gesture, briefly, at various complications we have not yet considered that push further against the idea that we can avoid preference reversals.

There are many changes that take place as time passes, as Callender (2021: 245) remarks:

> As time elapses from t to t', many things may change. Tastes change. New infor mation arrives, e.g. learning. Old information departs, e.g. forgetting. One's

> underlying values or even self may change. For all these reasons, when making a choice at t', it doesn't seem at all incumbent upon me to look back and match what I earlier preferred at t.

Consider the popular song 'Father and Son' written and sung by Cat Stevens (1970) (now known as Yusuf Islam). In this song, we hear an emotionally powerful exchange between a father and son about whether the son should leave home and pursue a very different life. The father and son have very different perspectives on it. One of the successes of the recording is that one can empathise with both characters, even though they disagree. The fact we can empathise with contradictory positions is independently interesting, but what I'd like you to consider here is that one and the same person can hold both of those contradictory positions without fault simply as a result of *changing their mind*. A young man may naturally identify with the position of the son, and as the man ages and becomes more worldly, he may come to identify with the father. And there's no particular reason to think that this change is merely the result of increased knowledge; for knowledge may be lost as well as gained (and the memory's the first to go).[24] Nothing we have said so far gives us any reason to believe that someone who changes from agreeing with the son to agreeing with the father has become wiser rather than more jaded (more plausibly, it will be some combination of the two). We can treat them as epistemic peers who value different things equally reasonably.[25]

Some experiences change us epistemically, and some change our values in such a way that they change significant or central narrative features of our lives. Some do both, since the change in our values might change what possibilities we can foresee or imagine. Following L. A. Paul (2014) we will call these *transformative experiences*. If we're interested in our attitudes towards other times, transformative experiences must be contended with, precisely because they make it so hard to know how to evaluate attitudes to other times.[26]

I have already rejected the approach of merely totting up the different experiences at times to assess a life, in favour of a more holistic, narrative-based treatment. But given how quickly the narrative may change, and how quickly we may value that narrative differently, there is an inherent instability in relying on these narrative properties. The same person, at different times, can (predictably) value different things and insist that

24 I had a reference for this, but I've forgotten it.

25 The problem of changing values as we age was discussed by Hume (1757/2008:110–11):

> A young man, whose passions are warm, will be more sensibly touched with amorous and tender images, than a man more advanced in years, who takes pleasure in wise, philosophical reflections concerning the conduct of life and moderation of the passions. At twenty, Ovid may be the favourite author; Horace at forty; and perhaps Tacitus at fifty. Vainly would we, in such cases, endeavour to enter into the sentiments of others and divest ourselves of those propensities, which are natural to us.

Of course, as Hume goes on to point out, the popularity of Ovid amongst twenty-year-olds may also be subject to 'continual revolutions in customs and manners' (p.111).

26 Though Paul writes as though transformations tend to be one-way, I will allow that any change significant enough to count as transformative counts, even if it is later reversed. Indeed, one can imagine cyclical cases, such as Dr Jekyll and Mr Hyde transforming into one another at intervals. Thanks to Julia Janczur for alerting me to this.

you ignore their preferences at other times. Odysseus can ask to be bound to the mast and for his cries to be released to be ignored, and later, when under the influence of the siren song, have equally strong or stronger preferences to be released. In this case, we have few qualms about ignoring Odysseus' cries, as we have reason to believe that he is not normatively competent.

It is not always so easy to dismiss transformative experiences as changes of normative competence.[27] For one thing, it presupposes some standpoint from which we can assess normative competence, and we often find that we are judges in our own case. This, for good reasons, is a practice that is avoided in good justice systems but is often unavoidable in deciding how to author our own lives.

If we know that we *will* change our preferences, must we now change our preferences to the ones we will have? Somewhat flippantly, the answer seems to be that we should if our future preferences will be better than our current ones, and we should not if they will not. But the problem may well lie in working out whether the change in preferences is progress or corruption.

Johanna Thoma (2018) discusses 'temptation cases' where we temporarily change our preferences (e.g. we are much more inclined to watch a further episode of TV just as the previous one finishes, but disinclined to spend the whole evening watching TV at other times).[28] In such cases, we can see a case of overruling our temptations because they are comparatively brief and incompatible with preferences we hold strongly at other times. But not all cases are so easily overruled.

In our 'Father and Son' case, the two seemed to be epistemic peers, and we can imagine a life evenly split between the two. It is hard to know what advice to give someone in such circumstances. Thomas Nagel (1970:40 fn.1) suggests it is important not to frustrate your future self: 'One does not have reason now to ensure the frustration of what it will be rational to do in the future.' But in these cases, we don't have a clear view of what will be rational in the future. We have, as it were, peer disagreement about what reasons should motivate action.

I propose treating such cases by analogy with those of peer disagreement in epistemology (e.g. Kelly 2005; Lackey 2008). A change in one's future values – where one has no evidence that the change is one of progress, or of corruption – fits well as a case of peer disagreement and so presents a similar dilemma. But although we are anticipating a transformative experience in which values that matter to us no longer motivate us as they do now, this is still our future self (for whom we have a special concern) that we are discussing.[29] As Beardman (MS) points out, if we didn't have a special concern for such future transformations, we wouldn't care that their values go so much against our own. But it is also, I claim, their anticipated special concern

[27] Pettigrew (2019) provides a thorough exploration of these difficulties.

[28] This is quite a nice example of TV producers exploiting *kairos* in creating their stories, by making you inclined to act in their favour at just the moment such action is required.

[29] Given that the reasons we have to do things might be shaped by our actual loving relationships to actual beloveds, what we have objective reasons to do may well change over time. Given such a commitment, we needn't think that we have to choose between a view that takes seriously the inescapability of the present as the time of deliberation and decision-making, and one that takes it to be the case that reasons are objective and not merely based on what we desire at particular times, as Stephanie Beardman (MS) suggests.

for us, as one of their past selves, that makes it more plausible we should treat them as epistemic peers. How, given that they apparently have access to (though may have overlooked) all the reasons we now have, can they have different values to us? Unless we have evidence that they are in a compromised epistemic state, such that it would justify our paternalism towards them, it's hard to know on what authority we should bind them (ourselves!) to act on the values we care about now.

There is some evidence from game-theoretic approaches to peer disagreement (e.g. Heesen and van der Kolk 2016) that sticking to one's guns (adopting a 'steadfast' strategy) or leaving one's options open (adopting a 'conciliatory' strategy) can each be rational in some circumstances, and can, in some circumstances, both be rational (though mutually incompatible). It may be that binding oneself in the future makes one predictably miserable (but has other properties that you now think more valuable), or it may be that keeping your options open allows the acquisition of more wisdom than it does smug complacency. The crucial thing to note about the analogy with peer disagreement is that we cannot step outside of our embedded perspective as decision-making agents and valuers, to compare values. For even our values about what to value are embedded in time with us and subject to change.

These cases of transformative experience combine the elements we have seen, where the narrative features of one's life change, the actualising of some things makes them valuable in ways we couldn't previously grasp, and produce preference reversals that confound forecasting principles because they (non-arbitrarily) undermine the kind of invariance in values that forecasting principles depend upon. That time passes, and ages us, and robs us of opportunities and forces us to become acquainted with what were previously hypotheticals, seems to be central to explaining the asymmetry in valuing that we began with.

There's a lot more that I could have done in this section; the literature on these issues is vast and intricate. Concerns about rationality, value theory, philosophy of action and moral psychology all interact with the issues in the philosophy of time that brought us here. It seems plausible to me, however, that these discussions about how we plan and choose and act over time need to be explicitly linked to questions about what time is, and in what ways the future differs from the past. I have argued that the non-fungibility of actuality makes the past quite different to the fungible potentialities of the future, and that these changes in time are changes that show up in our decision problems and make preference reversals reasonable and non-arbitrary.

8.2 Relief at last!

Moving away from Decision Theory, we can think more generally about the appropriateness of certain emotional responses (even if they don't involve choice). In Parfit's example, we encountered a feeling of relief (interpreted by decision theorists as an expression of preference) in a situation where there was no choice to be made. Relief has a significant history in the philosophy of time, in association with an argument that the appropriateness of relief is evidence for McTchange.

Imagine you had a dentist appointment at 2:30 p.m. You endure a root canal, which, though painless, involves sitting uncomfortably while listening to a series of unpleasant sounds (a drill, a blowtorch, the frustrated and reproving noises of a dentist) and smelling your own cauterised dental nerves. You come out of the dentist appointment and say (through a numb jaw) 'Thsank goothneth thsats over'.[30] You are thanking goodness for the dentist appointment being in the past, it seems! The argument for McTchange goes from this expression of relief to the conclusion that there must be some tensed fact for which you thank goodness – you are not thanking goodness that your root canal was at 2:30 p.m., but that it is *past*.[31]

8.2.1 Mellor and MacBeath's responses

The classic response to this argument is to separate out two distinct projects. One is about language, the other about metaphysics. This argument may show that tensed language cannot be reduced to tenseless language, but it does not yet show that there is McTchange, because whether there is McTchange is a matter of how the world is, not how we ordinarily speak. Hugh Mellor (1981;1983) and Murray MacBeath (1983) pushed a view that we could make sense of the expression of relief because how things are from the perspective of the speaker can be made true by an unchanging reality. We see here the familiar dispute between a view on which we find ourselves with a temporal perspective that can be described from an unchanging God's eye view, or as having a temporal perspective that we cannot escape from.

The question is not so much whether we can eliminate tensed claims from language; it is widely accepted that we cannot eliminate them from perspectival descriptions of the world. The question is whether our temporal perspective is inescapable in a way that our spatial perspective isn't. But I won't relitigate that debate, which was discussed at length in Part I. What concerns us here is a related question. If the difference between the past and future is the difference between the potential and the actual, why should we thank goodness that our root canal is actual?[32] Why would relief be the appropriate response to the *actuality* of a root canal?

It would be odd if the expression of relief that my dental surgery is over were in some way equivalent to gladness that my dental surgery is actual. This situation is unlike one in which I have experienced some pleasant thing and am glad of its actuality. I am unlikely to flip through the family photo album in years to come and relive the happy day of my root canal. It is not that I am glad I have had a root canal *simpliciter*. What my relief amounts to, in the dentistry case, if it is gladness of anything, is that I am not currently having a root canal and that I will not feel toothache in that tooth in the future. It is not merely the actuality that I thank goodness for but a change in what is potential.

30 'Thank goodness that's over!'

31 This argument is most strongly associated with Prior (1959), but can be found earlier, for example, in Broad (1938:267). Pearson (2018) defends a more recent version of the argument.

32 Thanks to R. A. Briggs for posing this question to me.

One concern you may have here is that the example is confounded by lots going on. People are often anxious before dental surgery that the anaesthetic won't work properly, that the surgery will go wrong, that the dentist will discover further problems requiring further appointments, and perhaps also, that they will acquit themselves poorly in front of their dentist. The relief at the surgery's completion is easy to explain as relief at a loss of potential loci of anxiety.

What we need, by way of an example, is some unpleasant event which is regular and predictable (so as to remove anxiety due to uncertainty), is temporary and doesn't affect one's subsequent health (so as to rule out anxiety about future problems), with respect to which one is passive (so as to rule out anxiety over one's own performance), and during which one is not closely observed (so as to remove anxiety about others' judgements). It has been suggested to me that under some circumstances a woman's menstrual cycle can furnish us with an example of such an event; for some women, it is regular and predictable, it is merely to be endured rather than requiring an intervention to conclude, and on some occasions, there is the option of curling up on the sofa with a hot water bottle without fear of the judgement of others. This is not something on which I have first-personal expertise, but I am given to understand that connoisseurs do often express relief that it's over. If you prefer that example, feel free to substitute it for the dentist appointment example.

There still seems to be some residual difficulty in fitting in our relief that an unpleasant event is over, even if we account for the loss of potential of various kinds. If to be past is to be actual, we can still generate cases on which it seems we are glad that the unpleasantness is actual. The incongruity of this thought is less striking if we consider that this relief is not in a situation where there was an option for the unpleasantness not being actual; we can take drugs that will prevent us from remembering, but as Aristotle observed, even God cannot make the past not have been.[33]

The worry might continue, however, that describing the past as actual gets things wrong; an unpleasant event ceases when it becomes past; how does that square with its becoming actual? To return, for a moment, to the case of pain, Mellor (1981:23) points out past pains differ from present pains:

> The headache which has just stopped . . . is really no longer a headache at all, because it is no longer painful. . . . So far as actual pain goes, the event is merely a retired or Emeritus headache, not something still in business as the genuine experiential article. And that is why I thank goodness for its pastness: by ceasing to be present it has ceased to be the unpleasant experience it was.

If the distinction between the past and future is that between actuality and potentiality, we need to be able to make sense of how being actual relates to having a 'retired or Emeritus status'.

We do, fortunately, already have the resources available for doing this. In §3.1.2, we discussed the progressive/perfective distinction. Processes and activities, we said, were marked out as being goers-on in time, whereas events (understood as mereological

[33] *Nicomachean Ethics* 1139b10, as quoted in the epigraph to Chapter 7.

Table 8.3 Potentiality and Actuality for the Tenses

	Potential	**Actual**
Future	✓	×
Present	✓	✓
Past	×	✓
?	×	×

sums of time-slices) are not. The relation between the potentiality/actuality distinction and the progressive/perfective distinction, once spelt out, can explain why *wholly* actual pains have a retired or Emeritus status.[34]

When contrasting the past as actual and the future as potential, it is important not to neglect the present. Indeed, one of the big puzzles in the philosophy of time is how one can use binary distinctions to get three tenses. If the past is actual and the future is potential, is the present neither or both? Provided we don't treat the present as a durationless boundary stretching across space-time (which our discussions in §5.2 suggest is difficult to give physical interpretation to), we should think that the present is a combination of potentiality and actuality. Processes and activities are ongoing changes of potentiality into actuality.

As we discussed in §7.4.1, for something to have potential, it must have some actuality for the potential to be the potential *of* that thing. Equally, there must be some unactualised potential for the thing to be *ongoing*. If a thing has no potential to be other than it is, it can't undergo change, and neither can it stay the same, for it doesn't have the potential to remain as it is for longer. So, given they have no more potential to hurt, past pains, in virtue of being *wholly* actual, have a retired or Emeritus status. And so for goings-on more generally.

We can consider the relationship between past, present and future, and the potentiality and actuality distinction in Table 8.3. As you can see from Table 8.3, the future is characterised by being wholly potential, and the past by being wholly actual. The present is a mixture of potential and actual, in transition from the former to the latter; there is something, but it has unspent potential in the process of becoming actual. This leaves a combination left unlabelled: what is *neither* potential nor actual.

What of the remaining combination? We might think of this cell in the table as the *counterfactual*. It picks out the possibilities that are either 'missed potentialities' – potentialities that can no longer be actualised – and/or possibilities that never were potential.[35] Just as the passage of time is a change from potential to actual, it is also a

[34] I do not mean to propose that some things have parts in addition to their actual parts, while others don't. I use 'wholly actual' to suggest that they are complete (whole!) or have exhausted their potential.

[35] Thanks to Taymaz Azimi for the phrase 'missed potentialities'. Normally, counterfactuals aren't considered in discussions of time but are part of a separate subfield of philosophy: modality. I find myself quite attracted to the view of these counterfactual possibilities as being the fourth tense determination, as it were. If the distinction between past, present and future is to be a modal one, counterfactuals are a natural extension of the modal distinctions we are using. My work with Ray Briggs (Briggs and Forbes 2019; MS), attempts to flesh out the counterfactual from a starting point of a dynamic view of time.

change from the potential to the no longer potential. The present is, in some sense, always haunted by the potentialities that never get actualised.[36]

Relief is not the only emotion we could have run this example with. Cochrane (2018) uses the temporal directedness of an emotion as one of its individuating features. Fear, grief, nostalgia, melancholia, hope, wistfulness, regret and pride all seem to involve some temporal directedness. Fear, for example, seems to be directed at something future and potential, grief at something past becoming wholly actual (though also at something potential ceasing to be potential, and becoming something that merely might have been). Nostalgia seems to be a longing for something that presents itself as past, but is usually an idealised past that was never actual. Melancholia, according to Hans Maes (2023), is a future-directed feeling associated with a sense of future potential and, unlike the idealism of nostalgia, tends to be based on a realistic sense of what could happen, rather than a confected sense of how the past used to be. Banfi (2021) argues that we can only make sense of existential dread if we accept a metaphysical difference between past and future.

Kieran Setiya (2016:2) has written about a certain kind of retrospective sadness:

> Alive in my reflection are the lives I have not led. When I think about them, when I imagine them vividly, I can be gripped by something like regret: a sense of loss. I do not believe I would have lived a better life in poetry or medicine; most likely worse. And yet I feel dissatisfied with how things are. I look back at my younger self, options open, choices not yet made. He could be anything. But I am committed, path fixed, doors closed.

Setiya argues that this is not to be explained in purely temporal terms:

> Imagine I make an irrevocable decision at eighteen, to pursue philosophy. Nothing explains why it is the time before the decision I envy, not the time right after it. Either way, my life lies in the future, my projects incomplete. And either way, there is the fact that I will miss out on some of what I most desire. Why does the decision itself bring regret, a sense of loss about the past but not the future? (p.3)

Setiya thinks we can explain this asymmetry, at least in part, by the *specificity* of the reasons for regret. I know what opportunities I have turned down and feel the loss of them. But specificity won't explain cases of grief:

> I am less troubled by the death of a friend in prospect than after the fact, not because it is suddenly more specific, but because it has come to pass. (p.4)

In both of these cases, I want to appeal to the modal change that has occurred to explain the appropriateness of the relevant emotions. It is the change from having the potential to be a career poet to no longer having that potential that I feel loss with respect to. And it is the loss of more potential friendship that I mourn when I experience grief.

[36] Setiya (2017) offers useful philosophical therapy if you find yourself so haunted.

In many of these cases, as with wistfulness and regret, it is often the modal change that is important. One longs for the time when opportunities were open and directs oneself towards past futures. Value asymmetries abound when we start considering the richness of our emotional lives; a broadly causal/modal distinction between past, present and future seems to give us the resources to start making sense of them (though I don't claim to have done that work here).

8.2.2 Control, agency and belief

The thank goodness that's over argument was, as I have mentioned in §8.2.1, originally intended as an argument in favour of a tensed view of time and so is often taken to be a classic argument in favour of McTchange. As MacBeath and Mellor noted, and as has become widely acknowledged since, the argument shows persuasively that we have an embedded perspective in time.

We should be very cautious about going from particular temporal value asymmetries to a dynamic metaphysics. It is not by accident that my discussion of this topic is in Part II, on the distinction between past and future, rather than in Part I, in defence of McTchange. As Fernandes (2019:4003) points out:

> If the [temporal value asymmetry in question] were to support a tensed metaphysics, it would need to (i) not vary with temporal distance, (ii) apply equally to events concerning oneself and others, (iii) be rational and judged to be so.

The discussion from §8.1.2 suggests that although people often do in fact have cyclic preferences allowing them to be pain pumped, we can easily see that this is not good. Failures of diachronic rationality are commonplace, and it is rare, where our emotions are concerned, to avoid egocentricity and near bias. So, arguments for McTchange from temporal value asymmetries had better not be arguments for solipsism or a metaphysical parochialism.

We can do much explanatory work in explaining why, from an embedded perspective, we have the temporal value asymmetries we do. Fernandes (2019:4010), summarising the work of Horwich (1987), Maclaurin and Dyke (2002), Greene and Sullivan (2015) and Callender (2017), amongst others, gives the form that such explanations take:

1. Emotions felt now about contemplated events can motivate appropriate action towards achieving those events.
2. Emotional responses are (potentially) evolutionarily adaptive for action-guidance only if the contemplated events are under our control.
3. We control some future events, but no past events.
4. So emotional responses to contemplated future events, but *no* past events, are (potentially) evolutionarily adaptive for action-guidance.
5. So, we feel stronger emotions towards future events than past events.

Fernandes acknowledges that this summary simplifies things, such as the role of uncertainty, the social roles of permission and punishment, and the role of emotional responses to situations not under our control in informing our responses to other situations. But these explanations are meant to show how we can explain temporal value asymmetries without appeal to McTchange. Indeed, one needn't appeal to the past/future asymmetry at all, if, as Fernandes (2017) argues, the asymmetry of intervention is to be explained in terms of an epistemic asymmetry arising from the second law of thermodynamics.[37] I concede that this explanation is open to those who disagree with me about McTchange and the past/future asymmetry. But note that this explanation is also open to me and fits nicely with the modal/causal past-future asymmetry I have defended.

Fernandes (2019) is wary of metaphysical explanations where scientific ones are available. I share her enthusiasm for an exploration of psychological mechanisms in virtue of which these value asymmetries manifest. But just as metaphysical stories are no substitute for psychological mechanisms, psychological mechanisms are unlikely to settle the relevant metaphysical debate. Here, as elsewhere, we must see which theories best make sense of things.[38] My concerns about debunking arguments from §3.5.3 remain relevant here. Perhaps I am an incredulous fool upon whom nature imposes for thinking the reason we have evolved to have asymmetries of value is that there is an asymmetry between past and future that we are trying to cope with. We have some biases that demonstrably conflict with the world. We should be cautious about accepting short-term gains that we can foresee will lead to long-term losses and in treating the roulette wheel as if the spins are dependent on previous ones. But these all-too-human biases don't undermine the sense-making explanation of differentially valuing past and future that I have offered.

8.3 So it goes . . .

There's a broad consensus that our experiences of temporal value asymmetry are products of our embeddedness in time, and that some of these value asymmetries are less rational or evolutionarily adaptive than others. James Lenman (2002:262) describes our 'generation-centred' moral community; we take people contemporaneous with us into account morally in a different way than past or future generations. He also says that if generation-centred reasons are 'invisible from a timeless and impersonal moral perspective, so much the worse, it may plausibly be urged, for a timeless and impersonal moral perspective'. I agree.

I want to return to a concern of the previous chapter: fatalistic worries. Many, such as Cockburn (1997), have thought that if there were no past/future asymmetry, there would be no temporal value asymmetry. To anticipate my argument in this section, in the absence of a past/future asymmetry, we can *explain* our temporal value asymmetries, but we cannot *justify* them.

[37] Cf. §§6.4 and 2.5.2.
[38] Cf. §6.5; Moore (2012).

8.3.1 The Tralfamadorians and Billy Pilgrim

In §7.1.2 we considered a knower who sees the world *atemporally*, promising to revisit this version of fatalism later. I propose to do that here. I'm going to assume that you were persuaded, in Chapter 3, that we have a first-personal awareness of our own persistence through time by enduring – being the proper subject of McTchange.

I will revisit fatalism by revisiting Kurt Vonnegut's Tralfamadorians.[39] They are an alien race who experience themselves as spread out in time and do not share our embedded perspective. One of these aliens describes itself to Vonnegut's protagonist, Billy Pilgrim, as follows:

> I am a Tralfamadorian, seeing all time as you might see a stretch of the Rocky Mountains. All time is all time. It does not change. It does not lend itself to explanations. It simply *is*. Take it moment by moment and you will discover that we are all, as I've said before, bugs in amber. (1969:70)

I'm interested here in why we would move from an atemporal experience to fatalistic worries. One response is that the distinction between fixtures and options doesn't arise unless one has an embedded perspective. Indeed, Maclaurin and Dyke (2002), amongst others, argue for an evolutionary explanation of our lack of temporal equanimity, based on the evolutionary pressures to develop a temporally embedded perspective. It may be true that the fixtures/options distinction depends on an embedded perspective, and an evolutionary explanation can be given for the development of such a perspective. But my worry is unresolved. Suppose one accepts that one *has* an embedded perspective, from which things appear to change, but that one is *really* trapped in a moment in time. Then it becomes unclear why we should care about other times once we have been convinced that it is a mere evolutionary pressure; that there are causal explanations for our lack of equanimity, but no justifications. Future pains, we said earlier, don't hurt but are of concern because they *will* be present pains. If we are just a momentary perspective, trapped in this moment, then future pains will never hurt *us*. The justification for our special concern for the future goes if we are trapped at a moment in time.[40] So too do many of the more complex narrative concerns that we have about how our lives are going.

The Tralfamadorians don't have any temporal value asymmetries: they respond to every event the same way. They say 'so it goes'. Nothing is to be looked forward to, relieved by or grieved over. Everything is just what happens at that time. Billy Pilgrim informs us:

> The most important thing I learned on Tralfamadore was when a person dies he only *appears* to die. He is still very much alive in the past, so it is very silly for people to cry at a funeral. All moments, past, present, and future, always have existed,

[39] I introduced these in §3.4.2.

[40] Indeed Velleman (2006) explicitly links this Tralfamadorian view to the Buddhist claim that we must accept the Buddhist no-self thesis (discussed in §3.3) to overcome suffering.

> always will exist. The Tralfamadorians can look at all the different moments just the way we can look at a stretch of the Rocky Mountains. (1969:22)

The Tralfamadorians are fictional. They don't provide direct evidence for how things would be if there were no McTchange, and if there were no temporal asymmetry. But they are useful, I contend, in prompting us to think through the pragmatic consequences of believing in a temporal asymmetry that results from an ongoing change of potentiality into actuality.

8.3.2 Fatalism as equanimity

The Tralfamadorian 'so it goes' attitude is one of equanimity. They don't get upset by deaths, and they don't celebrate life. They just selectively pay attention to the interesting bits. It is striking that their response to seeing time like we see a panorama is one of equanimity. I think we should consider this a version of fatalism.

If it is a version of fatalism, it is a distinctive kind of fatalism. It doesn't deny the counterfactual dependence of some events on others (though history 'does not lend itself to explanations'), and it needn't suggest that the events would have happened as they did regardless of what anyone did. But there does seem to be a lack of agency involved. Certainly, there isn't a practice of offering praise or blame. The evaluative response to events is not that of a participant in them but of a detached '*que serà serà*'. Even if there are still sequences of events in which things vary over time, there is no sense of value or disvalue here.

Suppose, for the moment, that the attitude of equanimity to all events is the appropriate response to a world with no McTchange and no temporal asymmetry. We might put an argument, a *reductio ad absurdum*, against the denial of McTchange and the denial of a temporal asymmetry; it is absurd to suppose that equanimity is the correct response to all events – it is *pathological* to feel equanimity towards everything – therefore, there must be either a temporal asymmetry or McTchange, or both. I'll call this the *anti-Tralfamadorian argument*.

The anti-Tralfamadorian argument looks like it is an argument from wishful thinking. Given that I value differentially valuing things, I should conclude that the world has a certain structure. But why should the world be structured so as to be convenient for my preferences about valuing?! Any such argument from wishful thinking would indeed be terrible. That I prefer something a certain way is not, in general, a good reason to suppose that things *are* that way. The world is often structured in ways that are uninfluenced by my preferences. As C. S. Peirce (1902 [CP2.113]) observes:

> It may be indispensable that I should have $500 in the bank – because I have given checks to that amount. But I have never found that indispensability directly affected my balance in the least.

When we are dealing with issues of the nature of time, it would be rather disturbing if my preferences were in any position to be influential.

There is a version of the anti-Tralfamadorian argument that is less crude, however. We might call it a *pragmatic transcendental argument*.[41] The argument is similar to a (perhaps more familiar) transcendental argument:

1. A temporal value asymmetry is a condition on the very possibility of having reasons for action.
2. We have reasons for action.
3. Therefore, there is a temporal value asymmetry.

The transcendental argument faces the objection that we can easily explain how, given our embedded perspective, we would take ourselves to have reasons for action without an objective temporal value asymmetry.[42] For although there is a temporal asymmetry from our limited perspective, this is compatible with its being contingent and local. With 1 being false, the move from 1 and 2 to 3 doesn't follow.

The pragmatic transcendental argument works slightly differently. It replaces 1 with 1*:

1*. A temporal value asymmetry is a condition on the very possibility of having *appropriate* reasons for action.

This is more plausible (and I shall defend it in §8.3.3). But it still doesn't get us to 3. So, we must replace the transcendental argument with a pragmatic transcendental one. This changes from an argument that establishes that the conditions of possibility for an actual state of affairs hold, to one that commits us to hoping that the conditions of possibility for an indispensable commitment hold:

2*. A temporal asymmetry is a condition on the very possibility of having appropriate reasons for action.
3*. We are indispensably committed to reasons for acting.
4*. We are committed to hoping that there is a temporal asymmetry.

This pragmatic transcendental argument doesn't provide *evidence* that 3 is true. But it might commit us to acting as though it is. The appeal to 'hope' may be striking to many here. What have hopes got to do with anything? I am using 'hope' in a somewhat technical sense, taken from Peirce.[43] This is meant to be a belief-like cognitive attitude, but one that doesn't require epistemic justification in the way belief does (though it can

41 Cf. Hookway (1999).

42 Reasons for action, in this sense, are a concern of practical rationality, where we are concerned with acting rather than choosing or deliberating, as we were with decision theory.

43 Peirce (1901/1998:107) puts it:

> there is a fundamental and primary abduction, a hypothesis which we must embrace at the outset, however destitute of evidentiary support it may be. That hypothesis is that the facts at hand admit of rationalisation, and of rationalisation by us. That we must hope they do, for the same reason that a general who has to capture a position, or see his country ruined, must go on the hypothesis that there is some way in which he can and shall capture it. We must be animated by that hope concerning the problem we have in hand, whether we extend it to a postulate covering all facts or not.

nonetheless be undermined to the extent it contradicts the evidence). Just as belief can be more or less confidently held, hope can be more or less confident. An abseiler has no evidence that there are no problems with the rope of which they are unaware. But they really commit themselves to the rope holding when they descend using it.

In order to defend the pragmatic transcendental anti-Tralfamadorian argument, we need to consider whether the premises are true. I'll start with the claim that we are indispensably committed to having reasons for action.[44] The argument for this is simple: when we act, we act under the guise of the good. If we are acting intentionally, we must find something *valuable*. As Raz (2010:125) puts it, 'barring conceptual ignorance or mistakes, actions performed with an independent intention are performed in the belief that the action has some value.' When there is something we are trying to do (captured by a particular description), we must think that what we are trying to do (under that description) is worth doing, that is that there is some reason for so doing. This is meant to be a conceptual truth about intentional action, that has been defended notably by Aristotle in the *Nicomachean Ethics*, and Anscombe (1963).[45]

The guise of the good has been attacked, for example, by Velleman (1992) and Setiya (2007), amongst others, and is the subject of an ongoing debate. But even if there are some rare and specific counterexamples to the claim that when we act, we have reasons for so acting, there aren't enough counterexamples to undermine the claim I need: that having reasons for action is an indispensable commitment. Even if not *all* action is under the guise of the good, enough actions are that we are committed to making a distinction in at least *some* cases between those actions that are valuable and those that are not.

Given the ubiquity of intentional action we take to have some point, in life generally, and in philosophical or scientific inquiry more specifically, we are not going to escape from a commitment to valuing things. If we are asking the question, 'why should I have reasons to act?', it is a presupposition of the question that we think there is some reason to ask it. If we ask the question, 'why should I inquire?' then we are already enquiring (and presuppose there is some reason to find out the answer to our question). Perhaps there are ways of avoiding a commitment to inquiry and having reasons for action, but Peirce (*CP* 1. 405; 1890) cautions us against this kind of scepticism about the value of these central human projects:

> The sole immediate purpose of thinking is to render things intelligible; and to think and yet in that very act to think a thing unintelligible is a self-stultification. It is as though a man furnished with a pistol to defend himself against an enemy were, on finding that enemy very redoubtable, to use his pistol to blow his own brains out to escape being killed by his enemy. Despair is insanity. . . . We must therefore be guided by the rule of hope.

[44] See Misak (2011) for a discussion of the history of indispensability arguments.

[45] See Moss (2010) for a discussion of Aristotle's position.

8.3.3 Action, passion, actuality and potentiality

Why think that we need a past-future asymmetry in general (and a causal/modal account of it in particular) in order to have appropriate reasons for action?[46] Why think 1* is true? The answer, I claim, relies on claims about the connection between our emotional responses and the account of activity as involving a change from potentiality to actuality. It is the modal situation that the agent is in; the fact that they are an agent rather than a patient. Not only must we avoid being a mere node in a network of counterfactual dependence, but even when we are undergoing processes over which we have no control, what emotional responses are appropriate is shaped by awareness of the change in our situation. If I am stuck in a meeting, knowing that I could still just catch my train if I hurried appropriately produces a very different emotional response to knowing that I have now missed my train.

The Tralfamadorians are (or think of themselves as) *patients* in a causal network. They are not agents, because they can't act intentionally. We saw in §7.5 a story about how the asymmetry of intervention could be understood in terms of a change from potentiality to actuality. Supposing that story is right, if there is no change from potentiality to actuality, there can be no interventions. Thus, believing themselves to simply be nodes in a network of counterfactual dependence, without the potential to intervene on the world, the Tralfamadorians don't attempt to intervene, and don't see any value in doing so. The question of intervention simply doesn't arise for them. The Tralfamadorians can't separate fixtures from options; everything is a fixture. They don't carry out intentions because they don't believe in reasons for acting, since that would require the possibility of changing options into fixtures. Since they can't change options into fixtures, they have no reasons for acting. So it goes.

The same is not true for Billy Pilgrim and us. We can't view time in the same way as he views a panorama of the Rocky Mountains. We *can* differentiate between options and fixtures. Why should the Tralfamadorian inability to separate fixtures and options affect *us*? Let us think about what it is to separate fixtures from options. This is to put a partition on the circumstances of some subject, such that some circumstances are not appropriate objects for intervention, and others are appropriate objects of intervention. My argument depends on thinking that although we can explain that we in fact partition things in this way, once we come to accept that we are merely nodes in a network of counterfactual dependence, or points on an entropic gradient, we can no longer understand why any particular partitioning makes sense. Our treating things as objects of intervention can be no more than conventionally appropriate.

You might look at the diversity of agential and affective responses and think that linking them to a modal change from open potentiality to fixed actuality doesn't work. But there is no especial problem with different subjects having different such partitions between fixtures and options. Some people are in a position to intervene where others aren't, and it is appropriate for some people to consider contingencies that other people need not. In a crisis, policymakers might hold fixed that the scientific advice is correct

[46] The word 'appropriate' is doing some work here; arguments, such as those in Maclaurin and Dyke (2002) which appeal to evolutionary considerations, would explain why we, in fact, take ourselves to have reasons for action, but by rejecting the need to treat them as *appropriate* reasons.

and work out how to respond, but the scientists hold the policymakers' responses as fixed and work out what advice to give. There is frequently a division of labour in making interventions, and there are vast inequalities in what interventions are available to different people, even in similar circumstances.

The problem comes if the Tralfamadorians are simply better placed to partition the circumstances into fixtures and options than Billy Pilgrim is (or we are). Once Billy becomes persuaded that the Tralfamadorians' partitioning of circumstances entirely into fixtures is *superior* to his previous one, he defers to them and becomes a fatalist. 'So it goes' becomes his response to everything too.

So, if *we* come to believe that *our* partitionings of things into fixtures and options presents *no options at all* (not even a single potential option that will become actual, as in §7.5), because we are mere patients in a network of counterfactual dependence, the appropriate response would be for us to view everything with equanimity too.

Consider a spatial analogy: John Bull thinks their compatriots are more morally valuable than people in other countries, in virtue of being compatriots. If Bull comes to see that geography doesn't provide a reason to separate people as having distinct moral worth, Bull ought to revise their evaluation of relative moral worth.[47] We might justly criticise Bull of being parochial in morally valuing his compatriots over others. Bull might respond that he is embedded within a political community, and from his embedded perspective, this view is the natural one to hold. Indeed, he can point out that it is shared amongst his political community and taken to be common sense! And we can explain various psychological and sociological reasons why communities like that exist. Here we should say that the naturalness of such parochialism is not a defence of its appropriateness.

Similarly, if Billy Pilgrim, or we ourselves, come to see our embedded perspective in time as parochial, it is undermined. Just as Billy Pilgrim adjusts his worldview, so would we have to. It might be natural that we divide the world into fixtures and options so as to think we have some options in the future, but its naturalness is no defence of its appropriateness, once we come to see how parochial it is. The appropriateness of partitioning our circumstances into options and fixtures depends on some things *really* being options, and others *really* being fixtures.

This is true even if lots of fixtures are only fixed relative to the context. When my grocer supplies me with potatoes, I hold it fixed that it is appropriate to pay for goods received in a commercial exchange.[48] The conventions and cultural expectations need to appeal to something beyond themselves to make sense. The intervention of paying makes sense as part of a larger system of practices that prevent me from going hungry. Though I am in a position where in practice I can hold it fixed that I won't go hungry, the practices that this rests upon depend on ongoing processes that ensure I don't need to.

Perhaps this is an uncharitable reading of the view that the distinction between fixtures and options arises from our embedded perspective. Defenders of this view

[47] We may, of course, have distinct moral duties towards compatriots in virtue of having a distinct political relationship with them.

[48] Cf. Anscombe (1957).

might insist that they think the reasons from an embedded perspective are real, but depend on the context to make sense. Because the distinction only makes sense within an embedded perspective, it doesn't arise otherwise. We might think that an umbrella (*qua* umbrella) is instrumentally valuable when it's raining, but not instrumentally valuable outside that context (or a context where it might potentially rain). The question of the usefulness of an umbrella (as such) only arises, you might think, when one is trying to keep rain off. Likewise, reasons for acting make sense within an embedded perspective, but not outside of it. But that doesn't mean they aren't appropriate within the embedded perspective.

I can accept that reasons for action only make sense within an embedded perspective, but still think there is a worry here. We can accept that reasons for action are response-dependent, but nonetheless distinguish between detectivist and projectivist proposals.[49] The marks of realism that we discussed in §6.5.1 are still relevant here. In particular, *evidence-transcendent truth* and *mind-dependence*. What are fixtures and what are options isn't simply a matter of our partial epistemic grasp of the situation, but something that transcends our evidence. If the response to the anti-Tralfamadorian argument is to make the fixtures/options distinction totally relative to our evidence, then that concedes anti-realism about options.

The reality of the fixtures/options distinction is compatible with a large number of things treated as fixtures that are only fixtures relative to a context; we can have options we are unaware of. We can simply not have the thought that lying or breaking our promises are amongst the options open to us; this is commendable, even. Likewise, we can falsely believe we have options that we do not, in fact, have. A detectivist realism allows an openness to the possibility that we need to get new evidence, rather than being assessed only against the evidence we have. It is not enough to describe an embedded perspective and assess whether, given the evidence, the partition of circumstances into fixtures and options is reasonable; we must allow the possibility that the embedded perspective could be improved upon.

If what are options are *always* mind-dependent, then that is anti-realism about options. Only if we are realists about options can we think we have reasons for action that are *appropriate*. But some fixtures are not up to us, and to deny this is to deny realism about fixtures. Saying that reasons for action only arise in the context of an embedded perspective, without accepting realism about the fixtures-options distinction, such that we could potentially be made to revise our partitioning as Billy Pilgrim has, is to reduce reasons for acting to internal drives, rather than normative demands we find in the world. Our actions have, no doubt, causal explanations. But it is not merely that we wish to explain, physiologically or psychologically, what processes are at play, but to explain what someone was aiming at or trying to achieve in so acting. As Setiya (2011:133) says:

> I may write the book review precisely on the ground that I promised to; that would be my reason for doing it. In cases like this, an agent's reason is not a psychological state or a fact about one. Indeed, it is only in rare conditions that an agent's reason

[49] Cf. §§1.4, 4.5.3 and 6.5.1.

> for doing something – the ground on which he acts in doing it – is a consideration of his own psychology. . . . The doctrine of 'psychologism', according to which the reasons for which we act are always states of, or facts about, our own psychology, is hopeless.

Many people may like the story on which we have internal drives that arise within our embedded perspective, but limiting ourselves to such a story, without consideration of what agents' reasons are, and whether they are good, loses us the guise of the good – the claim that (at least for the most part) we act in order to bring about something we take to be valuable.

If we take such a view seriously, we are no longer doing things because we think they are worthwhile, but merely because we are subject to internal drives. We lose the sense that things are up to us, that there are any options (even one inevitable one) because we are no longer interested in reasons, but merely in regularities and counterfactual dependence. We no longer intervene in order to achieve things we think worthwhile, but instead are nodes (or sequences of nodes) in a web of dependence relations. To all this, we might as well say 'so it goes' as any Tralfamadorian, except that from our perspective this doesn't tend to occur to us.

This connection between reasons for acting and the fixtures/options distinction links back to the argument of §8.2. Olley Pearson (2018) distinguishes between motivating reasons and justifying ones. He argues that Prior's Thank Goodness That's Over argument discussed in §8.2 works because it shows us that relief would only be *justified* if the root canal were over (to continue the example). I have tried to use the Tralfamadorian case to make a similar point with respect to action. Our action might be *motivated* from within our embedded perspective, but it can only be *justified* if our embedded perspective tracks a real past/future asymmetry. Because of this concern about justification of our action, rather than mere motivation of action, we have a dependence of appropriate reasons for action (i.e. justifying reasons) on a temporal asymmetry. We have reason to accept 1*, then: a temporal asymmetry seems to be a condition for the very possibility of *appropriate* reasons for action.

Reasons for acting only apply to agents. But even if we are like the Stoic dog attached to the cart, either moving with it or being dragged but exerting no agency over it,[50] we might still find some emotional responses appropriate and others not, as Pearson argues.

Even if everyone agrees that we are stuck in an embedded perspective, the difference between a past/future asymmetry that justifies various temporal value asymmetries and apparent past/future asymmetries that merely motivate temporal value asymmetries may be quite significant. This chapter has not primarily been concerned with defending a past/future asymmetry, as the previous two chapters were, but rather with fitting the account of it developed in the previous chapter with various temporal asymmetries of value. It is nonetheless helpful to have concluded with a pragmatic transcendental argument to the effect that we have to hope that our indispensable commitments to temporal asymmetries of value are justified.

[50] Cf. Long and Sedley (1987:I:386).

The questions we have considered in this chapter – of how we narrate our lives, how we regulate our emotions and how we weigh up actions – are ones about which there is much more to say. The links I have defended between the modal asymmetry between past and future, and these questions should be treated as somewhat tentative. Nonetheless, one of the central achievements of Part II has been to go from an account of asymmetry in physics and metaphysics, and show how it might contribute to concerns about how we get by in the world.

We have a picture of the nature of the past/future asymmetry that has the present as neither wholly past nor wholly future in terms of an ongoing change from potentiality to actuality. This picture fits with various ways in which we relate differently to the past and future, and to the present differently from both. With relief, we can say, that's Part II over, thank goodness!

Part III

The Growing Block

9

Temporal ontology

> *'Is it your opinion, Winston, that the past has real existence?'*
> *Again the feeling of helplessness descended upon Winston. His eyes flitted towards the dial. He not only did not know whether 'yes' or 'no' was the answer that would save him from pain; he did not even know which answer he believed to be the true one.*
> *O'Brien smiled faintly. 'You are no metaphysician, Winston,' he said. 'Until this moment you had never considered what is meant by existence.'*
>
> —Orwell (1989) part III.2

In Part I, I defended the claim that there is an asymmetry between space and time; time passes or flows in a way space does not. Things in time McTchange. In Part II, I defended the claim that there is an asymmetry between past and future. The past is actual, and the future is potential. There is a causal process of turning potentiality into actuality that is definitive of the present. We have now begun Part III. Part III assumes that I have persuaded you in Parts I and II. We have a dynamic view of time with an actual past and a potential future. You might think this means that the metaphysics is all over bar the shouting. But in fact, a major preoccupation of philosophers of time has received scant mention: whether the future/present/past *exist*.

The focus of this chapter will be on taking us from the considerations of Parts I and II to a particular view of what times exist: the Growing-Block view. This is the view that the past exists and the future doesn't, with the passage of time being what C. D. Broad (1923) called 'Absolute Becoming': the coming into existence of events. In §9.1, I will discuss how we should understand this word as neutral with respect to tense. §9.2 will survey different accounts of what times exist, showing for each how they account for the McTchange and a modal/causal past/future asymmetry of the kind defended in Parts I and II. One such view will be the Growing-Block view, which I endorse. §9.3 will examine why we should think that the past exists. There are many truths about the past, and we need to commit to an ontology that will make them true. In particular, a commitment to the existence of the past provides an appropriate way of differentiating between history and historical fiction. §9.4 will be concerned with reasons not to accept that the future, mere possibilia, or impossibilia exist. I will argue that the future resembles mere possibilia and impossibilia more than it does the past. Finally, in §9.5, things take a metaontological turn, and we consider what we're doing when we ontologically commit to things. I will argue that the kind of pragmatism I favour allows us to resist a deflationism or quietism in which we give up on metaphysics in favour of just giving an account of our discourse, but equally avoids thinking that the 'real world' is in some sense hidden from us rather than partially revealed to us in the practices that allow us to get by in it. Given that the Growing-Block view plausibly allows us to best make sense, in ontological terms, of the difference between history and historical fiction, and between the actual and potential, it is the most plausible temporal ontology to account for McTchange and the various asymmetries between past and future.

9.1 Tenseless, timeless and tensed 'is'

The curious thing about the ontological problem, as Quine (1948:21) famously remarked, is that it can be summed up in three Anglo-Saxon monosyllables: 'what is there?'. But for the philosopher of time, the word 'is' is one of the most controversial technical terms in philosophy. This brief section will clarify what the meaning of 'is' is as I use it, clearing away some obstacles to getting the debate going.

I wish to differentiate between the tensed sense of 'is' and the tenseless senses of 'is'. By the tensed sense of 'is' I mean the ordinary sense in which we might say 'it is raining' to communicate something about what goes on at present, viz. the raining. Equally, one might replace 'is' with 'was' or 'will be' to assert something about what went on previously or what will go on subsequently. 'Is', 'was' and 'will be' are all tensed locutions since they tell us something about the relation of a fact or event to the present.

There are a number of tenseless senses of 'is' which we might contrast with the tensed sense. This contrasting sense of 'is' is intended to be one that is neutral with respect to the relation of a fact or event to the present. There are, unfortunately, a number of different ways of understanding the tenseless 'is', as Oaklander (2004:338) observes. One that is 'literally without tense'. Oaklander's example is 'Oaklander's working on a paper on time and freedom has the property of presentness on October 13, 2003'. I don't find Oaklander's gloss on the literally without tense sense helpful; attributing the property of presentness on some date neither seems required to indicate tenselessness nor incompatible with reading it as a present-tensed claim. I would suggest that we think of it, instead, as being *neutral with respect to tense*, such that saying something *is* does not yet provide any information about whether it is past, present or future.[1] We might read the tenseless is, then, as 'is (at some time or other)' with further specification as to the time and tense optional. The remaining two comprise one that is 'timeless' (with the example of two plus two is four) and one which is omnitensed (so we read 'is' as 'is, was and will be').

I think the distinction here is useful, though some may take the view that there is no sense of 'is' which is neutral in this first sense. I don't think much hangs on this

1 Given my account of the *nature* of the past, present and future distinction, of course, if a process of writing is going on on that date, it will be the case that 13 October 2003, is present. But that is very much a substantive claim (with which Oaklander would substantively disagree). Forbes (forthcoming) explores the need not just to account for tense but also for verb aspect when considering the difference between past and future.

here, since on my view what is the case in the tense-neutral sense turns out to be co-extensive with what is presently the case.[2] I suspect that framing things in terms of whether there is McTchange, rather than whether there is an irreducible property of presentness, takes the pressure away from the need to make a dispute about the felicity of a tense-neutral sense of 'is' a major source of contention. If we insisted on a tensed 'is', rather than a tenseless one, it would be trivially true that there are only present things/stuff/events/objects/processes. Given that McTchange is the crucial difference between dynamic and static theories, there is no reason in principle why the property of presentness can't be reductively analysed in modal or topological terms, and the other tenses defined by being earlier or later than the present.

It is, you may have spotted, compatible to say 'Bob is (at some time or other) eating a sandwich' and to say 'Bob is (at some time or other) not eating a sandwich' provided the times in question are different. This should be unproblematic since it is equally unproblematic to consistently say 'Bob is tanned (on his face)' and 'Bob is not tanned (on his leg)'. Usually, it is left implicit that relevant places and times are constant throughout an argument unless otherwise stated. The important thing about the 'sometime or other' qualification is that it leaves open the question of the pastness, presentness or futurity of the event or fact in question. There are a couple of distinct uses of the verb to be that we may wish to note. We use the 'is' of predication to indicate a *way* something is (e.g. grass is green) and 'is' of ontological commitment to indicate *that* something is (e.g. there is grass).

I'm mostly concerned here with the 'is' of ontological commitment. I think we should read the 'is' of ontological commitment as neutral with respect to tense. Having a neutral understanding of ontological commitment is useful for ensuring that debates aren't merely verbal.[3] With these clarifications out of the way, we can proceed to consider various accounts of what, temporally speaking, there is.

9.2 A survey of temporal ontologies

Assuming that we have this neutral sense in which we can ask what there is, and we have accepted the central claims of Parts I and II – that there is McTchange and that there is a modal/causal asymmetry between past and future such that the future is potential and the past actual – we will now survey a number of temporal ontologies to see how they account for these claims.[4]

[2] This is because the role of the change from actuality to potentiality means that wholly past things exist but are not doing anything in either the present-tensed or tense-neutral sense. I'll revisit this in §10.2. See Briggs and Forbes (2012) for a detailed semantics, and Forbes (2023a) for a discussion of the interaction between verb aspect and tense.

[3] Recall our discussion in §1.5 about showing how different theories were distinct.

[4] Other temporal ontologies are available. Most notably, the 'static block' view on which there is no McTchange, and there is no ontological difference between past, present and future. Given that Parts I and II have argued against those positions, we shall not discuss that view, or other similar views, further here.

9.2.1 The Growing Block

The Growing-Block view, according to which the past exists and the future does not, will get its own chapter: Chapter 10. The intuitive idea is that spacetime is growing as more time comes into existence. On this view the McTchange from potentiality to actuality is an ontological change that Broad describes as 'absolute becoming'. Tim Maudlin (2007:109) glosses the view that the future is unreal as the view 'all will end, I will not exist tomorrow, I have no future'. This, to forestall any misunderstandings, is *not* what is claimed by the Growing-Block view. It need not deny that there *will* be a future to accept that the future does not (tenselessly) exist. For to say that there *will be* a future is to commit to change in what is present as time passes, on this view, not to commit ontologically to something that (tenselessly) exists later.

This view received its first explicit and unambiguous articulation in the 1920s, with the *locus classicus* being C. D. Broad (1923)[5] There was a small resurgence in defence of the Growing-Block view in the 1970s,[6] and the first monograph-length defence of it with Tooley (1997). More recently, it has received renewed defence, including by Forrest (2004), Button (2006; 2007), Correia and Rosenkranz (2013), Rosenkranz and Correia (2018) and Perović (2021).[7]

I will save further discussion of this view (and the different varieties in which it is defended) until later. For now, I will merely comment that it appears to be a fairly straightforward way of capturing the asymmetry between past and future; it's an asymmetry of ontological status. §§9.3 and 9.4 will examine the case for using ontology to capture this asymmetry. The Growing-Block view also captures McTchange in the form of absolute becoming; the coming into existence of events. Although I think it is the best, as we shall see, this picture isn't the *only* way we can capture McTchange and the modal/causal asymmetry between past and future.

9.2.2 Presentism

An alternative to the Growing-Block view (and one that has received more attention in the literature) is Presentism.[8] This is the view that only present things/stuff/processes/events exist. This view won't, as the Growing-Block view does, associate existence with the actual, but rather with the phenomenal or the ongoing. I will briefly survey a number of varieties of Presentism, though in a self-consciously coarse-grained way. I will consider, briefly, versions of Presentism that disagree with the considerations of Parts I and II, only to set them aside. I will then survey different ways Presentists deal with the past.

[5] Thomas (2015); (2019) draws our attention to Samuel Alexander (1920) and Henri Bergson (1911) as precursors of the view, and Hilda Oakeley (1927; 1932) as a defender contemporary with Broad.

[6] Cf. J. L. Mackie (1980) and R. C. Jeffrey (1979).

[7] My (2016) and my work with R. A. Briggs (2012; 2017; 2019) have also contributed.

[8] E.g. Adams (1986); Crisp (2003; 2004; 2007); Bigelow (1996); Bourne (2006); Dorato (2015); Ingram (2016; 2018); Keller (2004); Lombard (1999); McKinnon and Bigelow (2012); Markosian (2004); Mason (2006); Merricks (1999); Tallant (2018); Tallant and Ingram (2015; 2020); Zimmerman (1998; 2007).

John Bigelow (1996:35) claimed that 'Presentism was assumed by everyone everywhere' until the nineteenth century. This, at best, is a rhetorical exaggeration, but I don't think even a more modest claim that it is a common-sense view is true. As Ursula Coope (2005:4) warns us, often appearances of familiarity in the philosophy of time espoused by historical figures 'are largely deceptive'. It is not clear that philosophers, or people going about their lives, had an assumed ontology that committed them to Presentism over the Growing-Block view, nor indeed to any other of the views we will consider. Indeed, as we saw in §2.1, there is no precise temporal ontology that participants in experimental studies stably subscribe to. While the claims of Parts I and II might be, in some sense, presuppositions of common sense, the question of to what one is ontologically committed often seems anachronistic to impose on philosophers before the rise of analytic metaphysics, and a stretch to claim on behalf of the philosophically uninitiated.[9] The question of *existence* seems to lose traction outside some explanation of what contrast we are attempting to invoke. If the contrast is between a present which is *unlike* past and future in important ways, each of the various views discussed in this chapter can accept that. But there is room for debate over how dissimilarity between future, past, present and mere possibilia should be cashed out in the coin of existence.

Even if we wish to attribute Presentist inclinations to historical figures, some of these may produce Presentisms that aren't in the spirit of the modern view. Parmenides, for example, might be interpreted as a Presentist, but if he is, he does not seem to accept McTchange.[10] There is certainly a consistent position – we might call it *Parmenidean Presentism* – that accepts that only the present exists, but denies McTchange.[11] Let's set that aside, and henceforth assume Presentism involves both a commitment to McTchange and to the non-existence of anything non-present.

Not all versions of Presentism accept that there is an asymmetry between past and future of the kind defended in Part II. Dummett (1969; 2004) and Markosian (1995), for example, accept an 'open' past according to which any contingent truths about the past depend on the nature of the present (and any laws of nature), and so lack the modal status that actuality was intended to confer. Dummett's view, sometimes described as 'train carriage presentism,' is inspired by intuitionistic semantics, according to which any proposition p is true just in case that you have a proof of p. It is called train carriage presentism because if you look out of the window of your train carriage and fail to count the number of blades of grass in the field you are passing, there is simply no fact of the matter how many there were. Markosian's view, by contrast, doesn't rely on our actual knowledge of the world, but on what could in principle be derived from the directly observable state of the Universe and the laws of nature. If those laws of nature are indeterminate, what was the case in the past will be left open by them. For each view, it is possible, should all present physical traces be removed, to change the

9 Augustine of Hippo *Confessions* XI §22–4, is perhaps a case where an ancient philosopher does seem to commit to something recognisable as Presentism, but it's not clear that his arguments for this position – that there can be no secret place that things go to or come from – would get much traction today.

10 See Long and Sedley (1987) for fragments of Parmenides.

11 Parmenidean Presentism was discussed briefly in §2.6.2.

facts about what happened in the past. I describe these views as *Orwellian Presentism*, echoing a criticism of Dummett by Crispin Wright (1986) that alludes to Orwell's *Nineteen Eighty-Four* (quoted above), in which it is proposed that by controlling memory/records, we control the past.

There is a certain irony that philosophers, such as Markosian and Dummett – both of whom are noted for their personal commitment to egalitarianism – endorse a view of the open past more regularly associated with authoritarian regimes, fictional and real, who have conducted attempts to remove any trace of genocide or suppression of dissent. When I reflect on my attitude to intrepid journalists such as Jones (1933) or Cole (1990), who recorded acts of genocide and suppression of dissent respectively – at considerable personal risk – my response is not that if it weren't for them the events they recorded might not now have determinately happened. It is *because* the highly significant events have happened, whether any trace remains or not, it seems to me, that there was an imperative to attempt to record them.[12] Having defended the asymmetry between past and future in Part II, I will not dwell on it, or Orwellian Presentism further.

Assuming that the Presentist wants to accept the claims of Parts I and II, they have to give an account of what exists. We know what the Presentist says about the present things – they exist – but what do they say about the non-present things? A common approach for Presentists is to claim that truths that seem to be about the past and future are, in fact, truths about some other things. I will call these *Surrogate Presentisms*. There are two broad types of Surrogate Presentism: Ersatzer and Haecceitist Presentism. I'll start with Ersatzer Presentism before moving on to discuss the Haecceitist version.

Ersatzer Presentists include Thomas Crisp (2003; 2007) and Craig Bourne (2006).[13] Bourne (2006), for example, proposes that true propositions about Julius Caesar are present-tensed propositions containing the name 'Caesar', which are members of maximally consistent sets of present-tensed propositions (containing no embedded tense operators); these sets of propositions are related by the *E*-relation to further maximally consistent sets of propositions to create a timeline. One such set of propositions is concretely instantiated (by us). So 'Caesar fought at Bigbury Camp' is true iff the proposition 'Julius Caesar fights at Bigbury Camp' is a member of a set of propositions appropriately *E*-related to other sets of propositions forming a timeline, one set of which is concretely instantiated. These sets of maximally consistent propositions (containing no embedded tense operators) are known as *ersatz times*, and the *E*-relation is intended to be very much like the 'earlier than' relation, except that it relates ersatz times rather than anything concrete.[14] Rather than a commitment

12 The Orwellian may respond that, in practice, it will be very unlikely for an authoritarian regime to wipe out *all* traces of their crimes. However, this doesn't help, since it amounts to a criticism for not trying hard enough to change the past, not for attempting it in the first place. Machiavelli (1532, ch. 4) advises killing the previous ruler's whole family when overthrowing them, so there will be no one left to launch a rebellion. I can't believe Dummett or Markosian want to offer parallel advice, but a commitment to the openness of the past seems to undermine any objections to such advice if carried out thoroughly enough.

13 Crisp's view combines elements of Erstazer Presentism and Lucretian Presentism.

14 I'm not sure how useful the comparison to the 'earlier than' relation is, since according to Presentism nothing is ever related by such a relation, so we cannot claim familiarity with it in order

to Julius Caesar as a historic flesh-and-blood Roman with a distinctive nose, Caesar is, it seems, an ersatz Roman with an ersatz nose. Caesar is a set of descriptions, but there is nothing the descriptions are descriptions *of*. Insofar as we are persuaded that we can have singular thought about dead people, Ersatzer Presentism allows us only particularised thought.

Haecceitist Presentism is defended by Adams (1986), Mason (2006), Ingram (2016; 2018) and described (though not clearly endorsed) by Keller (2004). Adams appeals to singular thought as a reason to accept Haecceitist Presentism. According to this view, facts about Julius Caesar are about some remnant of Caesar, for example the property of being Caesar. The property of being Julius Caesar is also known as Caesar's *haecceity* or 'thisness'. Ingram (2018:70) describes the view as follows:

> my proposed thisness ontology invokes and involves a vast number of properties. I say that, for every entity x (that exists now or has existed), there is a primitive, particular, purely non-qualitative property: its thisness, the property being x . . . there are also myriad higher-level properties, viz. properties of properties. I assume that all (non-paradoxical) predicates express properties.

He continues, considering a major objection to the view:

> This ontology may strike the reader as unacceptably (perhaps problematically) vast. Indeed, those attracted to Quinean 'desert landscapes' (e.g., Quine 1948:23) will find little to entice them. But . . . accepting my thisness ontology allows presentists to solve several important metaphysical problems facing standard versions of their view.

The challenge for the Haecceitist isn't so much that it introduces too *many* things, but that the things it introduces are too *strange*. A case can be made for accepting haecceities, but they don't seem to be entities in their own right; they need to be thisnesses *of* something. But the thrust of the Haecceitist Presentist position is that the haecceities hang around after there has ceased to be anything for them to be haecceities of.

Although metaphysician's mystery is fools' gold, I can't make sense of uninstantiated properties generally, and in particular, I can't fathom *non-qualitative* uninstantiated properties. I find myself sympathetic to the view that David Seargeant (1985) expresses that properties are *ways of being*, and so there must be some entity of which the property is a way of that entity being. Properties existing without being properties of anything are most curious. As Lewis Carroll's (1897:59) eponymous Alice observed when the Cheshire cat disappeared, leaving only its grin remaining:

> 'Well! I've often seen a cat without a grin,' thought Alice; 'but a grin without a cat! It's the most curious thing I ever saw in all my life!'

to understand the *E*-relation.

The Haecceitist view, then, like Ersatzism, seeks to replace Caesar with something rather different, and Caesar's nose with something that lacks nostrils. The Haecceistist claims that this allows us to have singular thoughts about Caesar, but it only gives us a property, which is perhaps not substantial enough.

A third alternative, that accepts the past is not open, but does not appeal to surrogates for past entities are views which claim that past truths are made true by present things. Bigelow's (1996) *Lucretian Presentism* is perhaps the best-known example.[15] A recent defence of this view has also been given by Tallant and Ingram (2020), though it is not clear that they hold the view they defend, merely that they are critical of the objections to it. Lucretian Presentism is the view that there are *past-tensed* properties that truths about the past are true in virtue of. This can either come in the form that indestructible atoms have these *past-tensed* properties (e.g. this atom was previously part of Caesar's nose) or the world as a whole has past-tensed properties (such as previously having contained a battle at Bigbury Camp). The atomic view is more in keeping with the Lucretian inspiration of this view.[16]

This view initially seems to make the past dependent on the present (as Orwellian Presentism did), though because the properties in question are properties of necessarily eternally persisting entities, the relationship between what happens at various times and the properties of the atoms can't so easily be separated on this view. It makes as much sense to ask whether the world is how it is now because of the Lucretian properties some atoms used to have (e.g. that they *will* form part of your nose) as it does to ask whether the past happened the way it did because the atoms currently have (e.g. that they used to form part of Caesar's nose). Lucretian Presentism ensures that all properties are properties of something (avoiding my concerns about Haecceitistic Presentism), but the properties are of a rather strange sort. Sider (2001) describes them as 'ontological cheats', though this characterisation, by his lights, would also include most modal properties, such as dispositions, or capacities. While Lucretian Presentism, as presented by Bigelow (1996) and Tallant and Ingram (2020) is committed to a superabundant view of properties, where properties describe the entire past and possible future histories of an object in every detail, the relation between past and present does seem to be in question here.

Even if it is not straightforward to accuse Lucretian Presentism of agreeing with the Orwellian Presentism that the past depends on the present, it is equally unstraightforward to extract a claim that the present is a product of the past. Consider the issues of Chapter 6, where we attempt to make sense of records and can only do so by presupposing a past of which they are a product. The Lucretian Presentist faces a challenge because they do not accept the existence of a past event that was recorded, merely the existence of some properties that correspond to the veracity of the record. One can't say that this fossil depends on that ammonite, merely that this fossil is such that there used to be an ammonite. And this prevents us from having the referential resources to talk about the dependence of the present on the past or vice versa.

15 Although see also McKinnon and Bigelow (2012) and Crisp (2007).

16 See Lucretius 7A 1.445–82 in Long and Sedley (1987:32–3).

The fourth and final alternative version of Presentism I will discuss is one that denies the need to explain claims about the past and future in terms of any ontological commitments. Tallant and Ingram (2015) discuss 'Nefarious Presentism', which, as opposed to 'Upstanding Presentism' simply refuses to accept ontological commitment to a past of which the present is a causal product. This provides an elegant solution to the objections presented to the other versions of Presentism described above. If there is some question about what claims about Caesar's attack on Bigbury Camp amount to, the Nefarious Presentist will answer straightforwardly with an account of a camp at Bigbury and a Roman general who attacked it. But if you ask to inspect the list of Roman generals the Nefarious Presentist believes there to be, like all other Presentists, the Nefarious Presentist will not provide you with a list on which the name 'Gaius Julius Caesar' is to be found. Whereas an Upstanding Presentist would rifle through their pockets for alternatives to a Roman general that may satisfy your demand (the remains/records of Caesar, an ersatz general, the property of being Julius Caesar, a world which used to contain Caesar, etc.), the Nefarious Presentist will give you a cheeky wink and be on their way.[17]

We will revisit, in §9.3, whether any of these strategies to account for past truths will suffice. But here we can at least conclude that Surrogate views, Lucretian views and Denial views can plausibly accept the claims of Parts I and II. The question is whether they represent the *best* ways to make sense of those commitments.

In §3.2, we discussed the role of endurance in motivating McTchange. The role of Presentism in motivating endurance is a little murky. Haslanger (2003) and Hofweber and Velleman (2011) argue that one can be an endurantist with or without presentism, which is irrelevant to the 'innocent sense in which persisting objects are extended in time' (Hofweber and Velleman 2011:44). Tallant (2018), on the other hand, argues that presentism isn't compatible with *any* of the three views of persistence we discussed, because presently existing objects fail to depend on their pasts in the appropriate ways.

The view, defended by Merricks (1999), that endurantism is the only view of persistence compatible with presentism seems quite strong, as we common-sensically are happy with climbs, wars or lectures having earlier and later phases. We would normally contrast such events with things that endure. If the link between endurantism and presentism is as tight as that, the combination of both views seems like radical departures from common sense in excluding events with phases from our account of the world.

What's more, the alternatives presented in many of these discussions present a false dichotomy wherein one must either adopt eternalism, a static view, or Presentism, a dynamic view. The claim that one must adopt a dynamic view to make sense of persisting by enduring, which I defended in §3.4, is much weaker than the claim that one must adopt Presentism specifically. As we shall see, there are many other temporal ontologies available.

[17] They called the view 'Nefarious Presentism'. That's pretty strong evidence of a disposition to give a cheeky wink.

9.2.3 The Moving Spotlight

While Presentists claim that only the present exists, defenders of a Moving-Spotlight view claim that past, present and future all equally exist, and that the present is, in some metaphorical sense, 'lit up' in a way that other times are not. Cameron (2015) provides the most fully developed, though somewhat heterodox, treatment of this view, but Deasy (2015) and Miller (2019) have also offered recent defences of it. The view takes its name from a passage by Broad (1923:59), in which he uses the simile of a lantern with a focussing 'bull's eye' lens:

> We are naturally tempted to regard the history of the world as existing eternally in a certain order of events. Along this, and in a fixed direction, we imagine the characteristic of presentness as moving, somewhat like the spot of light from a policeman's bull's-eye traversing the fronts of the houses in a street. What is illuminated is the present, what has been illuminated is the past, and what has not yet been illuminated is the future.

The Moving Spotlight looks to accept the claims of Part I: it allows for McTchange. For Cameron (2015), every substance has a temporal distributional property – a property that specifies how its parts are arranged in time – and an age. From these properties, the present can be derived; if a substance is twenty years old, the present is twenty years later than the origin of the substance. The passage of time, on Cameron's view, is McTchange in the ages of substances.

At first glance, the Moving-Spotlight view has a case to answer on the asymmetry between past and future. It does not accept an asymmetry between past and future in terms of ontological asymmetry; the past and future are equally real. There is more than one way to accept a difference between past and future into one's metaphysics, however. Ross Cameron's (2015:195) defence of the Moving-Spotlight view builds on his work with Elizabeth Barnes (Barnes and Cameron 2009; 2011) on the open future. For them, it is a matter of indeterminacy of the future. The future is some way, but it is indeterminate what way that is. This contrasts with the Growing Block (§9.2.1) which claims there is no ontology of which we can say *it* is indeterminate:

> I think the moving spotlight view sits naturally with the thesis that if the future is open then it is a matter of future contingents being indeterminate: that for any prediction about how things will be, it is (determinately) either true that things will be that way or false that things will be that way, but it is (sometimes, at least) indeterminate which. The moving spotlighter agrees with the growing blocker that claims about what was the case are sensitive, in some respect, to past ontology and so, again due to considerations of parity, predictions about what will happen should be sensitive in the same way to future ontology. In that case, given that the moving spotlighter believes in future ontology, she believes in the relevant portion of reality that should speak to the truth or falsity of any claim about how things will be. Unlike the growing blocker, she believes that our temporal ontology is extensive enough to speak to the prediction, one way or another.

For Barnes and Cameron, and for Cameron *solus*, there is a sense in which the future is open. This sense, however, seems to capture there being multiple possibilities for how the future goes or vagueness. It doesn't capture the considerations of Chapter 7 regarding change from potentiality to actuality even in a deterministic Universe.

The Cresting Wave view that Kristie Miller (2019:113) explores is entirely compatible with the considerations of Chapter 7. The passage of time, on this view, is a cresting wave of causal efficacy, with an asymmetry between past and future in the form of a difference between potentiality and actuality, and the passage of time involving McTchange in what is causally efficacious:

> the only moment with causal efficacy is the present moment. Or, better, presentness *is* causal efficacy, which is like a cresting wave. . . . Once presentness has left a time its causal efficacy is spent.

The Cresting Wave, unlike Cameron's view, has resources to cope with a change from potentiality to actuality, but the Cresting Wave may have difficulty with multiple future possibilities. It could, of course, employ the ontic-indeterminacy approach of Barnes and Cameron.

9.2.4 The Shrinking Tree

If one thinks of there being multiple possible futures, a different model may be preferred. Storrs McCall (1994) proposes a model of the universe on which it forms a tree-like structure where the branches of the tree drop off as they cease to be possible, leaving the trunk-like past extending as possibilities drop away. Again, this view has an account of McTchange (the dropping away of future possibilities), though on McCall's view this is tied directly to quantum indeterminacy. There can be no passage of time without a reduction in the future possibilities. It strikes me as a cost of the theory that it can't allow change in what time it is to happen without there being change in what possible futures there are. It seems at least *possible* that the world could have a period of time in which it was really dull before the next indeterministic quantum event occurred.

There is a question about how the fully real regions of spacetime in the past and future differ from the present. The resources of the Cresting Wave theory could, of course, be used here. But there is also a question about how to think about the merely possible persons that currently exist, but most of whom will drop out of existence as they cease to be possible. Miller (2006:310) makes this point about McCall's view:

> there exist many persons on these future branches, only a small proportion of whom will become part of the present and the past, the remainder of whom will 'vanish' when the branch they are on drops off. Plausibly, regardless of our metaphysical views, most of us will hold that we have some duties with respect to future persons. It is usually held that we have a duty not to harm future persons, however exactly we understand that claim. Generally, though, while we might think that we have a duty not to beget children who will suffer lives that are beyond

> the threshold of decency, we do not think that we have a duty to beget children who would have fulfilling lives. The idea is that abstaining from creating such children is not to harm them in any way, since they are merely possible and do not exist. Thus, we might conclude, future persons do not have a right to existence, though whatever future persons will exist do have certain rights to have a decent life.

This is an argument linking the existence of future persons to our ethical duties. Recalling the discussion of Harman (2009) in §8.1.4, the difference between actual and merely potential people seems morally significant. *If* the Shrinking Tree can't allow us to distinguish between the moral status of begotten children and children that may never be begat, that seems like a reason to think the view is unable to capture the moral reality of the actual world. We'll revisit this worry in §9.4.

9.2.5 The Shrinking Block

While the Moving Spotlight (including the Cresting Wave variant) and the Shrinking Tree differ from the Growing Block and Presentism in an ontological commitment to the future, we might also examine a view which agrees with the Presentist about the non-existence of the past. The Shrinking Block accepts the existence of the future and denies the existence of the past. It has been defended by Casati and Torrengo (2011), Norton (2015) and Lam (2021).

John Norton (2015:103) likens the view to a burning fuse, in which the future is potential and the past actual:

> Please imagine a long fuse hanging down from the ceiling. We take a match, strike it and bring it to the end of the fuse. . . . Our attention is held by the flaming point that slowly advances upward. Those parts of the fuse ahead of the fire take their turn to be consumed by it and to disappear into the ashes. . . . The advancing point of fire is the now. It consumes the future, the unburnt fuse, which is converted into the past, the scattered nothingness of ashes. In the burning fuse model, the future states carry the capacity of being able to come to be present. That capacity is realized in momentary fires of the present after which all reality is extinguished. The future is real; the past is not. . . . The key novelty of the burning fuse model is the recognition that future events carry a property that is not carried by past events: they carry the capacity of being able to come to be present. Past events do not carry that capacity. Once they are past, they are spent. The capacity is lost. Indeed, they carry no potentialities at all. (Norton 2015:103)

It is obvious that this model is compatible with McTchange and with a view of the asymmetry between past and future in terms of a change from potentiality to actuality. But here potentiality is aligned with existence, rather than actuality. And why not? Common sense can be wheeled in to support this theory quite easily, as Casati and Torrengo (2011:242) argue:

> A [defender of the Shrinking-Block] has the unassailable belief that she would not be able to travel into the past; for you cannot travel to a destination that does not exist. On the other hand, her intuition does not foreclose travel into the future.

Perhaps, as Casati and Torrengo (2011:243) suggest, the comparative unpopularity of the Shrinking Block is a result of cultural bias (perhaps with a religious aftertaste):

> We suspect that a preference for the Growing Block over the Shrinking Block is a consequence of a cultural bias toward creation: 'if the universe exists, it is because it has been created; and of course it can only have been created in the past.'

Casati and Torrengo put the challenge of this chapter incredibly clearly. The move from accepting McTchange, and from the asymmetries of records, of potentiality and actuality, and of emotional attitudes like hope and regret, to *ontology* needs much more scrutiny than it has usually received. It is not a matter of simply taking any ontological story that's going, but thinking about why that temporal ontology does a better job than its rivals. Also, to echo the concerns of Callender (2012) from §1.5, about whether these views really involve any disagreement, either observationally, metaphysically or explanatorily.

9.2.6 The Glowing Shrub

Completists may note that there are more options. In §8.2.1 we considered the role of the counterfactual as being like a fourth tense: the neither potential nor actual. We could have a view that accepted the existence of future possibilities (like the Shrinking Tree), but also the existence of what could have happened but didn't. Such views might be thought of as Shrubs. They are like trees, but rather than having a trunk portion and a branched portion, we get branches from the very beginning. There are versions of this view, motivated by the Everett interpretation of Quantum Mechanics, defended by Saunders and Wallace (2008) and Wilson (2011), but with no room for McTchange. However, using the same asymmetry between potentiality and actuality that the Cresting Wave Moving Spotlight uses, we can have a cresting wave of causal efficacy sweeping through the shrub-like structure that the Everett interpretation of Quantum Mechanics provides.

Having surveyed this catalogue of temporal ontologies, represented diagrammatically in Figure 9.1, we have perhaps laboured the moral; a range of different temporal ontologies are compatible with the same set of metaphysical commitments about an asymmetry between space and time (time changes in a way that space does not) and a central asymmetry between past and future (the future is potential and the past is actual).

The question of which of these temporal ontologies we adopt, then, may have more to do with our views on temporal ontology than with settling pre-theoretical puzzles about time. The comparatively diminutive size of Part III compared to Parts I and II

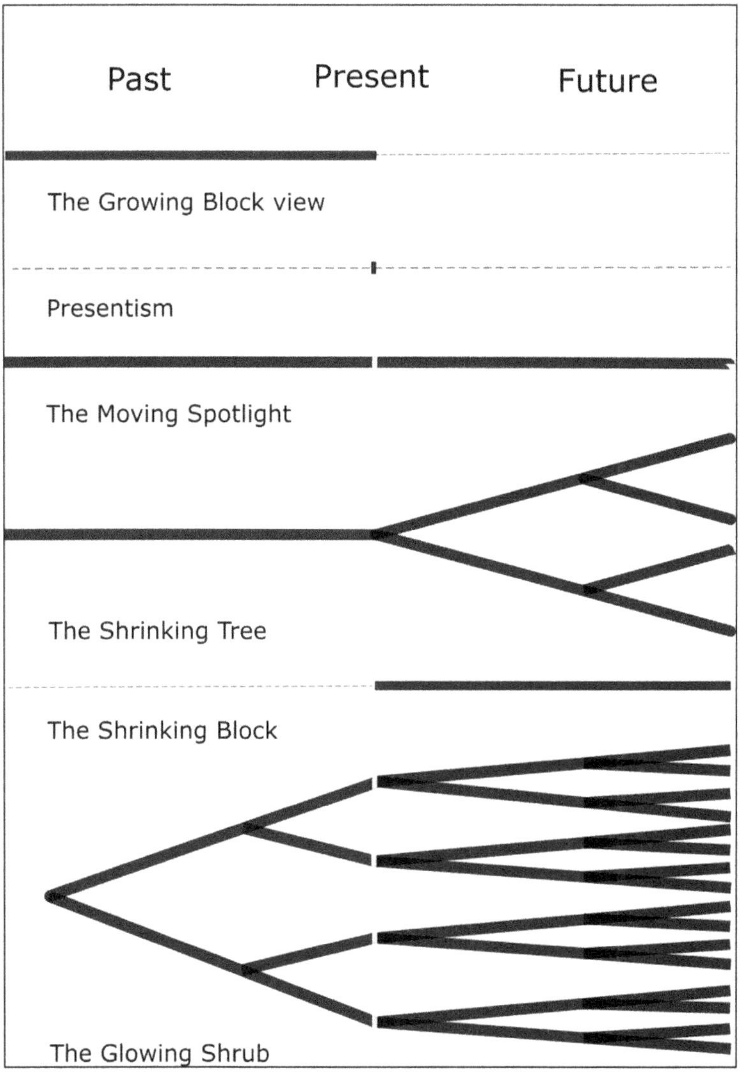

Figure 9.1 A range of temporal ontologies.

reflects my view on the comparative importance of engaging with the pre-theoretical puzzles about time rather than what seem like back-office functions of tidying up one's accounts. But Part III – for all it represents the theoretical back-office of the metaphysics business – is important, especially when the auditors come in. The remainder of the chapter, therefore, will be devoted to two tasks. First, it will outline the broad reasons for accepting the Growing-Block view over its rivals, in §§9.3 and 9.4, and then in §9.5, we will discuss more broadly still what temporal ontology (as distinct from the metaphysics of Parts I and II) is trying to do.

9.3 Making past truths true

In this section, I want to consider the case for thinking that the past *exists*. Why think the considerations about actuality of the past from Part II lead us to think that we should be ontologically committed to past states, events, things and stuff. As we discussed in §4.2, on the Quinean approach this amounts to a question of whether past entities are referred to by our best theory once it has been put into 'ontologese'; the canonical notation.

The central objection to those views that deny existence to the past is that we are committed to various truths about it, and the commitment to those truths requires a commitment to some ontology that *makes* those truths true. As Asay (2018:914) puts it:

> Truths have implications for our ontological commitments. So we can raise a question: for any given truth <*p*>, in what ways does <*p*>'s truth depend or not depend on ontology? What *constraints* on one's ontology does a particular truth make? I suggest that truthmaking investigations be understood in this light. To ask what the truthmaker is for <*p*> is to as Quine's (1948) ontological question in reference to it: What is there, with respect to <*p*>? (emphasis original)

The truthmaking demand, then, presents a trilemma to any theory. We must accept one of these three options:

- Commit to the things that the truths quantify over;
- Show how the truths avoid the commitments they appear to generate;
- Remove such truths from one's theory.

I will not engage with the view that there are no truths about the past;[18] neither will I engage further with the view that the past ontologically depends on the present. I'm pretty well committed to the view that the present is a causal product of the past.

There are two remaining strategies, then: one, which we have seen above, is to rely on present (or present and future) ontology to make truths about the past true (without accepting that the past depends on the present ontology) or to accept the existence of the past. Alternatively, one can reject the truthmaking demand entirely, as the Nefarious Presentist does. I will come back to the Nefarious alternative shortly; for now, I want to consider the position where one accepts truths about the past but denies that this involves a commitment to past things (events, processes, stuffs etc.).

There is a deflationary approach to truthmaking, where one provides instances of the 'because-schema' or 'B-schema' (cf. Asay and Baron 2020:5):

(B) <*p*> is true because *p*.

[18] This view has been defended, under the heading 'Hard Presentism' by Dawson (2020).

So, some truth-bearer, for example a proposition, is true because the world is as described by the proposition. <Grass is green> is true because grass is green, <Caesar fought at Bigbury Camp> is true because Caesar fought at Bigbury Camp, <Tractors replaced horses in agriculture during the 20th Century> is true because tractors replaced horses in agriculture during the twentieth century. These instances of the because-schema seem fine but leave the ontological problem untouched. The question remains of how to accommodate the descriptions of the world in *ontologese*. If truth is *constrained* by ontology, the descriptions of the world, for example 'tractors replaced horses in agriculture during the twentieth century', appear to commit us to tractors, horses and agriculture. But how are we to work out what commitments are required?

McFetridge (1990) offers two considerations that provide useful constraints in thinking of how we work out what commitments are generated from truths we accept. He illustrates one commitment with the example of things that appear in mirrors (1990:50–1):

> It would destroy the claims of mirrors 'reflecting' reality if, in attempting to explain why things appeared in the mirrors as they did, we were forced to invoke entities whose only role was to appear in mirrors. The 'theory of reflection' should not require, as it were, objects not required elsewhere in our science of nature. Allying this to the theory of truth (the 'theory of reflection in language') we might say: it would seem illegitimate to invoke, in explaining why certain sentences are true, categories of objects not otherwise needed in our explanatory projects. For, if the only ground for invoking such entities were to explain why certain sentences were true, this would undermine our view that the role of true sentences is to reflect a reality which would be there even if they were not.

We might call this constraint one of width of cosmological role (cf. Wright 1992; ch. 6.5). If we can make truth-bearers true in virtue of ontological categories to which we are otherwise already committed, we should. The test here is essentially a Quinean one of what is required by our best theory. But theory, here, is not merely to be understood as a set of sentences but as the product of an explanatory project which attempts to relate those sentences together to best describe the world.

McFetridge's second constraint is epistemic. If we could come to know that a proposition is true without any causal contact, however mediated, with the relevant commitments, that's a reason to think those commitments aren't needed. This seems to be related to a more recent constraint proposed by Baron (2013), that propositions ought to be true in virtue of what they are about.

These constraints look to pose problems for various views of truth about the past because we want to appeal to entities that have a wider role in our science of nature and explain why we might come to know those truths, such that our knowledge is non-coincidentally connected with our ontological commitments. More broadly, there is the thought that our account of what there is ought not to be pragmatically inert. Our commitments ought to show up in our practices. Even our commitment to the unknown, to the transphenomenality of the world, is pragmatically significant in that it drives us to further inquiry. Truthmakers should not be the sorts of things that pull apart from how we come to know things or how we regulate our inquiry.

We mustn't accept these constraints too wholeheartedly, however. Our knowledge <there are no pandas in Sheffield> is true does not require an ontological commitment to an anti-panda, and our knowing that there are no pandas in Sheffield does not require our standing in some causal relation to such an anti-panda. Indeed, we could know the locations of all the pandas and know that none of those locations are in Yorkshire, and know that Sheffield, if it is anywhere, is in Yorkshire. Thus, we can make <there are no pandas in Sheffield> true without ontological commitment to Sheffield, or to the existence of a lack of pandas or otherwise. Indeed, we can know <there are no Unicorns in the Unicorn Inn> if we know that there are no unicorns. This doesn't involve us knowing anything about the Unicorn Inn, or any commitment to unicorns outwith the Inn, or causal contact with unicorns. Rather, the commitment is to avoid a situation where we rack up ontological commitments that cannot be investigated and have an objectionably narrow cosmological role. It is precisely because unicorns only appear in fictions, and never in the wild, that we tend to eschew any ontological commitment to them.

Equally, there may be occasions where it is very difficult to know what we should be committed to. If, as discussed in §2.1, two hypotheses are empirically equivalent (and so our empirical evidence for them is equally good), but differently shape future inquiry, we may have a dispute about what truthmakers we can know. In such cases, at least, this can be a dispute about what we know as much as a dispute about what makes some truth-bearer true. As we discussed in §§1.5, 2.1 and again in 4.2, what commitments we have will not be determined by individual claims, but more holistically at the level of theory. It is important that we don't have commitments that don't fit with our theory, or become problematically disconnected from our capacities to know or talk of them.

Though I think, at the level of theory, attempts to provide surrogate truthmakers for past truths either don't work or are unsatisfying, I won't attempt to prosecute this point here. There are, as described above, sophisticated theories that allow one the ability to say all the things I think we ought to say, and provide a temporal ontology with no ontological commitments to the past when the theory is put into ontologese.

The reason for accepting the existence of the past that has the most intuitive pull for me is that it seems there is a clear distinction between how we talk of past people and fictional people. Let's start with a plausible thought about fictional people: that they are indeterminate in a way real, existing people are not. Anthony Everett (2013:92) puts it this way:

> if you tell me a story beginning 'Once upon a time a Hobbit lived in a hole . . .' you certainly do not have a specific creature in mind and you are most definitely not trying to refer to anything. But I will treat your utterance *as-if* you had a specific real creature in mind and coin a new singular mental representation accordingly.

When we talk about fictional people, although we might talk *as if* there are such people, we don't have specific people in mind.[19] By contrast, I submit, when we are

[19] Thomasson (1999), amongst others, thinks that we successfully refer to abstract artefacts when we refer to fictional people. One can accept that while thinking the relation we stand in to such

thinking about historical people, the people about whom we think outrun our ability to think of them. They are determinate and specific in ways such that we can always potentially (in principle) discover more about them. The ontological commitment to past people I favour is meant to reflect this *transphenomenality*. There is (always) more to them than our evidence of them.[20] What's more, as I argued above, the 'direction of fit' is such that we ought to change our beliefs about the past to match what happened (insofar as evidence allows), rather than changing the evidence so it matches our beliefs. Even if, as Danto (1985) supposes, *history* is a narrative created by humans, the past provides external reality that constrains such history.

Shakespeare wrote a number of plays. In *A Midsummer Night's Dream,* he wrote about a hobgoblin called 'Puck'. In *Julius Caesar*, he wrote about a Roman dictator called 'Julius Caesar'. Drama critics might criticise the characterisations in terms of their 'dramatic truth'; were these relatable, psychologically realistic characters, driven by complex motives and offering an artistic exploration of the human condition?[21] But there is another sense of truth in which, in a less literary spirit, we might criticise Shakespeare. *A Midsummer Night's Dream* is a fantastical comedy. Puck is a fictional character. We are not likely to criticise Shakespeare for *historical* inaccuracy in his portrayal of Puck.[22] But, with his Julius Caesar, we might wish to make very clear that although Caesar was assassinated, his funeral did not take place on the same day. Shakespeare was in a position to depict Caesar inaccurately in a way he was not in a position to depict Puck inaccurately.[23]

The suggestion, previously discussed in §7.4.2, that we are capable of singular thought about the past has been a staple of objections to temporal ontologies that deny this. Plantinga (1983), Markosian (2004) and Crisp (2005) discuss this in the context of Presentism, for example. Singular thought about the past requires the existence of the past, because it is existence-entailing. The Haecceitist strategy is intended to allow for a haecceity to be the object of thought when we think about past things, rather than the things themselves. I have argued above that uninstantiated haecceities don't seem well-placed to act as surrogates in this way. Particular accounts of singular thought differ wildly. While Markosian (2004:53) says 'It seems pretty

artefacts is still importantly different to the relation we stand in to the dead relatives that get pointed to in old photographs. Dead relatives are not artefacts that we create in talking about.

20 Cf. Polanyi (1983:32–3):

> This capacity of a thing to reveal itself in unexpected ways in the future I attribute to the fact that the thing observed is an aspect of a reality, possessing a significance that is not exhausted by our conception of any single aspect of it. To trust that a thing we know is real is, in this sense, to feel that it has the independence and power for manifesting itself in yet unthought of ways in the future.'

Even when things have been destroyed, there is a sense that they had such a capacity to have revealed themselves in ways that they did not.

21 Perhaps dramatic truth is the kind of thing discussed by cognitivists about the arts, such as Gibson (2008).

22 We might point out the anachronism of an English fairie in ancient Athens, but that's not quite the same.

23 Russell (1919:170) said, drawing a similar contrast:

> If no one thought about Hamlet, there would be nothing left of him. If no one thought about Napoleon, he would have soon seen to it that someone did.

clear to me that there are singular propositions about existing concrete objects', and Dickie (2010) argues that we are capable of singular thought about ordinary objects as a result of our acquaintance with them in perception, one could believe in singular thought as only applying to immediate sense-data, as Russell (1905) does. Ken Taylor (2010:79) made a helpful distinction between the referentially fit and the referentially successful:

> An objective representation is one that refers to a real existent (or expresses a real property). A representation is objectual, on the other hand, if it is 'fit' or 'ready' for the job of standing for a real existent. I will also say that objectual representations are *referentially fit* and that objective representations are *referentially successful.*
>
> Now a representation can be simultaneously both objectual and objective, both referentially fit and referentially successful. But some representations are *merely* referentially fit, without being referentially successful. The class of merely fit representations is the class of empty or non-referring singular terms. On the view that I shall outline here, it is crucial that merely fit singular representations are still, in one sense, fully singular. They are fully singular in the sense that they still enjoy, in virtue of their form, singular referential purport. It is just that they purport to refer without succeeding in so doing.

The claim I am pushing is that singular thought about the past is often referentially *successful* because it is objective in this sense (especially in those cases where I have vivid perceptual memories or records of the individuals in question).

Let's recall the five marks of realism discussed in §6.5: evidence-transcendent truth, explanatory ground, cognitive command, width of cosmological role and mind-independence. Past things, I claim, meet all of these marks of realism. We have discussed, in §9.2.2, how the present depends on the past, rather than the other way around, is a constraint on our historical narratives, and that the past exceeds our evidence of it. That's explanatory ground, mind-independence and evidence-transcendent truth, respectively. Cognitive command is satisfied; we think that we can assert truths about the past and have disagreements about it; historical statements are not merely expressions of our current emotional state (even if they are often also that).

Width of cosmological role is slightly more complex to state; past things are largely of the same kind as present things, which suggests a wide cosmological role. They are, however, importantly different from present things, in that they are past. Nonetheless, most of our science of nature relies on relating previous things to present things, so the past does seem to be very important in our cosmology (as argued for in Chapter 6). There is a challenge from Miller (2018) to which we will return in Chapter 10, regarding what we say about the retired or Emeritus status of the past. For if we say past people exist, we still need to distinguish them from present people. Notwithstanding this challenge, the transphenomenality and the direction of fit that the past affords provide compelling reasons to accept ontological commitment to the past.

9.4 Potentia, mere possibilia and impossibilia

That historical figures outrun our evidence of them, and normatively constrain what we say about the past, gives us (defeasible) reasons to think that the past exists. But what of the future? What of the rest: the things that would have happened if the past had been different, or couldn't have happened at all? Once we have accepted the past into existence, why not let a whole lot more in?

The moderate view that the past exists and the future doesn't finds itself between extremes. On the minimal side, there is the view that only present things exist – which Russell (1948:181) derisively referred to as 'Temporal Solipsism', but which we have been calling Presentism and even Dummet's 'train carriage presentism' (discussed in §9.2.2), on which there are no truths about how things are that outrun the present evidence. On the maximal extreme, there is the view that all possible times exist, defended by Lewis (1986) in the form of an infinity of possible worlds. More extremely, there is a position I will call 'Allism' that affirms the existence of impossibilia as well. Given our interests in going from an asymmetry of potentiality and actuality to a temporal ontology and having shown that our ontology should at least include past things, this section responds to the worry that we have no principled basis to stop there. First, I will explain the case for thinking there is an ontological asymmetry between past and future, and then I will discuss ontological restraint.

9.4.1 Ontological asymmetry

The first reason to think the past exists whereas the future doesn't is Valberg's asymmetry of reference, discussed in §7.4.2. There is a strong intuitive contrast between referring to my grandfather (long since dead, but of whom I have many fond memories) and referring to my grandchildren (at the time of writing I have no children, let alone grandchildren). It is clear that there may be many true things I can say about my grandchildren, if I have any. For example, I can say things about who their grandfather is, and I can say things about what sorts of genes they are likely to have and so on. But this case seems much more like the fictional case than it does the past case. I can act as if I had a real specific person in mind, but I do not. The contrast between what I do when I talk about my grandfather and what I do when I talk about my grandchildren is the contrast between an actual grandfather and merely potential grandchildren.

Actuality lines up nicely with existence here: there is someone actual *of whom* I can speak. Merely potential people don't seem to be specific people I can bring to mind, and as such, we might naturally say this is because there are (tenselessly) no such people. Though we can have *objectual* thought about the future that is *referentially fit*, the claim is that thought about my grandchildren cannot be referentially *successful*, because it is not *objective*, in the sense Talyor (2010) has in mind. Broad (1923:79) wryly quips that

> we cannot stand in the relation of direct acquaintance to future events, for the same reason which prevents us from robbing a Highlander of his breeks.

Scottish Highlanders, famed for wearing tartan kilts, have no 'breeks'[24] to steal, and there are no future events for us to stand in relations of direct acquaintance with. We don't need to buy Broad's sense-datum view of perception to accept the thought that future events are not (yet) available for singular thought. In any case, the thought we need here needn't stand alone; in Part II, we defended an asymmetry between past and future. The thought here is that an ontological asymmetry is well-placed to capture the asymmetry between actuality and potentiality argued for there.

Someone like Ross Cameron believes in the existence of future things but thinks it is ontically indeterminate how they are. This is compatible with their having transphenomenality, for they can outrun our *evidence* of them and be indeterminate nonetheless. The thing I find most confusing about this picture is in grasping what it is that is ontically indeterminate. I can understand how something may indeterminately be bald or not bald, or indeterminately be a mountain or a valley, but if it is unsettled whether I have any grandchildren, I am not clear *of what* we are contemplating the indeterminacy.

In general, there are many truths about the future (here I disagree with Broad (1923), and also with Dawson (2020)). It seems true that either the milk in the fridge will be drunk, or it will go off. We rely on such quotidian truths to get by in life. One reason to drink the milk in the fridge is that I don't wish it to go off. And we must have adequate commitments in our ontology to make true such truths. If we assume that we can only commit ourselves to the bare minimum that is required to make our truths true, it is plausible we should reject the existence of future things. Since they plausibly depend on the past and the present, along with some elements of the nomological package, future truths can be made true without commitment to future truthmakers.[25]

9.4.2 Ontological restraint

The assumption that we should be frugal with our ontological commitments is not universally accepted, however. Ruth Barcan Marcus (1975:40) attacked Quine's (mis) use of Occam's razor with the quip, 'If nature abhors a vacuum, it may prefer a plethora.' Nicholas DiBella (MS) argues that the world is much, much more likely to be complex than simple, both in terms of the number of things that exist and the kinds of things that exist. We need reasons beyond a taste for desert landscapes to think that views like the Shrinking Tree and the Glowing Shrub aren't preferable to the Growing-Block view.

Eklund (2008:391) for example, defends 'decadent ontology':

> For instance, the deflationist about ontology can embrace a maximally decadent ontology: *maximalism*, as I will call it. What maximalism says is that for any type of object such that there can be objects of that type given that the empirical facts are exactly what they are, there are such objects. The qualification about the empirical

[24] A variant of the plural noun 'Brekis', meaning breeches, or trousers (*Dictionary of the Scots Language* 2004).

[25] Cf. Briggs and Forbes (2017).

> facts being exactly what they are is there to rule out that maximalism should be committed to the existence of phlogiston or Vulcan. The ontological decadence of maximalism consists in its willingness to countenance all sorts of metaphysically weird sorts of objects – like incars, or like unintuitive abstract objects – as existing.

This decadent ontology can get us new objects such as 'incars' (objects composed of cars only when they are in garages, such that they cease to exist as the cars leave the garage) where a more austere ontology would not accept that a new object is created by putting a car in a garage.

Once we accept ontological decadence, we might accept new types of objects, as Eklund suggests. But we could go even further. There are lots of ways of maximalising one's ontology. Accepting the existence of the future is widespread. Lewis (1986) defends the existence of other possible worlds. Indeed, once we have given up on restraint, why stop at possible worlds? Yagisawa (2010) defends the reality (but *not* the existence) of impossible worlds. Lewis (1990)) discusses a position he calls 'Allism', which he attributes (perhaps unfairly) to Richard Sylvan (*né* Routley), which is similarly committed to impossibilia.[26] I will use the term 'Allism' to refer to the extreme position that accepts the existence of incars, abstract objects, the future, possible worlds, impossible worlds; all of it.

Do the empirical facts rule out future objects, mere possibilia or impossibilia? It seems to be an empirical fact that sea levels will be higher than they are at present. We appear to be committed to future higher sea levels. It seems to be an empirical fact that perpetual motion machines are impossible. Don't we appear to be committed to perpetual motion machines, of which we have claimed the impossibility?[27] A sufficiently decadent philosophy can offer ontological commitments to make true these various empirical facts about the future and about the impossible.

For any case where the existence of an entity is disputed, the Allist is committed to its existence. This is on the grounds that, in order to have a debate about whether an entity exists, there must be something about which we dispute the existence. Allism has not been a popular view. But in settling the issue of why we should be committed to the existence of past things but not future things, we find ourselves facing the issue of what principled grounds there are for avoiding either the extreme *Allist* position or something like Dummett's 'train carriage presentism' on the other hand. While both views warrant incredulous stares,[28] understanding why existence should give us the contrast I propose between past and future requires understanding what is wrong with the extreme positions, rather than merely noting their absurdity. In the next section, I will discuss various approaches to ontology to see how they can help us avoid either extreme.

26 Witherall (2000) argues that this attribution is unfair.

27 For someone committed to ontological restraint, this will be easy to resist; we can come up with a predicate that nothing satisfies. But the question under consideration here is, 'why resist?'.

28 Allism garnered one from Lewis, who more often found himself in receipt of them.

9.5 Metaontology: The story of '∃'

9.5.1 Quinean orthodoxy

In §4.2.1 we discussed the Quinean view that one is ontologically committed to something when it is a commitment on one's best theory. We put the theory in the canonical notation known as 'ontologese'. This Quinean approach has a simple story to tell about what there is: 'Everything'. Perhaps more illuminatingly, for Quine, to be is to be the value of a bound variable. Quine's canonical notation contains sentences in first-order logic that quantify over variables: '$\exists x\, Fx$' says there is something (an x) which is a certain way (the predicate 'is F' applies to it). The sentence '$\forall y\, Gy$' says every y is a certain way (the predicate 'is G' applies to it). 'F' and 'G' here are schematic letters that can be replaced by any predicates to create sentences that potentially describe the world. These sentences can be combined with various sentential operators to create new sentences, and from these sentences we craft competing theories, the merits of which we can compare.

The discussions of ontology so far have very much been within the Quinean tradition. We work out what the ontological commitments of our best theory are, and then accept those things into our ontology. Our survey of temporal ontologies has shown us that there is a dispute about what those commitments are; a dispute that seems largely independent of the distinctions that we might want the theory to draw.

The ontology I defend, the Growing-Block view, comes from an asymmetry of reference between past and future and between historical and fictional cases. But it is not, as we have seen, that one is forced to accept ontological commitments to Caesar but not to Puck, or to my grandfather, but not to my grandchildren. Such a move depends on metaontological considerations about what circumstances hold in order to generate commitments. I could appeal to my taste for temperate landscapes, eschewing both Quinean deserts and Sylvanian jungles.[29] But I hope the differences between me and my rivals are more than merely variations of taste.

What we need is a theory of how to choose the best theory. And choosing which of the various theories of theory choice is best is difficult; we will inevitably presuppose elements of such a theory in deliberating amongst the theories on offer. Indeed, the job of turning the work of Parts I and II into a temporal ontology gets us bogged down in this murkiest of territories very quickly. It is tempting to stipulate the rules by which I work out my ontological commitments and leave things there. I will attempt to go a small step further; I will articulate why I find the rules I use attractive, but without claiming to have shown that they are the only game in town.

9.5.2 Pragmatic arguments

If we wish to allow for the situation where ontological commitments outrun our knowledge, we need to have room for some non-occurrent aspects to our ontology. Dummett's intuitionistic 'train-carriage presentism' which we have already ruled out,

[29] Quine (1948) describes himself as having a taste for desert landscapes. Richard Sylvan (published as Routley (2018)) offers, in contrast to Quine's desert, a neo-Meinongian 'jungle'.

is ruled out on this basis. The motivating thought here is that in order to be in the business of finding out about the world, we need to be committed to something we don't already completely know. For the goal of enquiring into the nature of things to make sense, we need to think that things are not exhausted by our current acquaintance with them. The nature of the world, and its correct description, must be more than meets the eye, or there's no point in taking a closer look. This is a pragmatic argument, based on the conditions for the possibility of inquiry.

Such pragmatic arguments presuppose that certain practices are well-founded. It is open to scepticism in any individual case about whether such a practice is; there are some inquiries that are not going anywhere. Trying to explain one's destiny by the relations between constellations and planets, as seen from Earth at various times, just doesn't seem to be in a position to predict future human behaviour. We are not permitted, on the pragmatic grounds that it would be useful if true, to accept the veracity of astrological predictions, *even though the practice of astrology presupposes the well-foundedness of astrology*. Not all inquiries are well-founded. But surely *some* are. As long as some such inquiries are well-founded, there is a reason to be committed to more than we currently are aware of.

Equally, at the other extreme, we want to avoid a situation where we can generate ontological commitments too easily. Talk is cheap. Just because we can describe some scenario does not mean that we are ontologically committed to some scenario, *out there* somewhere, that our description can be about. We must make it possible for inquiry to find things out by allowing it to outrun our current awareness of it, but we must also allow inquiry to do work by providing a check on loose talk. The idea that we are committed to what our *best* theory talks about presupposes the possibility that our theory will have to be revised because we were *wrong*. The Allist can, of course, accommodate this; the way we are wrong, on such a view, is about how we relate to the various entities there are, or how we categorise them. This is to shift the role of the existential quantifier from cleaving the wheat of reality from the chaff of mere thought to some other role. Insofar as existence has a technical meaning in the Quinean philosophical tradition, such a move is to change the notation used to capture what there is (and how it is), rather than to reflect genuinely different ontological commitments.

Avoiding the harsh anti-realism of the train carriage presentist and the ontological profligacy of Allism doesn't make much progress amongst the more mainstream temporal ontologies: the Growing-Block, Presentism, the Shrinking Block, the Cresting Wave and so on. Here, perhaps, the way we demarcate things is a matter of giving prominence to some asymmetries at the expense of others; the disagreement between the views is more a matter of variation in emphasis than that one and only one view is compatible with the evidence. The Growing-Block view, unlike the others, captures Valberg's asymmetry of reference, on which past things exist and future things do not. This does the best job of relating existence to the possibility of inquiry. We are committed to things that are potentially required, in the course of our inquiries, to explain the phenomena. They are potentially required because they serve as the phenomena, the causes of the phenomena, or are otherwise indispensable to our explanations.[30]

[30] The 'we' here are those with whom we could collaborate in such inquiries, where such collaboration itself requires causal contact.

It is non-trivial, on such a view of ontological commitment, that the past exists and the future does not. One might take the view that, as on the Shrinking Block view, it is only future things that are required in our explanations; after all, all evidence is future evidence, and records only count as records if they are records when we come to examine them. The problem with such views is that it is hard to see how we are to fix reference on the persisting things of which we claim the potential to come into causal contact. The advantage of lining up ontological commitment with actuality is that it's much easier to refer directly to things that we have memory of than to attempt to directly refer to merely potential things with which we are yet to be causally related. The Growing-Block view is not the only way of generating ontological commitments from the asymmetries between space and time and between past and future, but it does a good job of giving us a theory where we can have inquiry as investigation into a world that outruns us, where we can also out-talk the world.

9.5.3 Metaphysical Realism, realism and anti-realism

I think the Growing-Block view does the best job of outlining the ontological commitments of our best theory. I have been coy about claiming that it is the One True Theory, before which all other theories must concede defeat. That's not because I am unconvinced by it (I think it *is* true), but because I have concerns about what sort of things we think the practice of giving a temporal ontology should achieve.

There's a view of metaphysics according to which the world contains a certain structure that it is the job of metaphysicians to describe. Our job, on such a view, is to come up with a theory (and perhaps a language in which to express that theory) which makes distinctions only where the world does, entirely independently of how we think about it. To recall the butchery metaphor from §4.5.3 (to which I will return), we must 'carve nature at its joints'.

This view faces some immediate problems. The first is that it is plausibly incoherent. The second is that if it is not incoherent, it must be false. This is well-trodden territory, so I will summarise the key argument on which I rely, from Button (2013).

The External Realist, as I will, following Button, call them, is committed to three principles.

The Independence Principle

> The world is (largely) made up of objects that are mind-, language-, and theory-independent.

The Correspondence Principle

> Truth involves some sort of correspondence relation between words or thought-signs and external things and sets of things.

The Cartesianism Principle

Even an Ideal Theory might be radically false.

We might contrast the External Realist with some other (groups of) views. Anti-realism rejects at least one of the Independence Principle or the Correspondence Principle. The pragmatist realist accepts both the Independence Principle and the Correspondence Principle but rejects the Cartesianism Principle; an Ideal Theory, for the pragmatic realist, is simply one in which words or thought-signs make the best possible sense of the external things or sets of things.

The objection to the External Realist is that if we accept the Correspondence Principle, we need some way of working out what that correspondence relation is. We need the Independence Principle to have something for our words or thought-signs to correspond to. But any way we have of grasping what the correspondence relation between our words or thoughts and the world they attempt to describe is inconsistent with the possibility of the radical falsity of an ideal theory.

For inquiry to get going, it not only must be the case that there is a world that outruns us for us to enquire into (i.e. the Independence Principle) but that inquiry is in a position, in principle, to succeed. Truth (of a kind that obeys the Correspondence Principle) has to be something that mere mortals can grasp if they inquire long enough and well enough.[31] If we allow that the world might, not in specifics but wholesale, be unknowable to us, then *this very sentence* does not communicate anything. Why? Well, the sentence can only succeed in communicating something if it provides a relationship to the world. If we accept the Cartesianism Principle, we need some method of connecting words or thought-signs to the world, such that there's no method of telling whether it has worked. And if the words in that sentence don't provide a method of telling how words or thought-signs relate to the world, they don't communicate anything. The world, on such a view, does not merely outrun us but is untethered from us; we are in 'ineffable ignorance' with respect to it.[32]

It is incoherent to think of there being a *real world* – the 'noumenal' world – hidden behind the world we experience – the world of phenomena. As Richard Rorty put it: "no phenomenal term can describe the phenomenon-noumenon relation, and we have no other terms" (1979:89). Whatever the 'real world' is, for us to talk about it it must be revealed by the phenomena, not hidden by them. For us to enquire about the world, we must be connected to the world, such that it is at least possible, some of the time, to find out about it. Again, this is a pragmatic argument. We can't make sense of inquiry that isn't just introspective unless this is presupposed.

This view might be contrasted with an anti-realist view, where the world is created by our talk about it; that all inquiry can be is introspection into what we can imagine. Time, on such a view, is either a projection onto the world by human language users, or, on a more global anti-realism, the world too is a projection of human language users.

31 There may be cases where our inquiries cannot succeed because, for example, evidence has been lost; these will count as cases where we are unable in practice to enquire well enough to succeed, rather than an in principle inability to inquire.

32 Cf. Lewis (2009).

Time, causation, modality and other concepts that I have used freely in this book are not, for the anti-realist, features of the world but merely part of grand narratives that humans have come up with. We have seen this move before. It is the move where we reject the world as a constraint on a theory because we can project whatever theories we like onto the world. It is the move where we fail to draw the appropriate contrast between history and fiction.

There is a sophisticated version of anti-realism, where the world-as-constraint-on-theory is also part of the theory. Such a view aims to undermine the ability to articulate the complaint that on that theory, the world isn't acting as a constraint. But expressing the constraint of the world serves a functional role purely internal to the theory, much as saying 'pleased to meet you' serves a social role that doesn't actually require that you are pleased, merely that you say so, and deny any suggestion that you are merely saying so.[33]

The ancient butchery metaphor used by Plato and Zhuangzi is a good one because it acknowledges, unlike the anti-realist, that the world is an external constraint on how we may successfully navigate it. But it is misleading in the hands of more recent users. For External Realists, like Sider (2009), fail to acknowledge that the butcher is carving for a purpose to achieve some end. The tools that the butcher uses are ones which are created to fulfil those ends, and at which the butcher can be more or less skilled. Their skill is determined, at least partly in an objective way, by whether they succeed in achieving those ends. What's more, new tools can be created to allow the butcher to do things in ways they couldn't envisage before. But even for the skilled butcher, they must make choices and trade-offs about which joints to carve at, and which cuts to leave on the bone. Some cuts are better than others (splinters of bone are best avoided, I understand), but there is no single best way to carve a carcass, independent of how you want to cook the meat.[34]

In contrast to External Realism and anti-realism we might opt for a realist [lowercase 'r'] pragmatism. The world is one that we can enquire into, sometimes successfully, and our best theory is constrained by that inquiry and by the presuppositions of that inquiry. Perhaps the *ontology* we get depends on the particular purposes for which we are carving up the world, but the cuts we get must avoid splinters of bone and be suitable for human consumption.

9.5.4 Pragmatism and deflation

The view I have outlined is a kind of pragmatism, inspired largely by the views of C. S. Peirce, Hilary Putnam and Christopher Hookway.[35] What the commitments of our best theory are depends on our purposes. But even purposes are not all equal;

[33] Such barriers to effective communication turn out not to be a substitute for genuine affection, however much they emulate the mannerisms of it.

[34] There are, of course, traditions that develop, that allow people to name standard cuts and expect the same cuts from different butchers. There are great advantages to having a shared standard, even if it is not the uniquely best one.

[35] Hookway (1999; 2000; 2012), Putnam (1990; 1994). My own defence of the approach is in Forbes (2023b)

some purposes seem frankly eccentric (e.g. drinking a can of paint or placing all the green books I own on my roof), whereas other purposes seem to be ones that are themselves presupposed by almost any purpose I might have. Inquiry into the world is going to be a pretty general purpose that creatures like us have, and it seems enquiring into the world is something that we do over time (through McTchanges in time) and with asymmetries between past and future. Intervening on the world is going to be presupposed by any activity I may engage in, even if the intervention is small and comparatively insignificant. Where my views appeal to such inescapable interests, my pragmatic arguments are going to apply not just for humans, but also for any potential collaborators in inquiry we may come across.

A view I have considered throughout this book is a deflationary pragmatist one, associated with Huw Price and others. In this section, I will consider how my pragmatism differs from the more deflationary kind. The answer, in brief, is that Price (2013) is a metaphysical quietist, whereas I embrace the metaphysical project. Price thinks that rather than doing metaphysics, we must look at the practical differences that make certain distinctions matter for us and understand how best to use those distinctions while not making any metaphysical commitments.

I object to the quietist on the grounds that they seem to accept the possibility of maintaining a distance from having metaphysical commitments. If our commitments are the presuppositions of projects that we all have, this involves a kind of ironic distance to all of our projects. The quietist can dismiss all of the things that they say as talk with the vulgar and then maintain a studied silence when talking ontologese. I would be sympathetic to such a view were it not for all the examples where they deflate away the very world we live in. The world needs defence against those External Realists who will thrust it behind the veil of experience, and those anti-realists, who deny it is anything more than a story we tell each other. It also needs defence from those who watch over this without stepping in to defend the world because they think telling the stories is all that matters. I think that the quietist can't get out of taking a stand on which stories matter and are better, and should be told. Quietism, here, amounts to losing the distinction between history and fiction. Both may be narratives, but one is constrained by the past in a way that the other isn't. The quietist can accept that there is a difference *according to the fiction*, but in so doing they can't tell the difference between history and fictional works like *Don Quixote*, that claim to be true histories.

We have seen the deflationary pragmatist appear throughout this book. In §1.5, they argued that that whether or not there is McTchange is a matter of verbal disagreement (but there isn't any from a scientific point of view). In §3.5.3, they argued that our beliefs in the persistence of the self are a product of our embeddedness in time (which we should not generalise from). In §6.5.2 (and later in §8.2.2), they argued that the asymmetries between space and time were due to the entropic gradient (and were contingent and local) rather than due to a causal/modal asymmetry. I agree with them on their pragmatist methods, in some sense. My approach, after all, has been to highlight asymmetries we find in our encounters with the world and look at how they work.

What we have seen from the deflationary pragmatist is that they think that if you can explain circumstances in which it would *seem* to us to make sense to make such distinctions, then there is no need to commit metaphysically to a world that is as it seems. There is a separate, 'naturalistic' metaphysics that accounts for the world as described by science. The world of lived experience is a curious by-product of this naturalistic world. The quietist, here, does not deny that the passage of time is *real-from-our-point-of-view*, and that asymmetries between the past and future are *real-from-our-point-of-view*, and that Valberg's asymmetry of reference is *real-from-our-point-of-view*. But none of these things are *real-from-a-scientific-point-of-view*. The understanding of science here, as I argued in §4.5, seems to be a rather austere one that seems metaphysically loaded in its own way. Where the External Realist thinks the world can come totally apart from our talk about it, and the anti-realist thinks that all there is to the world is our talk about it, the deflationary pragmatist says 'let me tell you how we talk about it, and why we talk about it like that rather than some other way'. Progress in a sense, but missing the final step. I think we should be in the business of working out whether to believe our stories about the world. That doesn't require comparing them to a noumenal world beyond our reach, hidden behind experience, and it doesn't mean merely deciding as a community whether the stories have a truthy ring to them. The deflationary pragmatists don't think these stories are ones for believing, just for telling.[36] But why tell them, unless we believe them? *We are not mere nodes in a network of counterfactual dependence to which it seems as if there is change, and hope, and potential, and love, and loss, doomed to tell stories as if there are.*

Of course, if I were a mere node in a network of counterfactual dependence, that is just what you would expect a collection of my time-slices to utter. My hope that there is hope does not become true just because I hope it. But as gambles go, we've got nothing to lose and everything to gain.

In answer to Callender's question in §1.5 about whether the various views are explanatorily distinct, the distinction between my view and the deflationary pragmatist's is not an empirical one, but a metaphysical one (insofar as one can disagree with a quietist). This metaphysical difference produces an explanatory difference. If we take the gamble that inquiry *is* possible, that we *can* intervene on the world to produce effects, that the world outruns our evidence of it, and that there really is McTchange and a modal/causal distinction between past and future, then we *can* believe many of the stories we tell ourselves about time, change, and persistence, and make sense of our lives in a way that is not possible if these are just stories that we've ended up with.

Where does that leave us with temporal ontology? In §9.2, we discussed a range and found that we could use many of these to describe the features of time that matter; McTchange and a modal/causal directionality to time given by a difference between potentiality and actuality. The arguments for accepting the Growing-Block view are that it does the best job at getting the contrast between histories and fictions, and

[36] One gets the impression that the 'naturalistic' one is for believing, but that would make them realists about that story and anti-realists about the others, which is not the Quietism they claim.

between referentially successful singular thought about my grandparents and merely referentially fit objectural thought about my grandchildren. The realist pragmatism I advocate entitles us to believe in the commitments of our best theory, and to constrain our best theory to commit only to those things that we have reason to believe in or are entitled to hope. I have argued that some version of the Growing-Block view is our best theory of time.

10

The Growing-Block view

Attempts have been made to account for the obvious difference [between the past and the future] by a theory of the universe as an infinite midden or rubbish-heap, in which the outworn states of the present are conserved: the past is thus real and the future unreal.

—Collingwood (1925–6:146)

I have looked at various temporal ontologies on offer and argued, albeit not conclusively, that the Growing-Block view is the best. But I have so far said little about the details of the view. Here, I will explain how the view works and to defend it against some objections. Principally, I defend it against the charge that its treatment of the past is not as straightforward as I supposed in §9.3. I will outline how causation and persistence work on my version of the Growing-Block view. Finally, I will contrast it with some rival versions of the view.

10.1 Recap

Lots of the work in spelling out the view has already been done in previous chapters.

The Growing-Block view I defend is

- a dynamic view (i.e. one committed to McTchange);
- committed to one-way, irreversible modal/causal change from potentiality to actuality;
- which is realised by a change in what exists;
- such that the passage of time is characterised by things/events/stuff coming into existence, but not going out of existence.

The ontological change, following C. D. Broad (1923), might be described as 'absolute becoming'. Absolute becoming is a term for coming into existence *simpliciter*, though it is important to note the present continuous verb aspect here. Absolute becoming is an ongoing process of coming to be; it is not merely that there is variation in what exists over time, but that the passage of time is an ongoing change in what exists.

This change – absolute becoming – is not just a change in what exists, and does not happen inexplicably. It is constrained, and caused, and it is often predictable. Though history never strictly repeats, it often rhymes. Once things are fully actualised, and so wholly past, they have no more potential, and so cannot be future. Nevertheless, what's going to happen often resembles what has happened, and the sorts of mechanisms that operate in the world can be very stable over time.[1] We can learn from the past to predict, usefully, how the future might go, and we can make interventions on the basis of our study of past events to make the future go how we intend.

As I present it, the Growing-Block view isn't simply a theory of what times exist. Important roles are played by modality (and causation understood so as to require commitment to modality) and by persistence. For example, there has to be something that persists through change, such that it can persistently constrain absolute becoming. In my (2015) I argued that we needed a theory of time to explain why time passes in an inescapable, orderly way, without skipping times or adding new times amongst past ones; it is the modal/causal constraint on what comes into existence that accounts for the orderliness of time. It is our first-personal awareness of enduring through change (and of being the proper subjects of change), and of intervening on persisting, ongoing processes, that provides the strongest abductive ground for the rich metaphysical picture that I have adopted.

I will come back to causation and endurance in §§10.3 and 10.4, respectively. Before then, we should consider one of the major objections to the Growing-Block view: that it faces a dilemma between being unmotivated and being untenable.

10.2 Past problems

Kristie Miller (2018) presents a dilemma for the Growing-Block view. The horns of the dilemma are that if the past is very different to the present, then treating them both as existing is unmotivated, and if the past is the same as the present, then the evidence for our own presentness is undermined. This section will be spent confronting this dilemma. I think one can avoid being impaled by the dilemma because of the distinction between actuality and potentiality that I have argued for in Part II. The present is different from the past because present things and processes are ongoing. But this response only focusses in on the challenges for explaining the Growing-Block view; Perović (2021) argues that my (2016) attempt to use the intrinsic/extrinsic distinction to explicate the difference between the ongoing present and the completed past leave something to be desired. And then, to round off the section, I consider the worry, from Sanson and Caplan (2010), that the past that the Growing-Block view is committed to is not the same as what was once present, and so unable to do the work required of a Growing-Block view to make past truths true.

1 Though there is a question about whether things worked in the distant past the way they have done more recently. Cf. Currie (2019).

10.2.1 Miller's dilemma

Let's imagine, for a moment, a view on which the past and present are alike in every respect, excepting that the present is later than the past. The present is defined not in terms of any mysterious primitive property, on the Growing-Block view, but can be thought of as the limiting case of the past; the time succeeded by nothing.[2] We can then ask you, dear reader, what grounds you have for thinking that you are an inhabitant of the present (succeeded by nothing), rather than the past (succeeded by something). As Craig Bourne (2006:21) argues, your grounds for believing you are present must be very strong:

> Is there anything more certain than the knowledge we have that we are present? It would be a scandal if our best theory of time could not guarantee such knowledge.

But if, *ex hypothesi*, present and past are exactly alike, then there are no grounds for such knowledge. Any evidence for your apparent presentness is no better than my evidence for mine, or Marie Curie's for hers, or Zhuang Zhou's evidence that he is a presently existing non-butterfly.[3]

One could respond by treating presentness as an indexical, that is treating it as an analogue to 'here' or 'I'.[4] But to do so is to rob the present of the distinction from space that motivated McTchange in Part I. No one should be in a position to ask if they are present and be wrong, because their only evidence for their presentness is that they are in a position to ask such a question! As we saw in Chapter 2, there is no special temporal quality of experience from which we can establish our objective presentness. If there is to be a non-indexical sense of presentness, it had better line up with the indexical sense. If it doesn't, given how very many past times there are, it would be exceedingly unlikely that we would get lucky and correctly identify that we were present in the non-indexical sense (i.e. succeeded by nothing).[5] This we can call the 'When am I?' problem, since the challenge is to answer how we can know when we are. It represents one horn of Miller's dilemma and, I think, gives us grounds to decisively reject the view of the Growing-Block on which the past and present are alike in every respect, excepting that the present is later than the past.

The other horn of the dilemma, I think, is more easily resisted. If one is going to accept that past and present are different in such a way that past people are not in a position to utter or even ponder the question 'When am I?' then one loses the motivation to treat the past and present as ontologically on a par. One should accept some version of presentism (surveyed in §9.2.2) instead.

[2] Strictly speaking, on the view I defend, presentness is going to be a property of enduring entities or processes that have some actuality and some potentiality, rather than being a property of moments of time. But we can treat the present as the limiting case of the past in this section for the purposes of posing the objection. Indeed, in previous work I have often described the present in precisely these terms.

[3] Cf. Zhaung *Zhuangzi*, 2.14.

[4] Cf. Mellor (1988).

[5] This point has been made by Braddon-Mitchell (2004) and Builes (2021).

My response is that the past is in some respects like the present and in other respects different. Both Presentists and defenders of the Growing-Block view must accept this. There is a challenge of getting a difference of ontological status to reflect the various differences between future, present, and past (cf. §8.2.1). The respect in which the past is like the present is that past things, like present things, are possible objects of singular thought, insofar as that thought is not only referentially fit, but also referentially successful. I can remember and reminisce about my grandfather, and I can remember and reminisce about my father. But in my father's case, he can contribute to the conversation, and in my grandfather's case, he cannot. I have no grandchildren, so there are no such people with whom I can fail to converse.

In saying the past exists, I am committing to this asymmetry between my grandparents and grandchildren, not to saying there are no differences between past and present. Is the position therefore less attractive because of its disunity? Perhaps. But imposing a false unity of the past and present seems at least as unattractive for the Growing-Block view, and imposing a false unity on the past and future seems rather unattractive for the Presentist. In §9.3, I gave arguments for accepting the existence of the past, and I don't see that they are undermined by acknowledging that there are respects in which the past differs from the present. Of the horns of Miller's dilemma, motivating the Growing-Block view in the face of some disunity of present and past is by far the least threatening. The bullet for the supporter of the Growing-Block view to bite is something that I defended in §§ 7.4.3 and 8.2.1 before we had even defended any specific *ontology* of time; presentness is a matter of having some actuality and some potentiality, so being like the past, which only has actuality, and like the future, which is only potentiality.[6] A temporal ontology, divorced from a richer picture, for example of modal change as time passes, isn't going to give us much illumination of what time *is*.

10.2.2 Intrinsic and extrinsic properties

This differentiation of past and present may seem to take me against the Growing-Block orthodoxy. Consider C. D. Broad's (1923:66) remark:

> Nothing has happened to the present by becoming past, except that fresh slices of existence have been added to the total history of the world. The past is thus as real as the present.

The divergence from Broad is easy to overstate. Broad, in this context, is clearly thinking about *intrinsic* changes. Intrinsic, rather than extrinsic, changes are changes in those properties that concern only the way something is (i.e. intrinsic properties),

[6] I also defended this view in my (2016), but in that context, it was a response to the 'When Am I?' problem. I want to emphasise here that something odd has gone on if we arrived at a Growing-Block view without already acknowledging *some* modal difference between past and present to be explained by such a view. Many objections to the view seem to end up in strange places because they don't allow room for a modal asymmetry between past and future, as if *existence* was all the view were concerned with.

rather than how it relates to other things (i.e. extrinsic properties). I agree that once things/stuff/events come into existence with some intrinsic properties, those intrinsic properties do not change.[7]

Broad is thinking of a sequence of momentary slices of existence being supplemented by further slices of existence. I am thinking of the present as being something with a duration that is ongoing, but incomplete. In §5.4 we discussed Zeno's arrow paradox, as it concerns thinking of times as instants or atoms. In summary, instants are too short for anything to happen, so if the difference between present and past is that the present is when things *happen* or *occur*, we can't think of the present as a maximally thin slice of existence. Presence isn't a matter of *slices of existence* having and then losing an additional intrinsic property, though it may be a matter of enduring objects or processes changing potentiality into actuality. Objects change in becoming past by ceasing to endure (i.e. by ceasing to have potential for further change) and processes change from processing to having processed. These changes are not well captured by a change in the intrinsic properties of maximally thin slices of existence as Perović (2021) has argued.

Perović (2021:638) presses me on my solution too, however:

> But saying that presentness better be extrinsic doesn't do anything to explain how the activity and life that go with it are to be extrinsic too. In fact, the air of mystery grows.
>
> To be fair, Forbes does appeal to the distinction between ongoing and finished events, thus suggesting that we associate the activity with the ongoing events and the lack thereof with events that are finished. But I worry that with this we end up just chasing our tail, for it is unclear what it might be for an event to be ongoing if not for it to be at the edge of the block.

It is not merely *disunity* that makes the version of the Growing Block I defend unattractive to some, but the *mystery* of how consciousness, process and ontology are related. Why is it that Zhuang Zhou isn't in a position to even contemplate whether he is a presently existing person or a butterfly dreaming of being a person, given he exists and is just as real as me? If the answer is that Zhuang Zhou has a 'retired or Emeritus status', some explanation is required of *what that means*. Whatever it means, it isn't an additional ontological status. I do not think that in addition to existing and not-existing, there is some other option! Although dividing past, present and future into two ontological statuses is difficult, using the asymmetry of reference to do so places present and past both as unambiguously existing.

[7] There are a couple of wrinkles here. One is that the distinction between intrinsic properties and extrinsic properties is notoriously difficult to give a satisfactory analysis of (see Marshall and Weatherson (2018) for a discussion). The other is that process philosophers will reject the idea that we have substances that are defined by intrinsic properties, rather than in terms of other processes. But if we grant Broad's acceptance of the intrinsic/extrinsic distinction, we can agree with him that *insofar as entities have intrinsic properties*, those properties will not change once those entities are wholly past.

I do think that there is a distinction between ongoing processes and spent ones, and that, as I remarked in §6.6.1, that neither the modal/causal picture where causation turns potentiality into actuality nor the temporal picture where the future becomes present and then past are liable to be reduced to the other. As Perović (2021:639) observes, 'What we are witnessing is a rock bottom of the [Growing-Block] theory and rather than looking at these concepts as *explaining* each other, we should see them as merely being intimately mutually related.'

She continues:

> But for this reply to work, there needs to be at least a sketch of how activity, consciousness, and ongoingness *can* all be treated as extrinsic. The first hurdle is that it doesn't look like the sense in which presentness and ongoingness are extrinsic is the same as the sense in which consciousness might be extrinsic. That is, relational features at the edge of the block would appear to pick out one sense of extrinsicness – something is present or ongoing due to there not being any succeeding slices of the block; but activity and consciousness, if they are to be treated as extrinsic as well, would have to be extrinsic to something else – not the slice at the edge of the block itself but perhaps to minds that are constitutive parts of those slices. But then consciousness would seem to be *intrinsic* to the slice of the block, while being *extrinsic* to the slice of the mind that is a constitutive part of the slice of the block. Would then extrinsicness of the latter somehow be derivative of the intrinsicness of the former? And if so, how exactly are the two related? There is certainly more work that needs to be done in this direction if the idea of consciousness being extrinsic in some sense is to convince at all.

Perović is right here; more work is to be done. More broadly, she is right that the problem in my (2016) was of my thinking in terms of 'slices of existence' as Broad did. Whether those 'slices' are part of a present thing/stuff/process will be context-dependent. It will be extrinsic to the slice of existence whether it is part of something active or not. The slices of existence, if they are sufficiently brief, change from potential to actual without ever being neither or both.[8] The change from not-existing to existing has no intermediate stages. But if, by contrast, we think not of slices but of a proper subject of change, it can undergo changes that are intrinsic to *it*. Not because it has some parts that change, but because it acquires new, later parts.[9] So a person who changes from being a presently existing person to a past person (i.e. by dying) changes in their intrinsic properties; it used to be an intrinsic property of them that they were alive, and now it is not. Nevertheless, they are still one and the same person they were; they have merely endured through a change from enduring to having endured but enduring no longer. Various emergent properties (such as consciousness) will be intrinsic properties for persisting subjects of change, but we cannot infer from something being

[8] Cf. Zeno's Stadium paradox (Aristotle's *Physics* 239b33), discussed in §7.4.4, fn.39.

[9] So, I think there is a sense in which enduring things have temporal parts! I will save detailed discussion of this until §10.4.

an intrinsic property of an enduring thing that it is a property (intrinsic or otherwise) of its parts.[10]

What the view does require, however, is that there is some fundamental persisting thing that endures. There must be some process or entity that persists through McTchange in order for the view to be one of a *growing* Universe. This enduring thing could be a quantum field, which increases as space-time atoms are added to it, or it could be the Universe as a whole that grows. Alternatively, it could be that the fundamental McTchanging thing is causal fizz, the process of change from potentiality to actuality itself. My view is compatible, I think, with various stories about fundamentalia, but there must be something that McTchanges fundamentally, because there must be some fundamental difference between the frozen present Fine (2005) discusses and a McTchanging universe.

To recall Leininger's example of a flip-book from §1.2, there must be something to our description of a flip-book in addition to enumerating the pages of the book, and where one's thumb is, that constitutes the flipping. That can't be the description of some further state, but must be something itself McTchanging. In order for something to be chang*ing* that something must persist through such change, and so must be the proper subject *of* change. This doesn't settle what the fundamental subject(s) of change are, merely that there must be some if the theory is to be true.

10.2.3 The once present and now past

If we've accepted that there is some sense in which the present differs from the past, one might worry, we seem to be committed to two things when we think of what was: the *once present* and the *now past*.[11] All the truths we care about with respect to the past seem to be truths about what was once present: truths about how changing things used to be when they were causally effervescent and partaking in activities, and undergoing various processes. But what my view gives us, it seems, is the now past: dead and lifeless sequences of states devoid of causal fizz, actuality devoid of potentiality.

The problem here isn't so much the mystery of the disunity of the past and future, but that the commitment the Growing-Block view has to the past isn't a commitment to the same thing that truths are about. In §9.2.2, I objected to Ersatzist and Haecceitist Presentisms on the basis that they offered poor substitutes for the past things that they appeared to be talking about. This objection has been prosecuted to great effect by Sanson and Caplan (2010). The worry that the once present is distinct from the now past is an extension of this worry to the Growing-Block view.

Back in §4.1 we made a distinction between events and processes. Events were taken to be temporally extended entities that have their temporal parts essentially. As a consequence, on the Growing-Block view, events will always be past; they cannot be incomplete, in the way processes can. Bacharach (2021) points out that this view has

10 I am assuming, rather tendentiously, that consciousness is emergent. Other views of consciousness are available. It is worth noting, in passing, that Samuel Alexander (1920) thought that the growth of space-time and the emergence of 'life' and 'mind' were linked.

11 Thanks to David Sanson for this distinction.

the counterintuitive consequence that events never *occur*; they are always such that they have occurred.

My view is not idiosyncratic. Consider Hornsby (2013:9):

> When a stretch of ongoing activity is over, an event is on the scene.

It is tempting to respond that 'event' as it is used in this sense is just a philosophers' term of art. Bacharach (2021:383) has an argument against so doing, however:

> something's substantially existing only as past is inconsistent with a basic feature of our thinking about the past: that whatever is now past was once present.

We have seen already that it would not do if the past were exactly like the present. We would encounter the 'When Am I?' problem. Now we return to the problem that, if the past differs from the present, it cannot be the thing we were interested in. We rejected the 'solution' of a second time dimension in §1.4.3, and we should stand by that rejection again here. It is not the case that the Growing-Block view is committed to two pasts: one long gone and another that exists now in a retired, Emeritus state. The Growing-Block view is committed to past events that previously were present. My late paternal grandfather is the very same one as the one whom I remember meeting during my childhood. He's not as active as he was, and I'm much taller now. Nonetheless, it was I and he who spent time together, and that time is now spent of its potential, he is spent of his potential[12] and I have some potential left (though less with each day!). The reason for this is that the change that constitutes the passage of time is a modal change. The modal connection between actuality with potential for more of it to come into existence and actuality without potentiality for more of it to come into existence doesn't undermine the self-identity of the actuality in question.

The concern that the now past can't be the same as the once present (since it has changed modal status by becoming past!) is our old friend McTaggart's argument (from §1.3) returning again. Indeed, Peter Tan (2020:10) argues that the solution is an appeal to hypertime, and a regress of hypertimes:

> The idea that proper initial segments of the block are the truthmakers for propositions about how big the block once was is analogous to making 'inner balls' do the truthmaking work for propositions about how big the rubber band ball once was. Both the block and the ball are concrete objects whose extent is added onto as time passes. And in both cases, the proposed alternative ontological strategy for propositions about how big that particular concrete object once was (the ball in one case, the entire universe in the other) is to accept a proper concrete part of that object, constituted as it currently is, as that proposition's 'truthmaker'. But the growing blocker already accepts that the proper strategy in doing ontology for propositions like that the rubber band ball was smaller is to locate a temporal

[12] Or at least his intrinsic potential, since he can still undergo extrinsic changes, he, *a fortiori*, has potential to undergo such changes.

> part of the rubber band ball at some time in the past. And if the right way to fill in the ontological details to account for the truth of past-tensed claims about the size of growing concrete objects like a rubber band ball is to do so, then the growing blocker ought to do the same when posed with true past-tensed propositions about another growing concrete object: the block itself.

Tan's worry is that claims like 'the growing block was once smaller' require truthmakers to make it true, and those truthmakers must be smaller blocks. The now past cannot make it true that things were once present. But smaller blocks that have been superseded by bigger blocks are not the same as the older parts of the bigger blocks.[13] So we must have a hypertime of increasingly bigger blocks to make true claims about how fewer things existed; that is, we must commit to 'hyper-time'. And then claims that hyper-time was smaller will lead to hyper-hyper-time, and so on *ad infinitum*.

The problem here is one of getting truthmakers for modal truths. Tan's argument would extend to claims about the ways the block could have grown differently, or indeed couldn't have grown differently. The move is to take us from the problems of §9.3 onto the problems of §9.4. The lesson to draw here is that McTchange in what time it is must be accompanied by a modal story, and that modal story has to allow that actuality can provide truthmakers for what was, what might have been and what couldn't be, just as it does for what will be.[14] Crucially, the idea that McTchange is modally constrained is key to allowing now past actuality to make true claims about the once present.

10.3 Causation

Given the central importance of the change from potentiality to actuality, I should say more about how I understand that change. A temporal ontology, divorced from a richer picture of modal change as time passes, isn't going to give us much illumination of what time is. Back in §7.2.3 and 7.4.3, I introduced causal fizz:

> The causal process whereby the potential becomes actual, in accordance with the laws of nature.

Causal fizz, on the Growing-Block view, is a causal process of coming into existence. A number of defenders of the Growing-Block view have thought of causation as being involved in the ontological change of absolute becoming.[15] On the Growing-Block view, as I defend it, causal fizz plays a crucial explanatory role, both accounting for the asymmetries between past and future, as outlined in Part II, and for the McTchange defended in Part I.

[13] This worry is perhaps related to the paradox of increase, discussed in Olson (2006), that things can't get bigger by growing because you end up with the original thing and some extra stuff, not the growth of the original thing.
[14] Cf. Briggs and Forbes (MS).
[15] E.g. R. C. Jeffrey (1979); Mackie (1980); Tooley (1997); Forrest (2004).

The reason to think of causal fizz as being a process of absolute becoming is that it helps solve a problem due to D. C. Williams (1951). Williams (1951:467) argued that it was unclear what 'passage' amounts to. It is something extra that doesn't seem to add anything to our theory:

> Let us hug to us as closely as we like that there is real succession, that rivers flow and winds blow, that things burn and burst, that men strive and guess and die. All this is the concrete stuff of the manifold, the reality of serial happening [i.e. temporal variation], one event after another, in exactly the time spread which we have been at pains to diagram. What does the theory [of passage] allege except what we find, and what do we find that is not accepted and asserted by the theory?

Williams was clearly of the opinion that the answer to his question was 'nothing'. But the challenge to someone who disagrees with Williams in this particular is to give content to what makes their theory different from his that seems to play some explanatory role in the theory. Causal fizz, though perhaps still heavily reliant on chemical metaphors of fizzing or effervescence, at least seems to play some explanatory role. Consider Dullworld:

> On Dullword, there is no biff, nothing ever interacts with anything else (perhaps there is just one thing, perhaps there is nothing at all, it doesn't matter). Dullworld has existed for five minutes. After another five minutes it ceases to exist. What explains its existing for ten minutes rather than five? Not biff (for there is none), but that after five minutes, more of Dullworld (five minutes more) was brought into existence.

Let's specify that, with respect to causal dependence, Dullworld never changed. Indeed, Dullworld, from its creation, had only one possible future. Nonetheless, Dullworld was different after five minutes than it was after ten. After ten minutes, ten minutes of Dullworld were actual, but after five minutes, only five minutes were. Causal fizz allows us to explain the causal/modal change in what is actual, even without a change in *what* is taking place, then. And since Dullworld is possible (or seems to be), we need to explain why, five minutes in, it lasts for another five minutes, and why, after ten, it stops. Causal fizz understood as a process of coming into existence allows us to explain this.

Perhaps we don't need causal fizz to explain the Dullworld case. Perhaps we could appeal to a fact about what time is the unique evanescent present – a fact that is left out by the otherwise fully described world. This would, after all, save us the heavy-duty machinery of causal fizz. The difference between Dullworld after five minutes and Dullworld after ten minutes would just be a difference in what time it was, which entailed a difference about what is the case, what was the case, and what will be the case (even though all these facts were settled in all but tense). The problem here is that the evanescent present seems to *dangle*, on such a view. D. C. Williams' worry that the passage of time doesn't add anything seems particularly pressing. In getting rid of a causal/modal asymmetry between past and future, the passage of time becomes a curious epiphenomenon that provides an evanescent present, but without it hooking up to the wider picture of the universe. Ontological change is a natural mechanism to realise such causal/modal change. On a view where causal fizz is a form of absolute becoming, the change from potentiality to actuality has clear substance.

Causal fizz also seems to be in a position to address a whole series of questions about how the passage of time works. Let's begin with my (2015) suggestion that we need explanations for why time continues to pass, why time passes without skipping times, and why there couldn't be past events that were never present. Causal fizz looks well set to help us with each of these. Time continues to pass because causal fizz brings the next moment into existence. Because of the intimate connection between what is nomologically necessary and what causal fizz will bring about, the next moment brought into existence will be a result of what the world is like now, and the laws of nature. And, if one accepts this view of causal fizz, it is natural to say that *what it is* for events to be past is to have been brought about by causal fizz. So the question of there being past events that were never in the process of being brought about doesn't arise.

One might still have concerns, however. How could causal fizz bring about the future when there is no future to bring about? Leininger (2021) argues that it cannot. If we think of causation as involving 'a necessary connection between distinct existences',[16] it is unclear what could be *connected* by causal fizz. For a causal relation to connect two things, both things must exist. But causal fizz is precisely meant to be a causal becoming. In this case, what is crucial is a change from potential into actuality, but there are no potential objects that causal fizz brings into existence. Just as the past doesn't have an intermediate status of not quite existing, the future doesn't have an intermediate status either. There isn't 'a limbo of unthings' (Williams 1951:462) that gets 'reified minute by minute'. When I say the future is potential, I mean that there is potential for things to come into existence, not that there are things with the potential to come into existence.[17]

Leininger (2021:14) argues that that causation cannot act as an 'enforcer' for truths about the future:

> appealing to an enforcer in a world without an existing future requires commitment to the idea that influence of what exists – in the form of this enforcer – extends over an unreal future. The idea that something has influence over an entity that doesn't exist is non-sensical.

The account of causal fizz changing potentiality into actuality seems to avoid this problem, however. For the potentiality already exists; it is contained within the actuality that exists; it is the potential for that actuality to become something else. Still, Leininger responds, why does a current potentiality limit what could happen in the future. Couldn't something interfere? Perhaps there is a sense in which there could be a miracle; but insofar as one takes potentiality to be a genuine modal notion, only

16 Cf Hume T1.3.14.15, SBN 162: 'the idea of cause and effect, of a necessary connexion of power, of force, of energy, and of efficacy.'

17 This amounts to a denial of the future-temporal version of the converse Barcan formula: $\neg(\mathbf{F}(\exists x Gx) \rightarrow (\exists x \mathbf{F} Gx))$.

potentialities can come to be actualised. Leininger insists that this requires a relation between an unreal thing and actuality, but I claim it doesn't. Either you have a relation between different bits of actuality (the past and the nomological package) that constrain what's potentially feasible, or a relationship between cause and effect, where both have happened, and you are attempting to explain why.

Backmann (2019) presses the point that it is not just causes as *relations* that are a problem on the Growing-Block view, but just the kind of neo-Aristotelian view of active causal 'powers' that claims to involve the inherently dynamic bringing about of potentiality that I have discussed.[18] Backmann proposes two problems for the Growing-Block view: if the properties instantiated in the past facts are supposed to be active powers, they are in tension with the view of the past as unchanging, and there is no way to make sense of activity in the present either.

Let's begin with the objection to there being active powers in the past. Backmann (2019:996) gives the example:

> 'My tacking the sailboat caused the boom to hit my partner's head' cannot be made true by a causal relation in the past block.

Because the past is actuality that is no longer ongoing, there is no longer the potential for a boom to hit anyone on the head. But this seems to be a feature, rather than an objection. Fortunately for Backmann's partner, this incident of domestic abuse is no longer any risk to her. The thing in the past lacks the potential or power to cause the boom to swing, but of course it is the actualisation of a previous potential. This objection is precisely the once-present/now-past objection applied to past causes. The mistake is to seek potential in the past, rather than actuality that is, as such, actualised past potential.

Friebe suggests a response on behalf of the Growing-Block view (2018:86fn19):

> Probably, the growing blockers must modify their view. If it is true that what exists simpliciter varies temporally, then – on the growing view – both the present and the past (but not the future) successively come into existence. Thus, on the growing block view the past is not analogous to eternalism but has its own characteristic status which does not make the past properties powerless.

It is not clear to me that the Growing-Block view must be modified to differ from eternalism. At least as I have presented it, the modal change from actuality to potentiality was a feature that the Growing-Block view was introduced to respect. The problem is to think of a Static-Block view, on which there is no change from actuality to potentiality, as providing the model of what the past is like. If we think of such a static view as the default, we end up thinking of the Growing-Block view as a botched compromise between the Static Block and Presentism. But the view is no compromise!

[18] 'Power' is an alternative translation of *dunamis*, which previously I have translated variously as 'capacity' or 'potentiality'.

It is the best attempt to make sense of two thoughts: that there is McTchange and that there is a modal asymmetry between past and future.

Backmann's (2019:1004) other criticism, that there is no activity in the present, rests on a similar view of the Growing Block as a messy compromise:

> Regarding the present, the GBT is identical to presentism, with the sole difference that there is no attrition of past facts.

Backmann's argument that there can be no activity in the present is one from Zeno's arrow paradox, which I raised in §5.4. Treating the present as a temporal atom *at which* things happen isn't going to work. But that's not what the Growing-Block view, as I defend it, does. Presentness is a property of things that endure through some duration and have some actuality and some intrinsic potentiality. The Growing Block isn't merely Presentism with the attrition of past facts, since what it is to be present is to be understood in terms of being in the process of changing from potentiality to actuality.

Backmann (2019:1000) worries about framing things in terms of this relation between actuality and potentiality, in a manner reminiscent of our concerns in §9.4:

> If this means that, parallel to Meinongianism, non-existing (future) things are in some way real, that they exist in some ontologically robust way, then the powers view is flat out incompatible with any temporal ontology like the [Growing-Block theory] or presentism in which the future is unreal. Alternatively, one could analyse the non-existing manifestation of a power, which the power stands in a relation to, in some ersatz way without positing its existence in any ontologically robust way.

There are two things to say here. One is that the potential for the future is contained within the actuality of the past and present (and the missed potentialities and actualised potentialities are also contained within the actuality of the past). The other is that we needn't think of causation as merely being a static relationship between cause and effect. The causal-counterfactual dependence between two existing, actual things results from the one being causally responsible for the other's coming into existence, not the other way around. Thinking of causation as being purely a matter of entities standing in unchanging relationships to one another is incompatible with an irreducibly dynamic picture of the world.[19] But that acknowledgement isn't an objection.

If the objection is meant to be not of the very possibility of one thing bringing about another, but, as it were, *how* causal fizz knows what to bring about, the fact that causal fizz is an unfolding of the potential inherent in the actually existing world should (if not illuminatingly) answer that question. (Recall the discussion of dormative virtue as an explanation in §6.5.2.) The view is realist about modality and claims modality is a feature of the world. Causation, being part of the nomological package, is just that sort of thing that can unfold the inherent potential of existence and couldn't bring about the impossible.

[19] This point is one that Leininger (2021) raises, rather than Backmann (2019).

To some extent, causal fizz is meant to be an unexplained explainer that earns its place in the theory because of its role in distinguishing a dynamic change from potentiality to actuality from a static network of counterfactual dependence that merely varies between different embedded perspectives. At this point, it remains a theoretical posit. But it plays the role of the enforcer, ensuring that only potentialities become actualised, that they do so in an orderly fashion and do so later than the actual stuff, without skipping times. There is likely to be more to say about it in the future, but progress is to be made by working through the details of empirically informed theories.

If the passage of time is the birth of space-time atoms, as Dowker (2014) proposes, then a number of live questions for research remain:

- What mechanisms result in new space-time atoms?
- What mechanisms constrain the properties of such atoms?
- How do such mechanisms interact with the curvature of space-time?
- How do those mechanisms lead to the expansion of the Universe spatially, as well as temporally?
- Can such a mechanism be intervened upon?

Answering questions about causal fizz won't be a straightforward empirical matter (for the reasons discussed in §2.2). The positing of a mechanism to be investigated will shape inquiry in different, and potentially fruitful, ways compared to the view that the change from potentiality to actuality stands in no need of explanation (either because there is no such change, or because it disconnectedly dangles in the wider theory of time). Given that I think there is such a change, it had better not dangle disconnectedly.

10.4 Endurance

Our first-personal awareness of our own persistence, as Chapter 3 made clear, is an important motivator for McTchange, and *a fortiori* the Growing-Block view. But how are we to make sense of enduring through change – of being the *proper subject of McTchange* – on the Growing-Block view with its commitment to the existence of the past?

People, on the view I defend, are themselves like mini-Growing Blocks. They have an actual past and potential for the future, and they grow as a result of changing potentiality into actuality. I should qualify that a little; *living* people are like mini-Growing Blocks. Dead people are not in the process of changing potentiality into actuality (though they may change properties as a result of extrinsic changes). There are, on my view, no wholly future people.

Because living people are like mini-Growing Blocks, it is both true of them that they are proper subjects of McTchange and that people have temporal parts. We may think of these temporal parts as 'histories'. The maximal temporal part of a living person is their entire life history: the temporal part that extends from their very beginning to the most recent thing that has happened to them. This is a temporal part of them in a similar way to someone's whole body being considered the maximal spatial part of them. But it is important to note that just as someone living isn't their body, someone living isn't their history. That's not to suggest that there's some further part not included in these things, just that people are not simply static arrangements of things in spacetime.

If the maximal temporal part of someone is their entire life history, the minimal temporal part, if there is one, will be one atom of proper-time thick. There will also be intermediate parts, which are sequences of minimal slices that are themselves proper parts of the entire life history. For such sequences that start at someone's very beginnings, those sequences will previously have been someone's entire life history and will have been superseded by a minimally longer sequence, which in turn became their entire life history. If there are no atoms of proper time – because time is continuous – then there will be no smallest sequence, and no next sequence, but each sequence will be superseded by a larger one of which it is a proper part unless it is the terminal sequence containing the last moment at which the relevant person is alive.

Notwithstanding issues about whether time is discrete, dense or continuous, none of the sequences will be identical with a living person. Although every person now coincides with a life history, the dynamic properties of persons differ from the dynamic properties of such histories. As time passes, living persons will grow to include more person-stages as parts. But the life history that coincides with me will persist without changing its parts; it will merely be superseded by a newer, longer life history.

This distinction between people and life histories is similar to Locke's (1689) distinction between a man and the matter that composes him. As we discussed in §3.2.2, one can think of someone as being constituted by their life history, appealing to constitution in an extended sense, where a thing can be constituted by temporal parts as well as spatial parts. Right now, you are composed of some particles, but are not identical to the collection of those particles, since they may pass in and out of you. Living persons are not identical to present slices of people, or present life histories, because living persons are not simply snapshots in time, but creatures which persist through change.[20]

At this point, people often ask me if I'm subscribing to *hylomorphism*, the Aristotelian view that substances (including people) are combinations of matter and 'form'. I am never sure how to answer this because, while there are readings of this that sound plausible to me, they depend on how one understands what 'form' means. Sometimes hylomorphism is precisely committed to form being an additional part, in addition to all the bits of matter. But I have just denied this. Whatever makes me different from a corpse isn't some additional part. If I had just been incredibly neatly decapitated, the issue, in this case, would be that I didn't have enough body parts. But if my head was put back on my body, I would have all the body parts and the issue would not be some further part, but that *the body was not alive*. It would be the *activity* of those parts that would be missing, and activity is not another sort of *thing*, but what things *do*. The same, *mutatis mutandis*, applies to my history. Being alive isn't simply

[20] Peter van Inwagen (1990) has the view that parts *only* compose when they participate in a life together. I'm not endorsing this view.

a matter of possessing all the right temporal parts, but that the thing which has those parts is *doing* the requisite things.

One might distinguish between personal and sub-personal processes (Drayson 2014). The brain processes information, but it is the person whose brain it is that does the thinking. There are processes going on in all or almost all spatial parts of the body, even if lots of the interesting ones must be attributed to the person, rather than to the parts of the body or the body itself. If we consider temporal parts, it won't be the case that all or almost all of the temporal parts will have processes going on *in* them. As we said in §4.1.2, processes are changes which McTchange. Processes involve an unfolding of reality, rather than merely being special temporal parts. If we are thinking of temporal parts as we described events in §4.1.2, as being filled regions of space-time, none of these will have processes going on in them, because they represent the spent remains of McTchange but are not McTchanges themselves. For both personal and sub-personal changes, it is not that the temporal parts change, but that the enduring entity changes which temporal parts it has. The processes of change from potentiality to actuality are irreducibly dynamic, but change isn't an additional type of thing that serves as an extra-temporal part.

Roberto Loss (2021) builds on the work I have done with Ray Briggs (Briggs and Forbes 2012) to offer a 'stage-theoretic' (i.e. exdurantist) account of the Growing-Block view. Loss shows how such an account can make sense of persistence on the Growing-Block view. When an exdurantist account is available, why insist on an endurantist one? A first answer is that in Chapter 3 our first-personal experience as of enduring through change was one of the motivations for accepting McTchange in the first place. Secondly, it looks like we need something to persist through change in order to make sense of how the future is constrained. Whatever it is that acts as an 'enforcer' for how the future turns out (causal fizz, as discussed in §10.3) must persist, in that it must be something that will (as a matter of necessity) continue to act as an enforcer in the future. In order to make sense of the modal constraint on the future, we need something that persists to enforce that constraint.

10.5 Rival views

We are near the end of our journey. We started by considering the difference between time and space, in Part I, and then considered the difference between past and future in Part II. In the previous chapter, we considered various ways of capturing those differences in a temporal ontology. In this chapter, I have outlined a version of the Growing-Block view. But one might accept that a Growing-Block view is the right *kind* of view while rejected my own version of it. In this section, we will consider rival versions of the Growing-Block view. I will confine myself to contemporary defences of the Growing-Block view, partially because the interpretative difficulties of comparing historic versions of the view are prohibitive: one inevitably compares Bramley apples to Bergamot oranges.

10.5.1 Dead past views

My view, according to Miller (2018), is a 'dead past approach' to the Growing-Block view. She jocularly refers to this as a 'shrivelling sausage' view. Forrest (2004; 2006) defended a similar view to mine, but in such brevity that useful comparison is hard to achieve. These views are distinguished from other varieties of the Growing-Block view because, according to these views, the past exists but doesn't contain any activity. A 'causal frisson',[21] which I have been calling 'causal fizz', distinguishes the present by providing the extra ingredient that is required for activities to take place. On any 'dead past' view, causation will need to be bound up with the passage of time, though this is not true of all Growing-Block views.

One area in which I differ from Forrest (2004) is that he explicitly claims that consciousness is confined to the 'hyperplane' that is uniquely present. I have avoided treating the present as a durationless location and, rather, as Kroll (2020) does, I treat it as a matter of things taking place. I have also avoided claims about the presentness of other things at scales where they fail to be contemporaneous with us. One can, for certain purposes, treat presentness as the limiting case of the past, but I reject the coherence of the God's Eye View from which we provide a hyperplane of presentness.

10.5.2 Whiteheadian views

Some Growing-Block views are continuations of a Whiteheadian Process Philosophy tradition, unmediated by C. D. Broad's influence. Perović (2021) is an exemplar of such a view. As discussed above, Perović is critical of Broad's tendency to think in terms of slices of existence, and her Whiteheadian view is characterised by an ontological holism more thorough-going that that I have presented here, and a commitment to processes as the fundamentalia. While I think the need to account for processes provides a strong motivation for the view I have defended (as witnessed by Chapter 4), I have, however, deliberately remained neutral on what the fundamentalia are. For all I know, it could be that the whole Universe is the fundamental process of absolute becoming, or it could be that there are fundamental processes and entities to be described by causal set theory. While I open myself up to criticism for failing to specify the view in the most detailed way possible, I have tried to remain neutral on those issues where I currently can't see reasons to settle on one alternative rather than another. For all that metaphysics has a reputation for being speculative, I'm wary of speculation that is unhinged from solutions to live philosophical puzzles. I can't see how to settle the question of the fundamentalia of a Growing-Block view, so I have suspended belief on the issue. Defenders of Whiteheadian views will criticise me for being in the grip of substance ontology, but given how revisionary abandoning substances would be, I prefer to supplement them with processes rather than replace them.

21 Cf. Forrest (2004).

10.5.3 Tooley's view

Tooley (1997) offered the fullest contemporary defence of the Growing-Block view. Like me, he has a role for causation on the Growing-Block view, but his arguments in favour of the Growing-Block differ interestingly from mine. Rather than appealing to our first-personal experience as of enduring through change or to the role of process, Tooley argues for the Growing Block from considerations about causation. His argument is that we can only explain certain otherwise implausible phenomena, such as diffuse collections of water molecules (clouds) forming an almost perfectly flat surface (the surface of puddles), if we accept a realist account of causation as making things happen. I think the link between causation and temporal ontology in his work is interesting, but I don't have much to say on it, in part because of other aspects of his view.

One curious feature of Tooley's view is that he has two notions of what there is – actuality-as-of-a-time and actuality simpliciter – both of which are, he says, primitive notions, not to be analysed in terms of any other. This leaves open questions about how we are to understand these apparently competing notions, and whether the resulting view is one on which the Growing-Block really grows. It seems actuality simpliciter McTchanges, but actuality-as-of-a-time does not. This isn't, by itself, a problem. On Tooley's view, however, actuality *simpliciter* and actuality-as-of-a-time seem to be in competition with one another. I can't tell whether the view is really committed to McTchange, or whether actuality *simpliciter* represents a changing embedded perspective, and we simply aren't in a position to articulate the unembedded reality that we capture partially with actuality-as-of-a-time.

10.5.4 Correia and Rosenkranz's semantic treatment

The other, more recent, monograph-length defence of the Growing-Block view, Correia and Rosenkranz (2018), takes its lead from Broad (1923), rather than from the tradition that provides a role for causation, following R. C. Jeffrey (1979) and J. L. Mackie (1980). Consequently, it doesn't have as prominent a role for the metaphysics of causation or endurance as my view does. Correia and Rosenkranz try to defend a Growing-Block view with much less metaphysical heavy-lifting.

Correia and Rosenkranz attempt to deal with a number of issues semantically, only using quantification, temporal operators, identity and the notion of being a time, where I deal with them metaphysically. They claim that, although the past exists, we cannot tenselessly predicate things of it.[22] They will, I suspect, find the account I offer far too rich for their blood. As Deng (2021:364) comments:

> They find it unnecessary to understand the debate as centering on whether what tenselessly exists varies over time. Instead they simply think of it as being about tensed existence.

[22] I respond to Correia and Rosenkranz (2013) in my (2016). As far as I can tell, my responses there apply to their (2018) too.

Correia and Rosenkranz (2018:90) argue against Forrest's claim that the past is dead. They claim such a view is unnecessary:

> The fact that Broad once was, but no longer is, having any conscious thoughts is no more important than the fact that he once was, but no longer is, indenting the cushion of his office chair. It is unclear what further work Forrest's hypothesis is meant to do.

This response seems to miss the mark. Forrest's claim is that life and sentience are activities, not states. He can count indent*ing* as an activity as much as he can thinking. As such, he doesn't need to distinguish between them, so the fact that Correia and Rosenkranz don't feel the need to is neither here nor there. The worry with Correia and Rosenkranz is that they attempt to rule out the 'When am I?' problem semantically, and their opponents feel like they are simply being robbed of the semantic resources to express their concerns. While Correia and Rosenkranz think the 'When Am I?' problem is incoherent, I think it is coherent, but not a problem.

Correia and Rosenkranz's view is set up in a rather different way from my own. Where McTchange is the central notion that gives me a dynamic view of time, for Correia and Rosenkranz (2018:15), 'temporaryism' is the relevant view. This is the view that sometimes there exists something that sometimes does not exist.[23] This is to be contrasted with 'permanentism', of which it is the negation. Permanentism says that always, for all things, they always exist.[24] Note that this contrast is distinct from the contrast Richard (1981) makes between temporalism and eternalism. Richard's distinction was about whether propositions change truth-value, and this one, from Williamson (2013), is about what exists.

There is an element of confusion about what temporaryism amounts to, however, as Deng (2021:369) remarks:

> If 'composing reality only temporarily' is to mean more than just composing reality relative to times (in which case even a B-theorists could hold that facts about what exists compose reality temporarily), at some point the perspective of the present time has to be given some metaphysical weight.

Because Correia and Rosenkranz deal only in tensed quantification, we don't get a clear sense of whether they are committed to McTchange. If they are not, their view collapses into a version of the purely tensed view, and if they are, it is unclear that the semantics will be sufficient for explaining how the McTchange works, and what constrains it. As Perović (2021:633) puts it: 'the metaphysical problem refuses to go away.'

[23] In formal terms: (TEM) Sometimes, $\exists x$ Sometimes, $\neg E!x$.

[24] In formal terms: (PER) Always, $\forall x$ Always, $E!x$.

10.5.5 Purely tensed views

Finally, let's briefly consider an alternative version of the Growing-Block view, defended by Button (2006; 2007). We might call this view the purely tensed Growing-Block view. What's notable about this is that it doesn't give us McTchange. For any time one cares to consider, that time will be treated as the latest time that exists. For the purely tensed Growing-Block view, the asymmetries between past and future are expressed in a Growing-Blockish sort of way, but such that we are simply taking up an embedded temporal perspective. There is no attempt to engage with questions about what is objectively the case (whether our embedded temporal perspective projects change onto the world, or detects McTchange in it).

Button's view, and to some extent Correia and Rosenkranz's view, give us semantic approaches that get the right sorts of claims to come out true but eschew anything metaphysically more heavyweight. If one were to adopt the deflationary pragmatism last discussed in §9.5.4, these semantics would suffice for a deflationary view that sounded like a Growing-Block view. One of my concerns about trying to get the semantics to do the work in defending the Growing-Block view is that it leaves the detectivist/projectivist debate about whether our embedded perspective genuinely changes or we just talk as if it does. There are multiple ways to get one's semantics to give one a Growing-Block view; the expressibility of the view is not, as such, in question. The biggest rival to my view isn't a different semantics; it's a deflationary approach that accepts the way of talking but denies (or refuses to assert) that there is McTchange, and denies (or refuses to assert) that there is a modal/ontological difference between past and future that is independent of our perspective on it.

As I argued in §9.5.4, the claims about McTchange, and about the modal/causal McTchange that grounds a series of asymmetries between past and future, provide a view that has distinct metaphysical commitments from its quietist competitor. It allows for different explanations to be offered for the phenomena, and allows for the possibility of fruitfully different ways for inquiry to develop. Callender's argument in §1.5.4 that these views all collapse into one another doesn't get going. Insofar as we're interested in fruitful inquiry, we should pursue, as abductive hypotheses, those theories that allow for inquiry to continue most fruitfully.

10.6 Conclusion

The defence of the Growing-Block view is a work in progress, as are we all. This book is unlikely to provide a definitive defence of the view, and no doubt, when I look back on this volume in a decade's time, my views will have matured. I may come to disagree with some of the things I have expressed here, and I may have developed much more specific views than I have presented. The relevant sciences are likely to have progressed, and the areas of philosophy upon which I have attempted to draw are likely to have developed in many directions. Indeed, the dynamism of the discipline of philosophy is such that while I have written this book, I have noticed new contributions to the relevant debates get published faster than I can write about them. The irrepressible tendency for things to happen, whether I would wish them to or not, is something I try to take as support for the view I defend, even though I am vain enough to want to produce a book that is unanswerable in its brilliance and comprehensiveness. I have often had cause to echo E. M. Cioran's (1973:113) remark: 'Each time Time torments me, I tell myself that one of us must back down, that it is impossible for this cruel confrontation to go on indefinitely.' I shan't apologise here for the defects of this book which I am yet to discover, I trust that task to the better-informed person in the future that I intend to become. For the defects that I already know about, I can only comment that Broad (1938:84) made a plausible claim when he described time as 'the hardest knot in all of philosophy'.

Having acknowledged that what I have said will inevitably not be the last word on the topic, I will conclude by considering those achievements I think contribute to the credit side of the ledger. In Part I, I have given reasons for thinking there is McTchange that don't rely on our empirically detecting it, are coherent and compatible with relativistic physics. These are our first-personal awareness of our own persistence and the need in scientific explanation to account for processes and opportunities for intervention.

In fact, having highlighted the Greek notion of *kairos*, at least as used in the context of Hippocratic aphorism, I think one of the tasks I have most obviously left undone is to say something general about the philosophical problem of how one identifies opportunities so that they can be grabbed by the forelock; they are notoriously bald behind. The epistemology of *kairos* is an area where I suspect the resources of analytic philosophy could be put to some significant practical use.

In Part II, I have given reasons for thinking that the asymmetries between past and future are grounded in a modal/causal asymmetry, which is a product of *causal fizz,* a process of changing potentiality into actuality, in accordance with the laws of nature, through absolute becoming. I have shown that this distinction can make sense of fatalistic worries, and that it can make sense of preference reversals over time as non-arbitrary. Indeed, we can, I have suggested, understand why our loving relationships – to others and ourselves – stop it from being arbitrary to prefer what we actually have to a merely possible more ideal alternative.

Having surveyed various options in temporal ontology, this chapter has attempted to summarise and flesh out how we are to understand the specific view I have defended. I've tried to show why my response to the 'When Am I?' problem naturally follows from the more general account of change from potentiality to actuality that I have defended, and shown how we can make sense of changes in what's true even in cases where we appear to be changing the facts of the past, or even necessary truths. Prominent features of my view that emerge from the discussions earlier in the book include metaphysically rich commitments to causation and persistence. As far as persistence is concerned, I am committed to things and processes that are the proper subjects of change but nonetheless have temporal parts. As far as causation is concerned, there is an ontologically productive causal process that brings effects into existence. Although there is more to say about endurance and causation (both being major topics in metaphysics in their own right), I have shown how they are part of the big picture of the passage of time I have presented.

In doing all these things, I have done a further thing: I have responded to a deflationary pragmatist challenge with a realist pragmatist view of my own. The methods I have used have sometimes been as significant as the conclusions that I have reached. I have paid close attention not only to our best science but to our lived experience, and the view I have offered is the result of clambering through difficult and sometimes boggy terrain. This journey has been achieved by starting with differences – between time and space, and between past and future – that we can notice in our lived experience of the world and our investigation of it. The view I offer is my best and most general attempt to make sense of these differences. My conclusion has been that there appear to be such differences from our perspective because we detect such differences in the world. But this world is one I take to be revealed by experience, not hidden by it. Time is neither beyond our grasp nor all in our heads; 'Withal it is the flow and go of very existence, nearer to us than breathing, closer than hands and feet.'[25]

[25] Williams (1951:461). Williams was being sarcastic.

References

Adams, R.M. (1986) 'Time and Thisness'. *Midwest Studies in Philosophy* 11: 315–29.
Albert, D.Z. (2000) *Time and Chance*. London: Harvard University Press.
Alexander, S. (1920) *Space, Time and Deity*. Basingstoke: Macmillan.
Anscombe, G.E.M. (1957) 'On Brute Facts'. *Analysis* 18(3): 69–72.
Anscombe, G.E.M. (1959) *An Introduction to Wittgenstein's Tractatus*. London: Hutchinson University Library.
Anscombe, G.E.M. (1963) *Intention*. London: Harvard University Press.
Anscombe, G.E.M. (1971) 'Causality and Determination'. In E. Sosa (ed.), (1975) *Causation and Conditionals*. Oxford: Oxford University Press.
Aristotle (1884) 'On the Heavens'. Trans J.L Stocks. In J. Barnes (ed.), *The Complete Works of Aristotle*. Guildford: Princeton University Press.
Aristotle (1996) *Physics*, Trans R. Waterfield. Oxford: Oxford University Press.
Aristotle (2002) *Nicomachean Ethics*, Trans. C. Rowe. Oxford: Oxford University Press.
Armstrong, D.M. (1980) 'Identity through Time'. In P. van Inwagen (ed.), *Time and Cause*. Dordrecht: D. Reidel, 67–78.
Armstrong, D.M. (1989) *A Combinatorial Theory of Possibility*. Cambridge: Cambridge University Press.
Armstrong, D.M. (1996) *A World of States of Affairs*. Cambridge: Cambridge University Press.
Asay, J. (2018) 'We Don't Need No Explanation'. *Philosophical Studies* 175: 903–21.
Asay, J. and S. Baron (2020) 'Deflating Deflationary Truthmaking'. *Philosophical Quarterly* 70(278): 1–20.
Attar, F. (1976) *The 'Ilāhī-nāma, Or, Book of God*, Trans. J.A. Boyle. Manchester: Manchester University Press.
Augustine of Hippo (2019) *Confessions*, Trans. T. Williams. Indianapolis: Hackett.
Augustine of Hippo. *City of God*.
Baber, H.E. (2021) 'Grounding Personal Persistence'. *Kriterion* 35(2): 113–33.
Bacharach, J. (2021) 'Are Events Things of the Past?'. *Mind* 130(518): 381–412.
Backmann, M. (2019) 'No Time for Powers'. *Inquiry: An Interdisciplinary Journal of Philosophy* 62(9–10): 979–1007.
Baldwin, J. (1966) 'Unnameable Objects, Unspeakable Crimes'. In The Editors of Ebony (ed.), *The White Problem in America*. Chicago: Johnson Publishing Company, 173–81.
Banfi, L. (2021) 'Existential Dread and the B-theory of Time'. *Synthese* 199: 14708–961.
Barbour, J. (1999) *The End of Time*. Oxford: Oxford University Press.
Barnes, E. (2010) 'Ontic Vagueness: A Guide for the Perplexed'. *Noûs* 44(4): 601–27.
Barnes, E. and R.P. Cameron (2009) 'The Open Future: Bivalence, Determinism and Ontology'. *Philosophical Studies* 146: 291–309.
Barnes, E. and R.P. Cameron (2011) 'Back to the Open Future'. *Philosophical Perspectives* 25: 1–26.
Baron, S. (2013) 'Talking about the Past'. *Erkenntnis* 78: 547–60.
Baron, S. and K. Miller (2015a) 'What is Temporal Error Theory?'. *Philosophical Studies* 172(9): 2427–44.

Baron, S. and K. Miller (2015b) 'Our Concept of Time'. In B. Mölder, Arstila and P. Ohrstrom (eds), *Philosophy and Psychology of Time*. New York: Springer, 29–52.
Baron, S., Miller, K., and Tallant J. (2022) *Out of Time*, Oxford University Press.
Barrie, J.M. (1902) *The Admirable Crichton*. London: Samuel French.
Barry, B. (1977) 'Justice between Generations'. In P.M.S. Hacker and Joseph Raz (eds), *Law, Morality and Society. Essays in Honor of H. L. A. Hart*. Oxford: Clarendon Press, 268–84.
Beardman, S. (MS) 'The Very Present Self'.
Beebee, H. (2000) 'The Non-Governing Conception of Laws of Nature'. *Philosophy and Phenomenological Research* 61(3): 571–94.
Belnap, N. and M. Green (1994) 'Indeterminism and the Thin Red Line'. *Philosophical Perspectives* 8: 365–88.
Belnap, N., M. Perloff, M., and M. Xu (2001) *Facing the Future*. Oxford: Oxford University Press.
Belot, G. (2013) 'Time in Classical and Relativistic Physics'. In Adrian Bardon and Heather Dyke (eds), *A Companion to the Philosophy of Time*. Oxford: Blackwell, 185–200.
Benovsky, J. (2013) 'Branching and (in)determinism'. *Philosophical Papers* 42(2): 151–73.
Bergson, H. (1911) *Matter and Memory*. London: Swan Sonnenschein.
Bernstein, S. (2020) 'Could a Middle Level be Most Fundamental?'. https://doi.org/10.1007/s11098-020-01484-1.
Besson, C. and A. Hattiangadi (2014) 'The Open Future, Bivalence and Assertion'. *Philosophical Studies* 167(2): 251–71.
Bigelow, J. (1996) 'Presentism and Properties'. *Philosophical Perspectives* 10: 35–52.
Blackburn, S. (2014) 'Truth, Realism, and the Regulation of Theory'. In *Essays in Quasi Realism*, Oxford: Oxford University Press, 15–34.
Blake, R.M. (1925) 'On Mr. Broad's Theory of Time'. *Mind* 34(136): 418–35.
Blake, R.M. (1926) 'The Paradox of Temporal Process'. *Journal of Philosophy* 23(24): 645–54.
Bobzien, S. (2001) *Determinism and Freedom in Stoic Philosophy*. Oxford: Oxford University Press.
Bobzien, S. (2021) *Determinism, Freedom, and Moral Responsibility*. Oxford: Oxford University Press.
Boethius. *The Consolation of Philosophy*.
Bourne, C. (2006) *A Future for Presentism*. Oxford: Oxford University Press.
Braddon-Mitchell, D. (2004) 'How Do We Know it is Now Now?'. *Analysis* 64(3): 199–203.
Braithwaite, R.B. (1928) 'Symposium: Time and Change'. *Proceedings of the Aristotelian Society, Supplementary Volumes* 8: 162–74.
Briggs, R.A. and G.A. Forbes (2012) 'The Real Truth about the Unreal Future'. In K. Bennett and D. Zimmerman (eds), *Oxford Studies in Metaphysics, volume 7*. Oxford: Oxford University Press.
Briggs, R.A. and G.A. Forbes (2017) 'The Growing-Block: Just One Thing After Another?'. *Philosophical Studies* 174(4): 927–43.
Briggs, R.A. and G.A. Forbes (2019) 'The Future, and What Might Have Been'. *Philosophical Studies* 176: 505–32.
Briggs, R.A. and G.A. Forbes (MS) Counterfeasibles.
Brink, D.O. and D.K. Nelkin (2013) 'Fairness and the Architecture of Responsibility'. In D. Shoemaker (ed.), *Oxford Studies in Agency and Responsibility*, Vol. 1. Oxford: Oxford University Press.
Broad, C.D. (1923) *Scientific Thought*. London: Routledge and Kegan Paul.

Broad, C.D. (1928) 'Symposium: Time and Change'. *Proceedings of the Aristotelian Society, Supplementary Volumes* 8: 175–87.

Broad, C.D. (1938) *Examination of McTaggart's Philosophy*, Vol II. Cambridge: Cambridge University Press.

Broad, C.D. (1959) 'A Reply to My Critics'. In P.A. Schlipp (ed.), *The Philosophy of C.D. Broad*. Tudor: New York, 711–830.

Bruekner, A. and J.M. Fischer (1986) 'Why is Death Bad?'. *Philosophical Studies* 50: 213–21.

Builes, D. (2021) 'Look at the Time!', *Analysis*, anab008, https://doi.org/10.1093/analys/anab008.

Burns, P., T. McCormack, A. Jaroslawska, Á. Fitzpatrick, J. McGourty, and E. M. Caruso (2019). 'The Development of Asymmetries in Past and Future Thinking'. *Journal of Experimental Psychology: General 148*(2): 272–288.

Button, T. (2006) 'There's No Time Like the Present'. *Analysis* 66(290): 130–5.

Button, T. (2007) 'Every Now and then, No-futurism Faces No Sceptical Problems'. *Analysis* 67(296): 325–32.

Button, T. (2013). *The Limits of Realism*. Oxford: Oxford University Press

Callender, C. (2000) 'Shedding Light on Time'. *Philosophy of Science* 67: 587–99.

Callender, C. (2012) 'Time's Ontic Voltage'. In A. Bardon (ed.), *The Future of Philosophy of Time*, London: Routledge, 73–98.

Callender, C. (2017) *What Makes Time Special*? Oxford: Oxford University Press.

Callender, Craig (2021). The Normative Standard for Future Discounting. Australasian Philosophical Review 5 (3):227-253.

Cameron, R. (2015) *The Moving Spotlight*. Oxford: Oxford University Press.

Campbell, J.K., M. O'Rourke, and M.H. Slater (eds) (2011) *Carving Nature at Its Joints*. London: MIT Press.

Carpenter, A. (2014) *Indian Buddhist Philosophy*. London: Routledge.

Carroll, L. (1897) *Alice's Adventures in Wonderland*, reprinted (2009). Oxford: Oxford University Press.

Cartwright, N. (1999) *The Dappled World: A Study of the Boundaries of Science*. Cambridge: Cambridge University Press.

Cartwright, N. (2020) 'Middle-range Theory: Without it What Could Anyone Do'. *Theoria: An International Journal for Theory, History and Foundations of Science* 35(3): 269–323.

Caruso, E.M. D.T. Gilbert, and T.D. Wilson (2008) 'A Wrinkle in Time: Asymmetric Valuation of Past and Future Events'. *Psychological Science* 19(8): 796–801.

Casati, R. and G. Torrengo (2011) 'The Not So Incredible Shrinking Future'. *Analysis* 71(2): 240–4.

Caws, P. (1993) *Yorick's World: Science and the Knowing Subject*. Oxford: University of California Press.

Chang, H. (2022) *Realism for Realistic People: A New Pragmatist Philosophy of Science*. Cambridge: Cambridge University Press.

Cicero (1991) *De Fato*, Trans. R.W. Sharples. Warminster: Aris and Phillips.

Cioran, E.M. (1973) *The Trouble with Being Born*, Trans. Richard Howard, 2012. New York: Arcade.

Cochrane, Tom (2018). *The Emotional Mind : A Control Theory of Affective States*. Cambridge, UK: Cambridge University Press.

Cockburn, David (1997) *Other Times: Philosophical Perspectives on Past, Present and Future*. Cambridge: Cambridge University Press.

Cole, C. (1990) [Photo] 'A Demonstrator Confronts a Line of People's Liberation Army Tanks on Chang'an Avenue, Beijing, during Protests for Democratic Reform on Tiananmen Square'. https://www.worldpressphoto.org/collection/photo/1990/33708/1/1990-charlie-cole-wy.
Collingwood, R.G. (1925-6) 'Some Perplexities About Time: With an Attempted Solution'. *Proceedings of the Aristotelian Society* 26: 135–50.
Confucius (2003) *Analects*, Trans. Edward Slingerland. Indianapolis: Hackett.
Coope, U. (2005) *Time for Aristotle: Physics IV.10-14*. Oxford: Oxford University Press.
Correia, F. and S. Rosenkranz (2013) 'Living on the Brink, or Welcome Back, Growing Block!'. *Oxford Studies in Metaphysics* 8: 333.
Correia, F. and S. Rosenkranz (2018) *Nothing to Come*. Cham, Switzerland: Springer.
Craig, H. (2018) 'The Psychological Effects of Solitary Confinement: A Systematic Critique'. *Crime and Justice* 47: 365–416.
Crane, T. (2009) 'Is Perception a Propositional Attitude?'. *Philosophical Quarterly* 59 (236): 452–69.
Crisp, T. (2003) 'Presentism'. In M. Loux and D. Zimmerman (eds), *The Oxford Handbook of Metaphysics*. Oxford: Oxford University Press, 211–45.
Crisp, T. (2004) 'On Presentism and Triviality'. In D. Zimmerman (ed.), *Oxford Studies in Metaphysics*, Vol. 1. Oxford: Oxford University Press, 15–20.
Crisp, T. (2005) 'Presentism and "Cross-Time" Relations'. *American Philosophical Quarterly* 42(1): 5–17.
Crisp, T. (2007) 'Presentism and the Grounding Objection'. *Noûs* 41(1): 90–109.
Crowther, T. (2018) 'Processes as Continuants, and Processes as Stuff'. In R. Stout (ed.), *Process, Action, and Experience*. Oxford: Oxford University Press.
Currie, A. (2019) *Scientific Knowledge and the Deep Past*. Cambridge: Cambridge University Press.
Dainton, B. (2000) *Stream of Consciousness: Unity and Continuity in Conscious Experience*. London: Routledge.
Dainton, B. (2011) 'Time, Passage, and Immediate Experience'. In C. Callender (ed.), *The Oxford Handbook of Philosophy of Time*. Oxford: Oxford University Press.
Dainton, B. (2012) 'Time and Temporal Experience'. In A. Bardon (ed.), '*The Future of the Philosophy of Time*'. London: Routledge.
Danto, A. (1985) *Narration and Knowledge*. Chichester: Columbia University Press.
Dasti, M.R. and E.F. Bryant (eds) (2013) *Free Will, Agency, and Selfhood in Indian Philosophy*. Oxford: Oxford University Press.
Davey, K. (2007) 'Aristotle, Zeno, and the Stadium Paradox'. *History of Philosophy Quarterly* 24(2): 127–46.
Davidson, D. (1980) *Essays on Actions and Events*. Oxford: Oxford University Press.
Dawson, P. (2020) 'Hard Presentism'. *Synthese* 198(9): 8433–61.
Deasy, D. (2015) 'The Moving Spotlight Theory'. *Philosophical Studies* 172(8): 2073–89.
Dehaene, S., V. Izard, P. Pica, & E. Spelke (2006) 'Core Knowledge of Geometry in an Amazonian Indigene Group'. *Science* 311(5759): 381–4.
Del Pinal, G. and S. Spaulding (2018) 'Conceptual Centrality and Implicit Bias'. *Mind and Language* 33(1): 95–111.
Deng, N. (2012) 'Fine's McTaggart, Temporal passage, and the A versus B debate' *Ratio* 26(1): 19–34.
Deng, N. (2015) 'How A-Theoretic Deprivationists Should Respond to Lucretius'. *Journal of the American Philosophical Association* 1(3): 417–32.

Deng, N. (2021) 'Plenty to Come: Making Sense of Correia's and Rosenkranz's Growing Block'. *Disputatio* 13(63): 363–72.
Dennett, D.C. (1984) *Elbow Room: The Varieties of Free Will Worth Wanting*. London: MIT Press.
Diamond, C. (2011) '"We Can't Whistle It Either": Legend and Reality'. *European Journal of Philosophy* 19(3): 335–56.
DiBella, N. (MS) 'The World Is (Almost Surely) a Strange Place'. https://drive.google.com/file/d/1mAO7-5NxO5sOe-75YK3GvltXvxkyPk5N/view.
Dickie, I. (2010) 'We Are Acquainted with Ordinary Things'. In R. Jeshion (ed.), *New Essays on Singular Thought*. Oxford: Oxford University Press.
Dictionary of the Scots Language (2004) 'Brekis n. pl.'. [online] Glasgow: Scottish Language Dictionaries Ltd. https://www.dsl.ac.uk/entry/dost/brekis (Accessed 28 December 2021).
Dixey, R.A. (1999) 'Fatalism', Accident Causation and Prevention: Issues for Health Promotion from an Exploratory Study in a Yoruba Town, Nigeria. *Health Education Research* 14: 197–208.
Dolev, Y. (2007) *Time and Realism: Metaphysical and Antimetaphysical Perspectives*. London: MIT Press.
Dorato, M. (2002) 'On Becoming, Cosmic Time, and Rotating Universe'. In C. Callender (ed.), *Time, Reality and Experience*. Cambridge: Cambridge University Press, 253–76.
Dorato, M. (2015) 'Presentism and the Experience of Time'. *Topoi* 34(1): 265–75.
Dowe, P. (2000) *Physical Causation*. Cambridge: Cambridge University Press.
Dowker, F. (2014) 'The Birth of Space-time Atoms as the Passage of Time'. *Annals of the New York Academy of Sciences* 1326: 18–25.
Drayson, Z. (2014) 'The Personal/Subpersonal Distinction'. *Philosophy Compass* 9 (5): 338–46.
Dummett, M.A.E. (1960) 'A Defence of McTaggart's Proof of the Unreality of Time'. In *Truth and Other Enigmas* (1978). London: Duckworth, 351–7.
Dummett, M.A.E. (1964) 'Bringing About the Past'. In *Truth and Other Enigmas* (1978). London: Duckworth, 333–50.
Dummett, M.A.E. (1969) 'The Reality of the Past'. In *Truth and Other Enigmas* (1978), London: Duckworth, 358–74.
Dummett, M.A.E. (2004) *Truth and the Past*. London: Columbia University Press.
Dupré, J. (2013) 'Living Causes'. *Proceedings of the Aristotelian Society Supplementary Volume* 87: 19–37.
Dupré, J. and M. Bertolaso (2018) 'A Processural Perspective on Cancer'. In D. Nicholson and J.A. Dupré (eds), *Everything Flows: Towards a Processual Philosophy of Biology*. Oxford: Oxford University Press.
Dupré, J. (2020) 'Life as Process'. *Epistemology and Philosophy of Science* 57 (2): 96–113.
Dyke, H. (2003) 'Temporal Language and Temporal Reality'. *Philosophical Quarterly* 53(212): 380–91.
Dyke, H. (2012) 'On Methodology in the Metaphysics of Time'. In A. Bardon (ed.), *The Future of Philosophy of Time*. London: Routledge, 169–87.
Eddington, A.S. (1929) *The Nature of the Physical World*. Cambridge: Cambridge University Press.
Effingham, N. (2020) *Time Travel: Probability and Impossibility*. Oxford: Oxford University Press.

Eklund, M. (2008). The picture of reality as an amorphous lump. In T. Sider, J. P. Hawthorne & D. W. Zimmerman (eds), *Contemporary debates in metaphysics*. Malden, MA: Blackwell, 382–96.

Emmet, D. (1984) *The Effectiveness of Causes*. Basingstoke: Macmillan.

Emmet, D. (1992) *The Passage of Nature*. Basingstoke: Macmillan.

Everett, A. (2013) *The Nonexistent*. Oxford: Oxford University Press.

Farr, M. (2018) 'Causation and Time Reversal'. *British Journal for Philosophy of Science*, axx025. https://doi.org/10.1093/bjps/axx025.

Farr, Matt (2020). C-theories of time: On the adirectionality of time. Philosophy Compass (12):1-17.

Fernandes, A.S. (2017) 'A Deliberative Approach to Causation'. *Philosophy and Phenomenological Research* 95(3): 686–708.

Fernandes, A.S. (2019) 'Does the Temporal Asymmetry of Value Support a Tensed Metaphysics?'. *Synthese* 3999–4016.

Fernandes, A.S. (2020) 'Freedom, Self-prediction, and the Possibility of Time Travel'. *Philosophical Studies* 177(1): 89–108.

Fernandes, A.S. (2023) *The Temporal Asymmetry of Causation*. Cambridge: Cambridge University Press.

Fine, K. (2005) *Modality and Tense*. Oxford: Oxford University Press.

Fitch, G. and M. Nelson (2018) 'Singular Propositions'. in Edward N. Zalta (ed.), *The Stanford Encyclopedia of Philosophy* (Spring 2018 Edition). https://plato.stanford.edu/archives/spr2018/entries/propositions-singular/.

Foot, P. (1967) 'The Problem of Abortion and the Doctrine of Double Effect'. *Oxford Review* 5: 5–15.

Forbes, G.A. (2015) 'Accounting for Experiences as of Passage: Why Topology Isn't Enough'. *Topoi* 34(1): 187–94.

Forbes, G.A. (2016) 'The Growing Block's Past Problems'. *Philosophical Studies* 173: 699–709.

Forbes, G.A. (2023a) 'The 2D Past'. In K. Jaszczolt (ed.), *Understanding Human Time*. Oxford: Oxford University Press.

Forbes, Graeme A. (2023). Critical Commonsensism in Contemporary Metaphysics. In Robert B. Talisse, Paniel Reyes Cárdenas & Daniel Herbert (eds.), *Pragmatic Reason: Christopher Hookway and the American Philosophical Tradition*. London: Routledge. 145-161

Forbes, G.A. and V.-P.K. Parkkinen (MS) 'Delayed Causation'.

Forrest, P. (2004) 'The Real but Dead Past: A Reply to Braddon-Mitchell'. *Analysis* 64(4): 358–62.

Forrest, P. (2006) 'Uniform Grounding of Truth and the Growing Block Theory: A Reply to Heathwood'. *Analysis* 66: 161–3.

Franklin, M.D., D.G. Schlundt, L. McClellan, T. Kinebrew, J. Sheats, R. Belue, A. Brown, D. Smikes, K. Patel, and M. Hargreaves (2007) 'Religious Fatalism and its Association with Health Behaviors and Outcomes'. *American Journal of Health Behavior* 31(6): 563–72.

Fraps, T. (2014) 'Time and Magic – Manipulating Subjective Temporality'. In D. Lloyd and V. Arstila (eds), *Subjective Time*. London: MIT Press, 263–86.

Friebe, C. (2018) 'Metaphysics of Laws and Ontology of Time'. *Theoria* 33(1): 77–89.

Frischhut, A. (2015) 'What Experience Cannot Teach us About Time'. *Topoi* 34(1): 143–55.

Frischhut, A. (2017) 'Presentism and Temporal Experience'. In Ian B. Phillips (ed.), *The Routledge Handbook of Philosophy of Temporal Experience*. Routledge.

Gale, R.M. (1968) 'Introduction to Section II'. In R. Gale (ed.), *The Philosophy of Time*. New Jersey: Humanities Press, 65–85.
Garfield, J.L. (2013) 'Just Another Word for Nothing Left to Lose: Freedom, Agency and Ethics for Mādhyamikas'. In M.R. Dasti and E.F. Bryant (eds), *Free Will, Agency, and Selfhood in Indian Philosophy*. Oxford: Oxford University Press, 164–83.
Garrett, B. (2016) 'Tim, Tom, Time and Fate: Lewis on Time Travel'. *Analytic Philosophy* 57(3): 247–52.
Geach, P.T. (1969) *God and the Soul*. London: Routledge and Kegan Paul.
Gibson, J. (2008) 'Cognitivism and the Arts'. *Philosophy Compass* 3(4): 573–89.
Gill, K. (1993) 'On the Metaphysical Distinction Between Processes and Events'. *Canadian Journal of Philosophy* 23: 365–84.
Godfrey-Smith, P. (2018) *Other Minds: The Octopus and the Evolution of Intelligent Life*. London: William Collins.
Gold, T. (1962) 'The Arrow of Time'. *American Journal of Physics* 30: 403–10.
Golding, W. (1954) *Lord of the Flies*. London: Faber and Faber.
Greene, Preston (2021). 'Pure' Time Preferences Are Irrelevant to the Debate over Time Bias: A Plea for Zero Time Discounting as the Normative Standard. *Australasian Philosophical Review* 5 (3):254-265.
Greene, P. and M. Sullivan (2015) 'Against Time Bias'. *Ethics* 125: 947–70.
Grünbaum, A. (1968) 'The Status of Temporal Becoming'. In R.M. Gale (ed.), *The Philosophy of Time*. New Jersey: Humanities Press, 322–54.
Grush, R. (2006) 'How to, and How Not to, Bridge Computational Cognitive Neuroscience and Husserlian Phenomenology of Time Consciousness'. *Synthese* 153(3): 417–50.
Gryb, S. and K.P.Y. Thébault (2016) 'Time Remains'. *British Journal of Philosophy of Science* 67: 663–705.
Hacking, Ian (1983) Representing and Intervening, Cambridge: Cambridge University Press.
Haack, S. (1992) '"Extreme Scholastic Realism": It's Relevance to Philosophy of Science Today'. *Transactions of the Charles S Peirce Society* 28(1): 19–50.
Halevy, Y. (2015) 'Time Consistency: Stationarity and Time Invariance'. *Econometrica* 83(1): 335–52.
Hall, N. (2004) 'Two Concepts of Causation'. In J. Collins, N. Hall, and L.A. Paul (eds), *Causation and Counterfactuals*. Cambridge: MIT Press, 225–76.
Harman, E. (2009) '"I'll be Glad I Did it" Reasoning and the Significance of Future Desires'. *Philosophical Perspectives* 23: 177–99.
Haslanger, S. (1989) 'Endurance and Temporary Intrinsics'. *Analysis* 49: 119–25.
Haslanger, S. (2003) 'Persistence Through Time'. In D. Zimmerman and M. Loux (eds), *The Oxford Handbook in Metaphysics*. Oxford: Oxford University Press, 315–54.
Hausman, D.M. and J. Woodward (1999) 'Independence, Invariance, and the Causal Markov Condition'. *British Journal of Philosophy of Science* 50: 521–83.
Hawking, S. and Mlodinow (2010) *The Grand Design*. London: Bantam.
Hawley, K. (2001) *How Things Persist*. Oxford: Oxford University Press.
Heesen, R. and P. van der Kolk (2016) 'A Game-Theoretic Approach to Peer Disagreement'. *Erkenntnis* 81: 1345–68.
Heller, J. (2011) *Catch-22: 50th Anniversary Edition*. London: Vintage Classics.
Hinchliff (1996) 'The Puzzle of Change'. In J.E. Tomberlin (ed.), *Philosophical Perspectives, x: Metaphysics*. Atasadero: Ridgeview, 19–36.
Hoerl, Christoph (2013) 'A Succession of Feelings, in and of Itself, is Not a Feeling of Succession'. *Mind* 122(486): 373–417.

Hoerl, Christoph (2014) 'Do we (seem to) perceive passage?' *Philosophical Explorations* 17 (2): 188–202.
Hoerl, Christoph (2023) 'The Flow of Time: Rationalism vs. Empiricism'. *Australasian Philosophical Review*, https://doi.org/ 10.1080/24740500.2022.2153069.
Hofweber, T. and D.J. Velleman (2011) 'How to Endure'. *The Philosophical Quarterly* 61(242): 37–57.
Hookway, C.J. (1988) *Quine: Language, Experience and Reality*, Cambridge: Polity Press.
Hookway, C.J. (1990) *Scepticism*. London: Routledge.
Hookway, C.J. (1999) 'Modest Transcendental Arguments'. In R. Stern (ed.), *Transcendental Arguments*. Oxford: Oxford University Press.
Hookway, C.J. (ed.) (2000) *Truth, Rationality and Pragmatism: Themes from Peirce*. Oxford: Oxford University Press.
Hookway, C.J. (2012) *The Pragmatic Maxim: Essays on Peirce and Pragmatism*. Oxford: Oxford University Press.
Hornsby, J. 2013. 'Basic Activity.' *Aristotelian Society Supplementary Volume* 87(1): 1–18.
Horwich. P. (1988) *Asymmetries in Time: Problems in the Philosophy of Science*. London: MIT Press.
Hull, D.J. (1988) *Science as a Process*. London: The University of Chicago Press.
Hume, D. (1739/1888) *A Treatise on Human Nature*, Ed. L.A. Selby-Bigge. Oxford: Clarendon Press.
Hume, D. (1757/2008) 'Of the Standard of Taste'. Reprinted In S.M. Cahn and A. Meskin (eds), *Aesthetics: A Comprehensive Anthology*. Oxford: Blackwell.
Hussey, E. (1972) *The Presocratics*. London: Duckworth.
Husserl, E. (1893-1917) *The Phenomenology of Internal Time-Consciousness*, Trans. J. Churchill (1966). Bloomington: Indiana University Press.
Husserl, E. (1900–01) *Logical Investigations*, Trans. J.N. Findlay, revised edition by D. Moran (2001). London: Routledge.
Iaquinto and Torrengo (2022) *Fragmenting Reality*. London: Bloomsbury Press.
Illari, P. McK., F. Russo, and J. Williamson (eds) (2011) *Causality in the Sciences*. Oxford: Oxford University Press.
Ingram, D. (2016) 'The Virtues of Thisness Presentism'. *Philosophical Studies* 173: 2867–88.
Ingram, D. (2018) *Thisness Presentism*. London: Routledge.
Ingthorsson, R. (2016) *McTaggart's Paradox*. London: Routledge.
Inwood, B. and L.P. Gerson (1994) *The Epicurus Reader*. Indianapolis: Hackett.
Ismael, J. (2012) 'Decision and the Open Future'. In A. Bardon (ed.), *The Future of Philosophy of Time*. London: Routledge, 149–68.
Ismael, J. (2016) 'How Do Causes Depend on us? The Many Faces of Perspectivalism'. *Synthese* 193: 245–67.
Jackson, F. (2004) 'Why We Need A-intensions'. *Philosophical Studies* 118(2-2): 257–77.
James, W. (1890) *Principles of Psychology*. New York: Holt.
James, W. (1896) *The will to believe*. New York: Longmans, Green and Co.
James, W. (1909) *A Pluralistic Universe*. New York: Longmans.
James, W. (1911) *Some Problems of Philosophy*. New York: Longmans.
Jeffrey, R.C. (1979) 'Coming True'. In C. Diamond (ed.), *Intention and Intentionality*. Ithaca: Cornell University Press, 251–60.
Jenkins, C.S.I. (2014) 'Merely Verbal Disputes'. *Erkenntnis* 79(S1): 11–30.
Johnson, W.E. (1921–4) *Logic*, Vol. I, 1921; Vol. III, 1924. Cambridge: Cambridge University Press.
Johnston, M. (1984) *Particulars and Persistence*. Dissertation, Princeton University Press.

Johnston, M. (1987) 'Is There a Problem About Persistence?'. *Proceedings of the Aristotelian Society* 61:107–34.

Johnston, M. (1989) 'Dispositional Theories of Value'. *Proceedings of the Aristotelian Society, Supplementary Volumes* 63: 139–74.

Johnston, M. (2011) *Surviving Death*. Oxford: Princeton University Press.

Jollimore, T. (2011) *Love's Vision*. Oxford: Princeton University Press.

Jones, G. (1933) 'Famine in Russia'. *Manchester Guardian*, 30 March.

Kajimoto, N., K. Miller, and J. Norton. (2021) 'Modelling Temporal Assertions for Global; Directional Eliminativists'. *Philosophers Imprint* 21(2): 1–16.

Kahneman, D., A. Tversky, and P. Slovic (1982) *Judgment Under Uncertainty: Heuristics and Biases*. Cambridge: Cambridge University Press.

Kalupahana, D.J. (1974) 'The Buddhist Conception of Time and Temporality'. *Philosophy East and West* 24(2): 181–91.

Kant, I. (2008) *Prolegomena to Any Future Metaphysics*. Cambridge: Cambridge University Press.

Kaplan, D. (1989) 'Demonstratives'. In Joseph Almog, John Perry, and Howard Wettstein (eds), *Themes From Kaplan*. Oxford University Press, 481–563.

Kauppinen, A. (2012). 'Meaningfulness and Time'. *Philosophy and Phenomenological Research, 84*(2), 345–377.

Kavka, G.S. (1983) 'The Toxin Puzzle'. *Analysis* 43(1): 33–6.

Kayani, A., M.J. King, and J.J. Fleiter (2012) 'Fatalism and its Implications for Risky Road Use and Receptiveness to Safety Messages: A Qualitative Investigation in Pakistan'. *Health Education Research* 27(6): 1043–54. https://doi.org/10.1093/her/cys096.

Keller, S. (2004) 'Presentism and Truthmaking'. In D. Zimmerman (ed.), *Oxford Studies in Metaphysics*, Vol. 1. Oxford: Oxford University Press, 83–104.

Kelly, T. (2005) 'The Epistemic Significance of Disagreement'. In J. Hawthorne and T.S. Gendler (eds), *Oxford Studies in Epistemology*, Vol. 1. Oxford: Oxford University Press, 167–96.

Kenny, A. (1963) *Action, Emotion and Will*. London: Routledge and Kegan Paul.

Khayyam, O. (1859) The Rubaiyat of Omar Khayyam. Trans. E. FitzGerald, London: Bernard Quaritch.

Kind, A. (2004) 'Personal Identity and Our Special Concern for the Future'. *Metaphilosophy* 34(4): 536–53.

Kroll, N. (2020) 'Passing Time'. *Erkenntnis* 85(1): 255–68.

Kon, M. and K. Miller (2015) 'Temporal Experience: Models, Methodology and Empirical Evidence'. *Topoi* 34(1): 201–16.

Korsgaard, C.M. (1989) 'Personal Identity and the Unity of Agency'. *Philosophy and Public Affairs* 18(2): 101–32.

Kouabenan, D.R. (1998) 'Beliefs and the Perception of Risks and Accidents'. *Risk Analysis* 18(3): 243–52.

Kuhn, T. (1962) *The Structure of Scientific Revolutions*. London: University of Chicago Press.

Kutach, D. (2011) 'The Asymmetry of Influence'. In C. Callender (ed.), *Oxford Handbook of Philosophy of Time*. Oxford: Oxford University Press.

Lackey, J. (2008) 'What Should We Do when We Disagree?'. In T.S. Gendler and J. Hawthorne (eds), *Oxford Studies in Epistemology*, Vol. 3. Oxford: Oxford University Press, 274–93.

Ladyman, J. (2008) 'Structural Realism and the Relationship Between the Special Sciences and Physics'. *Philosophy of Science* 75(5): 744–55.

Lam, D. (2021) 'The Phenomenology and Metaphysics of the Open Future'. *Philosophical Studies* 178(12): 3895–921.

Lam, V., and Michael Esfield (2013) 'A dilemma for the emergence of spacetime in canonical quantum gravity' *Studies in History and Philosophy of Science Part B: Studies in History and Philosophy of Modern Physics* 44(3): 286–293.

Latham, Andrew J., Kristie Miller, and James Norton (2020), 'An Empirical Investigation of Purported Passage Phenomenology'. *Journal of Philosophy* 117(7): 353–86.

Latham, Andrew J., Kristie Miller, and James Norton (2021) 'Is Our Naïve Theory of Time Dynamical?'. *Synthese* 198(5): 4251–71.

Lear, J. (1981) 'A Note on Zeno's Arrow'. *Phronesis* 26: 91–104.

Le Bihan, Baptiste & Linnemann, Niels Siegbert (2019) 'Have we Lost Spacetime on the Way? Narrowing the Gap between General Relativity and Quantum Gravity', Studies in History and Philosophy of Science Part B: Studies in History and Philosophy of Modern Physics 65 (C):112–121.

Lee, H.D.P. (1936) *Zeno of Elea*. Cambridge: Cambridge University Press.

Leech, J. (2017) 'Potentiality'. *Analysis Reviews* 77(2): 457–67.

Leininger, L. (2018) 'Objective Becoming: In search of A-ness'. *Analysis* 78(1): 108–17.

Leininger, L. (2021) 'Coordination and Coming to be'. *Australasian Journal of Philosophy.* https://doi.org/10.180/00048402.2021.1975138.

Lenman, J. (2000) 'Consequentialism and Cluelessness'. *Philosophy and Public Affairs* 29(4): 342–70.

Lenman, J. (2002) 'On Becoming Extinct'. *Pacific Philosophical Quarterly* 83: 253–69.

Leonelli, S. (2016) *Data-Centric Biology: A Philosophical Study*. London: University of Chicago Press.

Leonelli, S. (2018) 'The Time of Data: Timescales of Data Use in the Life Sciences'. *Philosophy of Science* 85: 741–54.

Lewin, K. (1922) *Der Begriff der Genese in Physik, Biologie und Entwicklungsgeschichte.* Berlin: Springer.

Lewis, D.K. (1973) 'Causation'. *Journal of Philosophy* 70(17): 556–67.

Lewis, D.K. (1976) 'Paradoxes of Time Travel'. *American Philosophical Quarterly* 13(2): 145–52.

Lewis, D.K. (1981) 'Are We Free to Break the Laws?', Theoria 47 (3): 113–21.

Lewis, D.K. (1986) *On the Plurality of Worlds*. Oxford: Blackwell.

Lewis, D.K. (1990) 'Noneism or Allism'. *Mind* 99: 23–31.

Lewis, David K. (2004) 'Void and Object' In John Collins, Ned Hall & Laurie Paul (eds.), Causation and Counterfactuals. MIT Press. pp. 277–290.

Lewis, D.K. (2004) 'Void and Object' In John Collins, Ned Hall & Laurie Paul (eds.), Causation and Counterfactuals. MIT Press. pp. 277–290.

Lewis, D.K. (2009) 'Ramseyan Humility'. In D. Braddon-Mitchell and R. Nola (eds), *Conceptual Analysis and Philosophical Naturalism*. London: MIT Press, 203–22.

Liggins, D. (2008) 'Quine, Putnam, and the "Quine-Putnam" Indispensibility Argument'. *Erkenntnis* 68: 113–27.

Lipman, Martin A. (2018) 'A Passage Theory of Time'. *Oxford Studies in Metaphysics* 11: 95–122.

Lipton, P. (1991). *Inference to the Best Explanation*. London: Routledge.

List, C. and P. Petit (2011) *Group Agency: The Possibility, Design, and Status of Corporate Agents*. Oxford: Oxford University Press.

Lloyd, Genevieve (1993) *Being in Time: Selves and Narrators in Philosophy and Literature.* London: Routledge.

Loewer, B. (2012) 'Two Accounts of Laws and Time'. *Philosophical Studies* 160: 115–37.

Locke, J. (1689) *An Essay Concerning Human Understanding.*
Lombard, L. (1999) 'On the Alleged Incompatibility of Presentism and Temporal Parts'. *Philosophia* 27: 253–60.
Long, A.A. and D.N. Sedley (1987) *The Hellenistic Philosophers.* Cambridge: Cambridge University Press.
Loss, R. (2021) 'Open Future, Supervaluationism and the Growing-block Theory: A Stage Theoretical Account'. *Synthese.* https://doi.org/10.10007/s11229-021-03419-7. Galperti, S., and B. Strulovici (2014) From Anticipations to Present Bias: A Theory of Forward-Looking Preferences, Working Paper, Northwestern University.
MacBeath, M. (1983) 'Mellor's Emeritus Headache'. *Ratio* 25: 81–8.
MacBride, F. (2005) 'Lewis's Animadversions on the Truthmaker Principle'. In H. Beebee and J. Dodd (eds), *Truthmakers: The Contemporary Debate.* Oxford: Oxford University Press, 117–40.
Machamer, P., L. Darden, and C. Craver (2000) 'Thinking About Mechanisms'. *Philosophy of Science* 67: 1–25.
Machiavelli, N. (1532) *The Prince.*
Mackie, J.L. (1980) *The Cement of the Universe.* Oxford: Clarendon Press.
Mackie, P. (2006). *How things might have been: individuals, kinds, and essential properties.* Oxford: Oxford University Press.
Mackie, P. (2003) 'Fatalism, Incompatibilism and the Power To Do Otherwise', Noûs 37 (4):672–689.
Maclaurin, J. and H. Dyke (2002) '"Thank Goodness that's Over": The Evolutionary Story'. *Ratio* 15(3): 276–92.
MacMurray, J. (1928) 'Symposium: Time and Change'. *Proceedings of the Aristotelian Society, Supplementary Volumes* 8: 143–61.
Maes, Hans (2023). Aesthetic Melancholy. Contemporary Aesthetics 21.
Makin, S. (2006) *Aristotle: Metaphysics Book Θ.* Oxford: Oxford University Press.
Makin, S. (2012) 'Energeia and Dunamis'. In C. Shields (ed.), *The Oxford Handbook of Aristotle.* Oxford: Oxford University Press.
Marcolis, J. (1993) *The Flux of History and the Flux of Science.* Oxford: The University of California Press.
Marcus, R. B. (1975). 'Dispensing with Possibilia'. *Proceedings and Addresses of the American Philosophical Association* 49: 39–51.
Markosian, N. (1993) 'How Fast Does Time Pass?'. *Philosophical and Phenomenological Research* 53(4): 829–44.
Markosian, N. (1995) 'The Open Past'. *Philosophical Studies* 79: 95–105.
Markosian, N. (2004) 'A Defense of Presentism'. In D. Zimmerman (ed.), *Oxford Studies in Metaphysics,* Vol. 1. Oxford: Oxford University Press.
Markosian, N. (2015) 'The Right Stuff'. *Australasian Journal of Philosophy* 93(4): 665–87.
Marshall, D. and B. Weatherson (2018) 'Intrinsic vs. Extrinsic Properties'. In E.N. Zalta (ed.), *The Stanford Encyclopedia of Philosophy* (Spring 2018 Edition). https://plato.stanford.edu/archives/spr2018/entries/intrinsic-extrinsic/.
Mason, F. (2006) 'What is Presentism?'. *Southern Journal of Philosophy* 44: 107–28.
Maudlin, T. (1993) 'Buckets of Water and Waves of Space: Why Spacetime is Probably a Substance'. *Philosophy of Science* 60(2): 183–203.
Maudlin, T. (2002) 'Remarks on the Passing of Time'. *Proceedings of the Aristotelian Society* 102: 259–74.
Maudlin, T. (2007). *The metaphysics within physics.* Oxford: Oxford University Press.

Maudlin, T. (2010) 'Time, Toplogy and Physical Geometry'. *Proceedings of the Aristotelian Society Supplementary Volume* 84: 63–78.
Maudlin, T. (2012) *Philosophy of Physics: Space and Time*. Oxford: Princeton University Press.
Maugham, W.S. (1933) *Sheppey* reprinted in *Plays: One*, London: Methuen.
McCall, S. (1994) *A Model of the Universe*. Oxford: Oxford University Press.
McDowell, J. (1985) 'Values and Secondary Qualities'. In T. Honderich (ed.), *Morality and Objectivity*. London: Routledge and Kegan Paul.
McDowell, J. (1994) *Mind and World*. London: Harvard University Press.
McFetridge, I. (1990) 'Truth, Correspondence, Explanation, and Knowledge'. In J. Haldane and R. Scruton (eds), *Logical Necessity and Other Essays*. London: Aristotelian Society, 29–53.
McInerney, P.K. (1991) *Time and Experience*. Philadelphia: Temple University Press.
McKinnon, N. and J. Bigelow (2012) 'Presentism, and Speaking of the Dead'. *Philosophical Studies* 160: 253–63.
McTaggart, J.McT.E. (1908) 'The Unreality of Time'. *Mind* 17 (68): 457–74.
McTaggart, J.McT.E. (1927) *The Nature of Existence*, Vol. II. Cambridge: Cambridge University Press.
Meincke, S.A. (2019) 'The Disappearance of Change: Towards a Process Account of Persistence'. *International Journal of Philosophical Studies* 27(1): 12–30.
Mellor, D.H. (1981) 'Thank Goodness that's Over'. *Ratio* 23: 20–30.
Mellor, D.H. (1983) 'MacBeath's Soluble Aspirin'. *Ratio* 25: 89–92.
Mellor, D.H. (1998) *Real Time II*. London: Routledge.
Menzies, P. and C. List (2010) 'The Causal Autonomy of the Special Sciences'. In Cynthia McDonald and Graham McDonald (eds), *Emergence in Mind*. Oxford: Oxford University Press, 108–29.
Merleau-Ponty, M. (1962) *Phenomenology of Perception*, Trans. C. Smith. London: Routledge.
Merricks, T. (1999) 'Persistence, Parts, and Presentism'. *Noûs* 33(3): 421–38.
Michelson, A.A. and E.W. Morley (1887) 'On the Relative Motion of the Earth and the Luminiferous Ether'. *American Journal of Science* 34 (203): 333–45.
Miller, K. (2006) 'Morality in a Branching Universe'. *Disputatio* 1 (20): 1–21.
Miller, K. (2008) 'Backwards Causation, Time, and the Open Future'. *Metaphysica* 9(2): 173–91.
Miller, K. (2016) 'Persisting Particulars and their Properties'. In Francesco Federico Calemi (ed.), *Metaphysics and Scientific Realism: Essays in Honour of David Malet Armstrong*. Berlin: De Gruyter, 139–60.
Miller, K. (2017) 'A Taxonomy of Views about Time in Buddhist and Western Philosophy'. *Philosophy East and West* 67(3): 763–82.
Miller, K. (2018) 'The New Growing Block Theory vs Presentism'. *Inquiry: An Interdisciplinary Journal of Philosophy* 61(3): 223–51.
Miller, K. (2019) 'The Cresting Wave: A New Moving Spotlight Theory'. *Canadian Journal of Philosophy* 49(1): 94–122.
Miller, K., A. Holcombe, and A.J. Latham (2020) 'Temporal Phenomenology: Phenomenological Illusion vs Cognitive Error'. *Synthese* 197(2): 751–71.
Misak, C. (2011) 'American Pragmatism and Indispensibility Arguments'. *Transactions of the Charles S. Peirce Society* 47(3): 261–73.
Misak, C. (2020) Frank Ramsey: A Sheer Excess of Powers, Oxford: Oxford University Press.

Mitova, V. (2021) 'How to Decolonise Knowledge Without Relativism'. In S. Khumalo (ed.), *Decolonisation as Democratisation*. Cape Town: HSRC Press, 24–47.
Mohammadian, M. (2019) 'Abduction – The Context of Discovery + underdetermination = Inference to the Best Explanation'. *Synthese*. https://doi.org/10.1007/s11229-019-02337-z.
Molière (1666) *Le Médecin malgré lu*, Théâtre du Palais-Royal. Paris: la Troupe du Roi.
Molière (1673) *Le malade imaginaire*, Théâtre du Palais-Royal. Paris: la Troupe du Roi.
Moore, A.W. (2012) *The Evolution of Modern Metaphysics: Making Sense of Things*. Cambridge: Cambridge University Press.
Moore, G.E. (1903) 'The Refutation of Idealism'. *Mind* 12(48): 433–53.
Moss, J. (2010) 'Aristotle's Non-Trivial, Non-Insane View that Everyone Always Desires Things under the Guise of the Good'. In Sergio Tenenbaum (ed.), *Desire, Practical Reason, and the Good*. Oxford: Oxford University Press, 65–76.
Mourelatos, A.P.D. (1978) 'Events, Processes and States'. *Linguistics and Philosophy* 2: 415–34.
Mourelatos, A.P.D. (1993) 'Aristotle's Kinêsis/Energeia Distinction: A Marginal Note on Kathleen Gill's Paper'. *Canadian Journal of Philosophy* 23(3): 385–8.
Nagel, T. (1970) *The Possibility of Altruism*. Princeton: Princeton University Press.
Nagel, T. (1971) 'Brain Bisection and the Unity of Consciousness'. *Synthese* 22: 396–413.
Nagel, T. (1986) *The View from Nowhere*. Oxford: Oxford University Press.
Nerlich, G. (1998) 'Falling Branches and the Flow of Time'. *Australasian Journal of Philosophy* 76(2): 309–16.
Neurath, O. (1935) 'Einheit der Wissenshaft als Aufgabe'. *Erkenntnis* 5: 16–22.
Nicholson, D. and J.A. Dupré (eds) (2018) *Everything Flows: Towards a Processual Philosophy of Biology*. Oxford: Oxford University Press.
Nolan, D. (2013) 'Why Historians (and Everyone Else) Should Care About Counterfactuals'. *Philosophical Studies* 163(2): 317–35.
Nolan, D. (2014) 'The Dangers of Pragmatic Virtue'. *Inquiry* 57(5-6): 623–44.
Norenzayan, A. and A. Lee (2010) 'It was Meant to Happen: Explaining Cultural Variations in Fate Attributions'. *Journal of Personality and Social Psychology* 98(5): 702–20.
Norton, J.D. (2015) 'The Burning Fuse Model of Unbecoming in Time'. *Studies in History and Philosophy of Science Part B: Studies in History and Philosophy of Modern Physics* 52(Part A): 103–5.
O'Neill, J. (1993) 'Future Generations: Present Harms'. *Philosophy* 68(263): 35–51.
Oakeley, H. (1927) 'The World as Memory and as History'. *Proceedings of the Aristotelian Society* 27: 291–316.
Oakeley, H. (1932) 'The Status of the Past'. *Proceedings of the Aristotelian Society* 32: 227–50.
Oaklander, L.N. (1990) 'The New Tenseless Theory of Time'. *Philosophical Studies* 58: 287–92.
Oaklander, L.N. (2001) 'Is there a Difference Between the Metaphysics of A- and B-Time?'. *Journal of Philosophical Research* 26: 23–36.
Oaklander, L.N. (2004) *The Ontology of Time*. Amherst: Prometheus Books.
Olson, E.T. (2006) 'The Paradox of Increase'. *The Monist* 89(3): 390–417.
Olson, E.T. (2009) 'The Rate of Time's Passage'. *Analysis* 69(1): 3–9.
Orwell, G. (1989) *Nineteen Eighty-Four*, London: Secker and Warburg.
Owen, G.E.L. (1958) 'Zeno and the Mathematicians'. *Proceedings of the Aristotelian Society* 58: 199–222.

Papineau, David (2016) 'Naturalism'. in Edward N. Zalta (ed.), *The Stanford Encyclopedia of Philosophy*. https://plato.stanford.edu/archives/win2016/entries/naturalism/.
Parfit, D. (1971) 'Personal Identity'. *The Philosophical Review* 80: 3–27.
Parfit, D. (1984) *Reasons and Persons*. Oxford: Oxford University Press.
Paul, L.A. (2010) 'Temporal Experience'. *The Journal of Philosophy* 107: 333–59.
Paul, L.A. (2014) *Transformative Experience*. Oxford: Oxford University Press
Paul, L.A. (2017) 'The Subjectively Enduring Self'. In Ian Phillips (ed.), *The Routledge Handbook of the Philosophy of Temporal Experience*. London: Routledge, 262–72.
Pearson, F.O.C. (2018) 'Appropriate Emotions and the Metaphysics of Time'. *Philosophical Studies* 175: 1945–61.
Peirce, C.S. (1901/1998) 'On the Logic of Drawing History from Ancient Documents'. In The Peirce Edition Project (eds), *The Essential Peirce, volume 2 (1893-1913)*. Bloomington: Indiana University Press.
Peirce, C.S. (1902/1934) 'Elements of Logic'. In C. Hartshorne and P. Weiss (eds), *Collected Papers of Charles Sanders Peirce*, Vol. 2. London: Harvard University Press.
Peirce, C.S. (1903/1934) 'Pragmatism and Abduction'. In C. Hartshorne and P. Weiss (eds), *Collected Papers of Charles Sanders Peirce*, Vol. 5. London: Harvard University Press.
Pereboom, D. (2001) *Living Without Free Will*. Cambridge: Cambridge University Press.
Perović, K. (2021) 'Three Varieties of Growing Block Theory'. *Erkenntnis* 86(3): 623–45.
Perrett, R. (2016) *An Introduction to Indian Philosophy*. Cambridge: Cambridge University Press.
Pettigrew, R. (2019) *Choosing for Changing Selves*. Oxford: Oxford University Press.
Phillips, I. (2009) 'Rate Abuse: A Reply to Olson'. *Analysis* 69: 503–5.
Pickering, A. (1995) *The Mangle of Practice*. London: The University of Chicago Press.
Plantinga, A. (1983) 'On Existentialism'. *Philosophical Studies* 44: 1–20.
Plato. *Cratylus*.
Plato. *Euthyphro*.
Plato. *Phaedrus*.
Polanyi, M. (1983) *The Tacit Dimension*. Gloucester: Peter Smith.
Pollock, J.L. (1983) 'How Do You Maximise Expected Value'. *Noûs* 17(3): 409–20.
Polt, R. (MS) '*Kairos* Punctuates *Chronos*'. Presented at online Workshop on the Phenomenon of Temporality, Organized by Tina Röck (University of Dundee) and Daniel Neumann (University of Klagenfurt), September 2021.
Power, S.E. (2018) *Philosophy of Time and Perceptual Experience*. London: Routledge.
Price, H. (1996) *Time's Arrow and Archimedes' Point*. Oxford: Oxford University Press.
Price, H. (2007) 'Causal Perspectivalism'. In H. Price and R. Corry (eds), *Causation, Physics, and the Constitution of Reality: Russell's Republic Revisited*. Oxford: Clarendon Press.
Price, H. (2011) 'The Flow of Time'. In C. Callender (ed.), *The Oxford Handbook of Philosophy of Time*. Oxford: Oxford University Press, 276–311.
Price, H. (2013) *Expressivism, Pragmatism and Representationalism*. Cambridge: Cambridge University Press.
Price, H. and R. Corry (eds) (2006) *Causation, Physics, and the Constitution of Reality: Russell's Republic Revisited*. Oxford: Oxford University Press.
Prior, A. N. (1958) 'Time after time' Mind 67 (266):244–246.
Prior, A.N. (1959) 'Thank Goodness That's Over'. *Philosophy* 34: 12–27.
Prior, A.N. (1962a) 'Changes in Events and Changes in Things'. Reprinted In P. Hasle, P. Øhrstrom, T. Braüner, and J. Copeland (eds), (2003) *Papers on Time and Tense*. Oxford: Oxford University Press, 7–20.

Prior, A.N. (1962b) 'The Formalities of Omniscience'. Reprinted In P. Hasle, P. Øhrstrom, T. Braüner, and J. Copeland (eds), (2003) *Papers on Time and Tense*. Oxford: Oxford University Press, 39–58.

Prior, A.N. (1968) 'Tense Logic and the Logic of Earlier and Later'. In P. Hasle, P. Øhrstrom, T. Braüner, and J. Copeland (eds). (2003) *Papers on Time and Tense*. Oxford: Oxford University Press, 117–38.

Prosser, S. (2007) 'Could Ee Experience the Passage of Time'. *Ratio* 20(1): 75–90.

Prosser, S. (2011) 'Affordances and Phenomenal Character in Spatial Perception'. *Philosophical Review* 120(4): 475–513.

Prosser, S. (2013a) 'Passage and Perception'. *Noûs* 47(1): 69–84.

Prosser, S. (2013b) 'Experience, Thought, and the Metaphysics of Time'. In K.M. Jaszczolt and L. de Saussure (eds), *Time: Language, Cognition, and Reality*. Oxford: Oxford University Press.

Prosser, S. (2016) *Experiencing Time*. Oxford: Oxford University Press.

Putnam, H. (1967) 'Time and Physical Geometry'. *Journal of Philosophy* 64(8): 240–7.

Putnam, H. (1990) *Realism with a Human Face*. London: Harvard University Press.

Putnam, H. (1994) *Words and Life*. London: Harvard University Press.

Quine, W.V.O. (1948) 'On What There Is'. *Review of Metaphysics* 2(5): 21–36.

Quine, W.V.O. (1950) 'Identity, Ostension, and Hypostasis'. In *From A Logical Point of View*. London: Harvard University Press, 65–79.

Quine, W.V.O. (1953) 'Two Dogmas of Empiricism'. In *From a Logical point of View*. London: Harvard University Press, 20–46.

Quine, W.V.O. (1969) *Ontological Relativity and Other Essays*. New York: Columbia University Press.

Quine, W.V.O. (1970) 'Truth and Disquotation'. In W.V.O. Quine (ed.), (1976) *The Way of The Paradox and Other Essays*. London: Harvard University Press.

Quine, W.V.O. (1981) 'Empirical Content'. In *Theories and Things*. London: Harvard University Press.

Ramsey, F. (1929) 'General Propositions and Causality'. In his (1931) *Foundations of Mathematics*. London: Routledge and Kegan Paul.

Rashbrook, O. (2013) 'An Appearance of Succession Requires a Succession of Appearances'. *Philosophy and Phenomenological Research* 87(3): 584–610.

Raven, M.J. (2011) 'Can Time Pass at a Rate of 1 Second per Second'. *Australasian Journal of Philosophy* 89: 459–65.

Raz, J. (2010) 'On the Guise of the Good'. In S. Tenenbaum (ed.), *Desire, Practical Reason, and the Good*. Oxford: Oxford University Press.

Reichenbach, H. (1938) Experience and Prediction Chicago: University of Chicago Press.

Reichenbach, H. (1949) 'The Philosophical Significance of The Theory of Relativity'. In Paul Arthur Schilpp (ed.), *The Library of Living Philosophers, Volume 7. Albert Einstein: Philosopher-Scientist(pp. 315–319)*. Open Court.

Reichenbach, H. (1951) *The rise of scientific philosophy*. Berkeley: University of California Press.

Reichenbach, H. (1956) 'The Primacy of Physical Time'. Reprinted In C. Sherover (ed.), (1975) *The Human Experience of Time*. New York: New York University Press, 315–29.

Reid, T. (1863) 'An Inquiry into the Human Mind on the Principles of Common Sense'. Reprinted In R.E. Beanblossom and K. Lehrer (eds), (1983) *Inquiry and Essays*. Indianapolis: Hackett.

Rennick, S. (2015) 'Things Mere Mortals Can Do, but Philosophers Can't'. *Analysis* 75(1): 22–6.

Rescher, N. (2000) *Process Philosophy: A Survey of Basic Issues.* University of Pittsburgh Press.
Resnik, M. (1995) 'Scientific vs. Mathematical Realism: The Indispensability Argument'. *Philosophia Mathematica* 3: 166–74.
Richard, M. (1981) 'Temporalism and Eternalism'. *Philosophical Studies* 39(1): 1–13.
Rideout, D.P. and R.D. Sorkin (2000) 'A Classical Sequential Growth Dynamics for Causal Sets'. *Physical Review D* 61. https://doi.org/10.1103/PhysRevD.61.024002.
Röck, Tina (2021) *Dynamic Realism: Uncovering the Reality of Becoming Through Phenomenology and Process Philosophy*. Edinburgh: Edinburgh University Press.
Rorty, R. (1979) 'Transcendental Arguments, Self-reference, and Pragmatism'. In P. Bieri et al. (eds), *Transcendental Arguments and Science.* Dordrecht: D. Reidel, 77–103.
Rosenkranz, S. and F. Correia (2018) *Nothing to Come: A Defence of the Growing Block Theory of Time*. Berlin: Springer Verlag.
Routley, R. (2018) *Exploring Meinong's Jungle and Beyond: The Sylvan Jungle,* Vol. 1. Berlin: Springer Verlag.
Russell, B. (1903) *Principles of Mathematics.* Cambridge: Cambridge University Press.
Russell, B. (1905) 'On Denoting'. *Mind* 14(56): 479–93.
Russell, B. (1912a) 'On the notion of cause' *Proceedings of the Aristotelian Society*, 13: 1–26.
Russell, B. (1912b) *Problems of Philosophy*, London: Williams and Norgate.
Russell, B. (1915) 'On the Experience of Time'. *The Monist* 25(2): 212–333.
Russell, B. (1919) *Introduction to Mathematical Philosophy.* London: George and Allen Unwin.
Russell, B. (1948) *Human Knowledge.* New York: Simon and Schuster,
Ryanasiewicz, R. (2011) 'Newton's Views on Space, Time, and Motion'. In E. Zalta (ed.), *Stanford Encyclopedia of Philosophy.* https://plato.stanford.edu/entries/newton-stm/.
Sainsbury, R.M. (1995) *Paradoxes,* 2nd ed. Cambridge: Cambridge University Press.
Salmon, W. (1998) *Causality and Explanation*. Oxford: Oxford University Press.
Sanford, D.H. (1968) 'McTaggart on Time'. *Philosophy* 43: 371–8.
Sanson, D. and B. Caplan (2010) 'The Way Things Were'. *Philosophy and Phenomenological Research* 81(1): 24–39.
Saunders, S. and D. Wallace (2008) 'Branching and Uncertainty'. *British Journal for Philosophy of Science* 59: 293–305.
Savitt, S.F. (2002) 'On Absolute Becoming and the Myth of Passage'. In C. Callender (ed.), *Time, Reality and Experience*. Cambridge: Cambridge University Press, 153–68,
Schaffer, J. (2010) 'Monism: The Priority of the Whole'. *Philosophical Review* 119(1): 31–76.
Schechtman, M. (1996) *The Constitution of Selves.* Ithaca: Cornell University Press.
Schellenberg, S. (2011) 'Perceptual Content Defended'. *Noûs* 45(4): 714–50.
Schellenberg, S. (2018) *The Unity of Perception.* Oxford: Oxford University Press.
Seargeant, D (1985) *Plurality and Continuity: An Essay in G.F. Stout's Theory of Universals.* Lancaster: Martinus Nijhof.
Sellars, W. (1962) 'Time and the World Order'. In Herbert Feigl and Grover Maxwell (eds), *Minnesota Studies in the Philosophy of Science, Volume III: Scientific Explanation, Space, and Time*. Minneapolis: University of Minnesota Press.
Setiya, K. (2007) *Reasons without Rationalism.* Princeton: Princeton University Press.
Setiya, K.(2011) 'Reasons and Causes'. *European Journal of Philosophy* 19(1): 129–57.
Setiya, K.(2016) 'Retrospection'. *Philosophers' Imprint* 16(15): 1–15.
Setiya, K.(2017) *Midlife.* Oxford: Princeton University Press.
Setiya, K. (2022) *Life is Hard.* London: Hutchison Heinemann.

Shakespeare, W. (1597) *Romeo and Juliet,* Reprinted (1981). London: Penguin.
Shakespeare, W. (1599) *Henry V*, Reprinted (1968). London: Penguin.
Shakespeare, W. (1604/5) *Hamlet,* Reprinted (2005). London: Bloomsbury.
Shardlow, Jack, Ruth Lee, Christoph Hoerl, Teresa McCormack, Patrick Burns, and Alison S. Fernandes (2021) 'Exploring People's Beliefs About the Experience of Time'. *Synthese* 198(11): 10709–731.
Shebilske, W.L. (1977) 'Visuomotor Coordination, Visual Direction, and Position Constancies'. In W. Epstein (ed.), *Stability and Constancy in Visual Perception: Mechanisms and Processes.* New York: Wiley.
Sider, T. (1997) 'Four-dimensionalism'. *Philosophical Review* 106(2): 197–231.
Sider, T. (2001) *Four-Dimensionalism.* Oxford: Oxford University Press.
Sider, T. (2009) 'Ontological Realism'. In D.J. Chalmers, D. Manley, and R Wasserman (eds), *Metametaphysics.* Oxford; Oxford University Press.
Sider, T. (2011) *Writing the Book of the World.* Oxford, England: Oxford University Press.
Siegel, S. (2009) 'The Visual Experience of Causation'. *The Philosophical Quarterly* 59(236): 519–40.
Siegel, S. (2010) *The Contents of Visual Experience*, Oxford: Oxford University Press.
Siegel, S. and A. Byrne (2017) 'Rich or Thin?'. In Bence Nanay (ed.), *Current Controversies in Philosophy of Perception.* New York: Routledge.
Simons, P. (2000) 'Continuants and Occurrents', I. *Aristotelian Society Supplementary Volume* 74(1): 59–75.
Simons, P. (2018) 'Processes and Precipitates'. In D.J. Nicholson and J. Dupré (eds), *Everything Flows: Towards a Processual Philosophy of Biology.* Oxford: Oxford University Press, 49–60.
Sklar, L. (1974) *Space, Time, and Spacetime.* Berkeley: University of California Press.
Skow, B. (2011a) 'On the Meaning of the Question "How Fast Does Time Pass?"'. *Philosophical Studies* 155: 325–44.
Skow, B. (2011b) 'Experience and the Passage of Time'. *Philosophical Perspectives* 25: 359–87.
Skow, B. (2015) *Objective Becoming.* Oxford: Oxford University Press.
Smart, J.J.C. (1949) 'The River of Time'. *Mind* 58: 483–94.
Smart, J.J.C. (1980) 'Time and Becoming'. In P. van Inwagen (ed.), *Time and Cause.* Dordrecht: D. Reidel, 3–15.
Smith, D.C. (2011) 'Mind-independence and the Logical Space of Wright's Realist-relevant Axes'. *The Southern Journal of Philosophy* 49(2): 164–91.
Smith, J.E. (1969) 'Time, Times, and the "Right" Time: "*chronos*" and "*kairos*"'. *The Monist* 53(1): 1–13.
Smith, N.J.J. (1997) 'Bananas Enough for Time Travel?'. *British Journal for the Philosophy of Science* 48(3): 363–89.
Sorabji, R. (1983) *Time, Creation, and The Continuum.* London: Duckworth.
Sorkin, R.D. (1991) 'Space-time and Causal Sets'. In J.C. D'Olivo, E. Nahmad-Achar, M. Rosenbaum, M.P. Ryan, L.F. Urrutia, and F. Zertuche (eds), *Relativity and Gravitation: Classical and Quantum, Proceedings of the SILARG VII Conference, Cocoyoc, Mexico, December 1990.* Singapore: World Scientific, 150–73.
Sorkin, R.D. (2007) 'Relativity Theory Does Not Imply that the Future Already Exists: A Counterexample'. In V. Petkov, (ed.), *Relativity and the Dimensionality of the World. Fundamental Theories of Physics.* Springer: Chicago.
Speaks, J. (2009) 'Transparency, Intentionalism, and the Nature of Perceptual Content'. *Philosophy and Phenomenological Research* 79(3): 539–73.
Stebbing, L. Susan (1937). *Philosophy and the physicists.* London: Pengiun.

Stein H. (1968) 'On Einstein-Minkowski Space-Time'. *Journal of Philosophy* 65: 5–23.
Stevens, C. (1970) 'Father and Son'. In *Tea for the Tillerman*. London: Island Records.
Steward, H. (2013) 'Processes, Continuants, and Individuals'. *Mind* 122(487): 781–812.
Steward, H. (2015) 'What is a Continuant?'. *Aristotelian Society Supplementary Volume* 89(1): 109–23.
Steward, H. (2020) 'Agency as a Two-Way Power: A Defence'. *The Monist* 103(3): 324–55.
Story, G.W., I. Vlaev, B. Seymour, A. Darzi, and R.J. Dolan (2014) 'Does Temporal Discounting Explain Unhealthy Behaviour? A Systematic Review and Reinforcement Learning Perspective'. *Frontiers in Behavioural Neuroscience* 8(76): 1–20.
Stout, R. (1997) 'Processes'. *Philosophy* 72: 19–27.
Stout, R. (2015) 'The Category of Occurrent Continuants'. *Mind* 124(496): 41–62.
Stout, R. ed. (2018) *Process, Action, and Experience*. New York, NY: Oxford University Press.
Sullivan, M. (2018) *Time Biases: A Theory of Rational Planning and Personal Persistence*. Oxford: Oxford University Press.
Swinburne, R. (1981) *Space and Time*, 2nd ed. Basingstoke: MacMillan.
Tallant, J. (2009) 'Time and Realism: Metaphysical and Antimetaphysical Perspectives by Yuval Dolev'. *Analysis* 69 (2): 372–4.
Tallant, J. (2018) 'Presentism, Persistence and Trans-temporal Dependence'. *Philosophical Studies* 175(9): 2209–20.
Tallant, J. and D. Ingram (2015) 'Nefarious Presentism'. *Philosophical Quarterly* 65(260): 355–71.
Tallant, J. and D. Ingram (2020) 'A Defence of Lucretian Presentism'. *Australasian Journal of Philosophy*. https://doi.org/10.1080/00048402.2019.1697709.
Tan, P. (2020) 'The Growing Block and What was Once Present'. *Erkenntnis*: 1–22.
Taylor, K. (2010) 'On Singularity'. In R. Jeshion (ed.), *New Essays on Singular Thought*. Oxford: Oxford University Press.
Thoma, J. (2018) 'Temptation and Preference-Based Instrumental Rationality'. In J. Bermudez (ed.), *Self-Control, Decision Theory, and Rationality*. Cambridge: Cambridge University Press.
Thomas, E. (2015) 'Hilda Oakeley on Idealism, History and the Real Past'. *British Journal for the History of Philosophy* 23(5): 933–53.
Thomas, E. (2018) *Absolute Time: Rifts in Early Modern British Metaphysics*. Oxford: Oxford University Press.
Thomas, E. (2019) 'The Roots of C. D. Broad's Growing Block Theory of Time'. *Mind* 128(510): 527–49.
Thomas, H.G. (1995) 'The Principle of Recombination and the Principle of Distinctness: A Puzzle for Armstrong's Theory of Modality'. *Australasian Journal of Philosophy* 73(3): 444–57.
Thomasson, A. (1999) *Fiction and Metaphysics*. Cambridge: Cambridge University Press.
Todd, P. (2016) 'Future Contingents are all False! On Behalf of a Russellian Open Future'. *Mind* 125(499): 775–98.
Todd, P. (2021) *Future Contingents Are All False: A Classical Open Future*. Oxford: Oxford University Press.
Tooley, M. (1997) *Time, Tense, and Causation*. Oxford: Oxford University Press.
Torre, S. (2009) 'Truth-Conditions, Truth-Bearers and the New B-Theory of Time'. *Philosophical Studies* 142(3): 325–44.
Torre, S. (2011) 'The Open Future'. *Philosophy Compass* 6(5): 360–73.

Torrengo, Giuliano (2017) 'Feeling the Passing of Time'. *Journal of Philosophy* 114(4): 165–88.
Torrengo, Giuliano (2018) 'Hyper-Russellian Skepticism'. *Metaphysica* 19(1): 1–17.
Uhlenbeck, G.E. and G.W. Ford (1963) *Lectures in Statistical Mechanics*. Providence: American Mathematical Society.
Ullmann-Margalit, E. and S. Morgenbesser (1977), 'Picking and Choosing'. *Social Research: An International Quarterly* 44(4): 757–85.
Urminsky, O. and G. Zauberman (2016) 'The Psychology of Intertemporal Preferences'. In G. Keren and G. Wu (eds), *The Wiley Blackwell Handbook of Judgment and Decision Making*. Chichester: John Wiley and Sons.
Valberg, J.J. (1992) *The Puzzle of Experience*. Oxford: Oxford University Press.
Valberg, J.J. (2012) 'The Irrevocability of Being'. *Philosophy* 87(339): 65–77.
van der Ham, I.J.M., Y. Hamami, and J. Mumma (2017) 'Universal Intuitions of Spatial Relations in Elementary Geometry'. *Journal of Cognitive Psychology* 29(3): 269–78.
van Inwagen, P. (1975) 'The Incompatibility of Free Will and Determinism', Philosophical Studies 27 (3): 185–99.
Van Inwagen, P. (1983) *An Essay on Free Will*. Oxford: Oxford University Press.
Van Inwagen, P. (1990) *Material Beings*. Ithaca: Cornell University Press.
Van Inwagen, P. (2002) *Metaphysics* 2nd edn. Boulder: Westview.
Velleman, J.D. (1991) 'Well-Being and Time'. *Pacific Philosophical Quarterly* 72(1): 48–77.
Velleman, J.D. (1992) 'The Guise of the Good'. *Noûs* 26(1): 3–26.
Velleman, J.D. (2005) *Self to Self*. Cambridge: Cambridge University Press.
Velleman, J.D. (2006) 'So It Goes'. *The Amherst Lecture in Philosophy* 1: 1–23. http://www.amherstlecture.org/velleman2006/.
Vendler, Z. (1967) *Linguistics in Philosophy*. Ithaca: Cornell University Press.
Vetter, B. (2015) *Potentiality*. Oxford: Oxford University Press.
Vihvelin, K. (1996) 'What Time Travelers Cannot Do'. *Philosophical Studies* 81(2–3): 315–30.
Vineberg, S. (2016) 'Dutch Book Arguments'. In E.N. Zalta (ed.), *The Stanford Encyclopedia of Philosophy*. https://plato.stanford.edu/archives/spr2016/entires/dutch-book/.
Vlastos, G. (1966) 'A Note on Zeno's Arrow'. *Phronesis* 11: 3–18.
Von Wright, G.H. (1963) *Norm and Action: A Logical Enquiry*. London: Routledge and Kegan Paul.
Vonnegut, K. (1969) *Slaughterhouse 5*, Reprinted 1991. London: Vintage.
Weaver, C.G. (2020) 'On the Argument from Physics and General Relativity'. *Erkenntnis* 85(2): 333–73.
Weyl, H. (1949) *Philosophy of Mathematics and Natural Science*, Trans. Olaf Helmer. Princeton: Princeton University Press.
Whitehead, A.N. (1929) *Process and Reality*. New York: Free Press.
Wiggins, D. (2016) 'Activity, Process, Continuant, Substance, Organism'. *Philosophy*, 91(2): 269–80.
Williams, C. (1996) 'The Metaphysics of A- and B-time'. *Philosophical Quarterly* 46(184): 371–81.
Williams, D.C. (1951). The myth of passage. *Journal of Philosophy* 48(15): 457–472.
Williams, R.G. (2008) 'Ontic Vagueness and Metaphysical Indeterminacy'. *Philosophy Compass* 3(4): 763–88.
Williamson, T. (2013) *Modal Logic as Metaphysics*. Oxford: Oxford University Press.

Wilson, A. (2011) 'Macroscopic Ontology in Everettian Quantum Mechanics'. *Philosophical Quarterly* 61(243): 363–82.
Wilson, D.S. (1990) 'Species of Thought: A comment on Evolutionary Epistemology'. *Biology and Philosophy* 5: 37–62.
Wilson, D.S. (1995) 'Language as a Community of Interacting Belief Systems: A Case Study Involving Conduct Towards Self and Others'. *Biology and Philosophy* 10: 77–97.
Wilson, D.S. (2002) *Darwin's Cathedral: Evolution, Religion, and the Nature of Society*. Chicago: University of Chicago Press.
Wilson, J. (2010) 'What is Hume's Dictum, and Why Believe It?' *Philosophy and Phenomenological Research* 80(3): 595–637.
Witherall, A. (2000) 'Lewis and Sylvan on Noneism'. *Grazer Philosophische Studien* 58/59: 181–202.
Wittgenstein, L. (1922) *Tractatus Logico-Philosophicus*, Trans. C.K. Ogden. London: Routledge & Kegan Paul.
Wittmann, M. (2014) *Felt Time,* Trans. E. Butler (2016). London: MIT Press.
Wolf, S. (1986) 'Self-Interest and Interest in Selves'. *Ethics* 96(4): 704–20.
Wolf, S. (2015) 'The Meanings of Lives'. In *The Variety of Values(pp. 89-106).* Oxford: Oxford University Press.
Woodward, J. (2003) *Making Things Happen: A Theory of Causal Explanation*. Oxford: Oxford University Press.
Wright, C. (1986), 'Anti-realism, Timeless Truth, and *Nineteen Eighty-Four*'. In *Realism, Meaning, and Truth,* 2nd ed. (1993). Oxford: Blackwell, 176–203.
Wright, C. (1992) *Truth and Objectivity.* Cambridge: Cambridge University Press.
Wüthrich, C. (2017) 'Raiders of the Lost Spacetime'. In D. Lehmkuhl, G. Schiemann, and E. Scholz (eds), *Towards a Theory of Spacetime Theories*. New York: Springer, 297–335.
Wüthrich, C. and C. Callender (2017) 'What Becomes of a Causal Set?'. *British Journal of Philosophy of Science* 68: 907–25.
Wylie, C.D. (2019) 'The Plurality of Assumptions About Fossils and Time'. *History and Philosophy of the Life Sciences* 41: 21. https://doi.org/10.1007/s40656-019-0260-3.
Yaghmaie, A. (2017) 'How to Characterise Pure and Applied Science'. *International Studies in the Philosophy of Science* 31(2): 133–49.
Yagisawa, T. (2010) *Worlds and Individuals: Possible and Otherwise*. Oxford: Oxford University Press.
Zagzebski, L.T. (2015) 'Omniscience and the Arrow of Time'. In J.M. Fischer and P. Todd (eds), *Freedom, Fatalism, and Foreknowledge*. Oxford: Oxford University Press.
Zahavi, D. (2007) 'Perception of Duration Presupposes Duration of Perception– or Does it? Husserl and Dainton on Time'. *International Journal of Philosophical Studies* 15(3): 453–71.
Zimmerman, D.W. (1998) 'Temporary Intrinsics and Presentism'. In P. van Inwagen and D. Zimmerman (eds), *Metaphysics: The Big Questions*. Oxford: Blackwell, 206–19.
Zimmerman, D.W. (2007) 'The Privileged Present: Defending an A-Theory of Time'. In T. Sider, J. Hawthorne, and D.W. Zimmerman (eds), *Contemporary Debates in Metaphysics*. Oxford: Blackwell.
Zhuang, Z. *Zhuangzi*.
Zupko, J. (2018) 'John Buridan'. In E.N. Zalta (ed.), *The Stanford Encyclopedia of Philosophy* (Fall 2018 Edition). https://plato.stanford.edu/archives/fall2018/entries/buridan/.

Index

Note: *Italicized* and **bold** page numbers refer to figures and tables. Page numbers followed by "n" refer to notes

abduction 34, 35, 42, 74, 75, 86, 137, 142, 152, 155, 157, 214 n.43, 254, 272
Absolute Becoming 13
absolute elsewhere 112, *112*, 113, 120, 121, 125
absolute present 113, 114
absolute simultaneity 103, 112, 116, 118
actuality 199–202
  asymmetry of reference 175–6
  causal fizz and 177–81
  distinguished from potentiality 173–5, 183, 185
  evolving universe 181–2
  Growing-Block view 226
  temporal value asymmetry 199–202
  for tenses **208**
Adams, R. 71, 226 n.8, 229
adaptive imaginary representation (AIR) 75, 76
adverbialism 72
affine transformation 107, 112, 115
agency 159–86
  asymmetry of 154–7
  thank goodness that's over argument 210–11
  unity of 69–71, 100
AIR, *see* adaptive imaginary representation (AIR)
Albert, D.Z
  on causation 132, 135–7, 139, 140, 142, 143 n.25, 150, 151, 155
Alexander, S. 226 n.5, 259 n.10
anaemic change 6
Anscombe, G.E.M. 23, 36, 37, 84 n.6, 165 n.17, 215, 217 n.48
anti-rate group 11–13
anti-realism 247–9
applied science 97–8
Aristotle 90, 115, 173
  De Caelo 34 n.18
  Metaphysics 83 n.5
  Nichomachean Ethics x, 86, 159, 207 n.33, 215
  Physics 181 n.42
  on substance ontology 63
armchair philosophy xi
Armstrong, D.M. 54, 56 n.8, 88
Arstila, V. 50
Asay, J. 237
asymmetry
  of agency 154–7
  causal 130, 138, 140, 155, 159, 167, 170, 176, 182, 225, 226, 273
  of control 138 n.18
  entropic 152
  epistemic 130, 132–7, 139, 211
  of intervention 137–9, 141, 184, 211, 216
  ontological 242–3
  past/future 133, 134, 138, 140, 169, 170, 172, 173, 175, 177, 183, 187, 188, 211, 216, 219, 220, 223, 233
  of records 138–41, 147, 148, 151–7, 235
  of reference 129 n.3, 175–6, 242, 245, 246, 251, 257
  temporal value 188–205
  time/space 5
Attar, F. 48 n.43
Augustine of Hippo 3
  Confessions 227 n.9
axes, conventional labelling of 131–2

Baber, H.E. 67
Bacharach, B. 56, 259–60

Backmann, M. 264, 265
backwards causation 157
Balanced Forecasting 199
Baldwin, J. 196 n.14
Banfi, L. 209
Barbour, J. 3 n.2
Barnes, E. 121 n.23, 167, 232, 233
Baron, S. xi n.3, 3 n.2, 5 n.9, 38 n.27, 123, 125, 134, 135, 238
Barrie, J.M
  Admirable Crichton, The 64
Barry, B. 74
Beardman, S. 204–5
Beebee, H. 8, 143
Belot, G. 116–18
Bergson, H. 226 n.5
Bernstein, S. 87 n.13
Bertolaso, M. 87
Besson, C. 167 n.20
biffy causation 179–81, 262
Bigelow, J. 226 n.8, 227, 230
Billy Pilgrim 212–13, 216, 217
biology
  activities in 89–91
  processes in 87–91
Blackburn, S. 151
Blake, R.M. 6 n.10
Bobzien, S. 164 n.12, 184 n.46
Bourne, C. 4 n.4, 20, 167 n.19, 226 n.8, 228, 255
boustrophedon 131
Braddon-Mitchell, D. 255 n.4
Braithwaite, R.B. 3 n.3, 6 n.10
branching time 169–70
Briggs, R.A. 61 n.15, 154, 165 n.16, 168 n.21, 173 n.28, 200 n.23, 208 n.35, 225 n.2, 226 n.7, 243 n.25, 261 n.14, 268, 1674 n.19
Brink, D. 166
Broad, C.D. 3–5, 6 n.11, 12–14, 18, 44, 167, 206 n.31, 223, 226, 242–3, 253, 256, 257 n.7
Brueckner, A. 200 n.22
Bryant, E.F. 65 n.23
Buddhist soteriological project 64–6
Builes, D. 255 n.4
Burns, P. 189E
Button, T. 226
Byrne, A. 41 n.32
Callender, C. 3 n.3, 4, 20–2, 26, 27 n.5, 37, 74, 75, 91, 101 n.33, 157, 195, 202–3, 210, 235, 251
  on evidence-transcendent truth 148
  on invariance 193
  on relativity 103, 110, 120, 121, 123, 129
  on stationarity 192–3 n.7, 193
Cameron, R.P. 4 n.4, 121 n.22, 167, 232, 233, 243
Campbell, J.K. 101 n.32
Canonical Quantum Gravity 123–5
Caplan, B. 254, 259
Carpenter, A. 65 n.23, 66 n.25
Carroll, L. 229–30
Cartesianism Principle 248
Cartwright, N. 87, 91 n.20, 97, 98, 180–1
Caruso, E.M. 189
Casati, R. 131, 234–5
causal asymmetry 130, 138, 140, 155, 159, 167, 170, 176, 182, 225, 226, 273
  presupposing 141–3
causal connexions, comparison with experiencing 35–9
causal determinism 163, 164, 166, 169, 180, 186
causal effervescence, *see* causal fizz
causal fizz 170–1, 177–83, 186, 259, 261–3, 265, 266, 268, 269, 273
causal set theory 104, 119–22, *120*, 123, 125, 157, 182, 269
causal theory of time 155, 156
causation 129–57
  backwards 157
  cognitive command 149, 241
  dependence 179
  detection of 143–5
  evidence-transcendent truth 146–7
  explanatory ground 147–9
  Growing-Block view 261–6
  against Humean projection 151–4
  immanent 170, 180 n.40
  mind-independence 150–1, 241
  ‘process’ views of 179
  productive 154–7, 179
  realism about 141–54
  transeunt 180 n.40
  width of cosmological role 149–50, 154, 238, 241

Caws, P. 93 n.24
Chang, H. 93 n.24
changing preferences, choosing for 202–5
chronometers 93
Cicero 162–3
Cioran, E.M. 273
circularity 155
Cochrane, T. 209
Cockburn, D. 211
cognitive command 149, 241
Cole, C. 228
Collingwood, R.G. 253
colour properties 9–10
common sense xi, 24, 30, 37, 56, 60, 65 n.23, 74, 76, 77, 79, 80, 103–5, 107–8, 110, 115, 119, 144, 217, 227, 231, 234
  and sense experience 33–5
Confucius, *see* Kongzi
consistency 192, 195, 202
contemporaneity 109–11, 119, 121
content externalism 40
content internalism 40
continuants-occurrents distinction 55–6
Coope, U. 227
Correia, F. 226
  Growing-Block view 270–1
correspondence principle 247
Corry, R. 150
Craver, C. 89
Crisp, T. 20, 226 n.8, 228, 230 n.15, 240
cross-temporal potentialities 173
Crowther, T. 56 n.9, 84
C-theory 135
Currie, A. 254 n.1

Dainton, B. 27 n.5, 48 n.46, 49 n.47, 49 n.50, 50
Danto, A. 240
Darden, L. 89
Darzi, A. 192 n.4
Dasti, M.R. 65 n.23
Davey, K. 181 n.42
Davidson, D. 56 n.6, 84 n.6
Dawson, P. 237 n.18, 243
dead past views 269
Deasy, D. 232
debunking arguments 75–7
decision theory 189–92, **190**, **191**, 205
deflation 249–52
Dehaene, S. 108 n.9
Del Pinal, G. 41 n.33
delayed two-slit experiment 140 n.22
Deng, N. 8–9, 15, 200 n.22, 270, 271
Dennett, D.C. 163 n.7
dependence causation 179
diachronic rationality 192–5
diachronic unity of agency 68–71
DiBella, N. 243
Dickie, I. 241
discovery 94–7
dogmatism 155
Dolan, R.J. 192 n.4
Dolev, Y. 20
Dorato, M. 50, 114, 226 n.8
Dowe, P. 92, 179
Dowker, F. 120–2, 125, 186, 266
Dullworld 262
Dummett, M.A.E. 8 n.15, 96, 139, 183, 227, 228
  intuitionistic 'train-carriage presentism' 245–6
Dupré, J.A. 80, 87–90
Dyke, H. 3 n.3, 73 n.35, 75, 76, 210, 212, 216 n.46

Eddington, A.S. 45 n.38
Effingham, N. 184 n.45
Einstein's simultaneity convention 111
Eklund, M. 243–4
eleutheria 164 n.12
embedded perspective 33, 76, 113, 114, 266, 270, 272
  of science 80, 94, 98–102
  of thank goodness that's over argument 205, 210, 212, 214, 217–19
Emmet, D. 68 n.30, 88, 180
endurance 37, **57**, 254, 266–8
  changing perspectives 73–5
  debunking arguments 75–7
  Growing-Block view 266–8
  incoherence of 59–66
  problems for 58–66
  proper subject for change 66–77
  responses to problem of change 58–9

special concern for the future, practical necessity of 68–9
through McTchange 53–77, 231
unity of agency 69–71
endurantism 54–6, **57**, 59, 60, 62–4, 65 n.33, 71, 72, 73 n.35, 231
energeia/kinesis distinction 84 n.7
entropic asymmetry 152
entropic B-theory 135
entropic gradient 129 n.2, 130, 135–43, 154, 159, 176, 195 n.10, 216, 250
entropy 134–7
epistemic, the 34, 114, 121, 163, 176, 188, 190, 199, 218, 238
asymmetry 130, 132–7, 139, 211
causation 140, 142, 144, 149 n.34, 151–4
fixtures/options distinction 183
goods 100
justification 35, 214
peers 203–5
value of science 96
equanimity, fatalism as 213–15
Ersatzer Presentism 228–9
Esfeld, M. 123–4
Eternalism 50
Euclidean geometry 108
Everett, A. 239
evidence-transcendent truth 146–7, 218
exdurantism 54–7, 59
expected utility 190, 191, 193, 197 n.15
experience 23–52
comparison with causal connexions 35–9
content of 39–43
indiscriminability argument 27–9
perceptual 23, 26, 30–2, 37, 41–3, 48–50
puzzles of 48–52
sense 33–5
temporal (*see* temporal experience)
unintelligibility argument 29–30, 32
explanatory ground 147–9
extentionalism 49–51
External Realism 249

Farr, M. 130 n.5, 134
fatalism 159–64, 166, 183, 186, 212
bad arguments for 161–2
as equanimity 213–15
logical 162
modality and 162–6
reasons for 162–6
fatalistic worry 160–6
Fernandes, A.S. 138 n.19, 139, 185, 210, 211
Fine, K. 7, 8, 11, 17–18, 50, 72, 186, 259
Fischer, J.M. 200 n.22
Fitch, G. 175–6 n.36
Fitzgerald, E. 129
Fitzgerald, F.S. xii
fixtures/options distinction 183, 185, 212, 218, 219
Fleiter, J.J. 160
flow 46–8
folk theory of time xii
Foot, P. 84 n.6
Forbes, G.A. 50 n.53, 61 n.15, 71 n.33, 81 n.2, 144 n.27, 154, 165 n.16, 167 n.19, 200 n.23, 225 n.2, 243 n.25, 249 n.35, 261 n.14
Ford, G.W. 134
forecasting principles 198–9
Forrest, P. 170, 226, 261 n.15, 269
fragmentalism 18–19
frame relativity 104, 105, 109
Fraps, T. 36 n.21
Friebe, C. 264
Frischhut, A. 27 n.5, 43, 49, 50
frozen present argument 186

Galperti, S. 202
Gale, R.M. 17
Garfield, J.L. 66 n.26
Garrett, B. 184 n.45
Geach, P.T. 6 n.10, 46
General Theory of Relativity 103, 114–20, *116*, *117*, 123, 124, 141
Gerson, L.P. 196 n.13
Gibson, J. 240 n.21
Gill, K. 83
'Glowing Shrub' view 169 n.23, 235–6, 243
Godfrey-Smith, P. 69
Gold, T. 136
Golding, W
Lord of the Flies 64
Greene, P. 189, 193–5, 210

Growing-Block view x, 168 n.21, 223, 224, 226, 227, 236, 243, 245–7, 251–74
causation 261–6
endurance 266–8
extrinsic properties 256–9
intrinsic properties 256–9
Miller's dilemma 255–6
once present and now past 259–61
past problems 254–61
recap 253–4
rival views 268–72
Grush, R. 48 n.46
Gryb, S. 124, 125

Haack, S. 21 n.27
Hacking, I. 96
Haecceitist Presentism 229
Haecceitist strategy 240
Hall, N. 179
Hamami, Y. 108 n.9
Harman, E. 200–1, 234
Haslanger, S. 58, 62 n.19, 72, 231
Proper Subject of Change 54, 55
Hattiangadi, A. 167 n.20
Hausman, D.M. 181 n.41
Hawking, S. 103–4
Hawley, K. 54, 58
Heidegger, M. 93 n.28
Heller, J
Catch-22 23–6, 28
Heraclitus 5 n.5, 53, 62 n.18
Hillary, E. 82
Hinchliff 59
Hippocrates 93–4
Hoerl, C. 27 n.5, 35 n.20, 42, 48 n.46, 49–50
Hofweber, T. 61, 67
Holcombe, A. 42
holism 85
Hooke's law 22
Hookway, C. 33–5, 86, 155 n.42, 214 n.41, 249
Hornsby, J. 260
horror separationis 90
Horwich, P. 210
Hull, D.J. 93 n.24
Hume, D. 203 n.25, 263 n.16
Humean projection, causation against 151–4
Husserl, E. 45, 49
Hussey, E. 122 n.24

Iaquinto, S. 18, 19
Illari, P.McK. 150
immanent causation 170, 180 n.40
independence principle 247
indiscriminability argument 27–9
indispensability claims 85
inference to the best explanation 31
Ingram, D. 20, 226 n.8, 229, 231
intentionalism 49 n.50
intervention 137–40, 142
asymmetry of 137–9, 141, 184, 211, 216
intuitive view, the x
invariance 193
Inwood, B. 196 n.13
Ismael, J. 74, 130 n.5, 183
isometry 107, 112, 115
Izard, V. 108 n.9

James, W.E. 6 n.11, 18, 44, 48 n.44, 49, 55 n.4, 75–6, 85 n.8, 122 n.24
Jeffrey, R.C. 170, 226 n.6, 261 n.15, 270
Jenkins, C.S.I. 19
jerk 46–8
Johnston, M. 9–10, 56 n.3, 60
Jollimore, T. 150 n.36
Jones, G. 228
Julius Caesar 228, 229, 231, 238, 240
justification 94–7

Kahneman, D. 24–5 n.1
kairos 93, 94, 95 n.29, 97, 98, 100, 101, 204, 273
Kajimoto, N. 130 n.5, 134
Kalupahana, D.J. 5 n.6
Kant, I. 24 n.1
Kaplan, D. 111 n.12, 175 n.36
Kauppinen, A. 198, 200
Kavka, G.S. 198
Kayani, A. 160
Keller, S. 226 n.8, 229
Kenny, A. 56 n.10, 83 n.5
Khayyam, O. 129
Kind, A. 70–1
King, M.J. 160
Kon, M. 48 n.46

Kongzi
  Analects 3.21 187
Korsgaard, C. 67, 70
Kroll, N. 177 n.38, 269
Kuhn, T. 93 n.24, 100
Kutach, D. 135

Ladyman, J. 87 n.13
Lam, V. 123–4, 234
Latham, A.J. 25 n.3, 42
Le Bihan, B. 124
Le Poidevin, R. 24, 37
Lear, J. 122
Lee, H.D.P. 122 n.24
Leininger, L. 6–7, 263, 265 n.19
Lenman, J 197 n.15
Leonelli, S. 95
Lewis, D. 56 n.3, 62 n.17, 82 n.4, 93 n.23, 163 n.8, 184 n.45, 244, 248 n.32
Liggins, D. 84–5 n.8
Linnemann, N.S. 124
Lipman, M.A. 18
Lipton, P. 85, 86, 105 n.4
List, C. 69, 87 n.13
Lloyd, G. 198
Loewer, B. 135
logical dominance 190
Lombard, L. 20, 226 n.8
Long, A.A. 162 n.5, 199 n.21, 219 n.50, 227 n.10, 230 n.16
Loop Quantum Gravity 123, 124
Loss, R. 268
Lucretian Presentism 230
Luther, M. 165 n.18

MacBeath, M. 188, 206–10
Machamer, P. 89
Mackie, J.L. 142, 163, 168, 170, 178 n.39, 181, 226 n.6, 261 n.15, 270
Maclaurin, J. 210, 212, 216 n.46
MacMurray, J. 24
Maes, H. 209
Makin, S. 83 n.5, 84 n.7, 165, 173
Marcolis, J. 93 n.24
Marcus, R.B. 243
Markosian, N. 4 n.4, 11, 56 n.7, 226 n.8, 227, 228, 240–1
Mason, F. 226 n.8, 229
Maudlin, T. 11, 12, 14, 16, 106, 117 n.18, 226
Maugham, S
  Sheppey 161
maximalism 243
McCall, S. 4 n.4, 169, 233
McDowell, J. 9, 10
McFetridge, I. 238
McInerney, P.K. 25, 31, 43 n.36
McKinnon, N. 226 n.8, 230 n.15
McTaggart, J.McT.E. 3 n.2, 4, 16–19
McTaggartian change 5–7, *6*
McTchange
  coherence 7–10
  endurance through 53–77, 231
  experience of 23–52
  relativity and 103–25
  response-dependent properties of 9–10
  science of 79–102
  temporal variation *versus* 3–22, 25, 29, 32, 33, 38, 44, 73, 80, 91–3, 97, 99, 101, 113, 262, 1229 n.1
  time's passage 10–16 (*see also individual entries*)
Meincke, A.S. 59, 62, 64–6
Mellor, D.H. 3 n.3, 8 n.16, 36, 73 n.35, 170, 188, 206–10, 255 n.4
memory 129–57
Menzies, P. 87 n.13
Merleau-Ponty, M. 47 n.41, 79
Merricks, T. 59, 226 n.8, 231
metaontology 223, 245–52
  deflation 249–52
  metaphysical realism 247–9
  pragmatic arguments 245–7
  pragmatism 249–52
  Quinean orthodoxy 245
metaphysical realism 247–9
Michelson, A.A. 105
Miller, K. x, xi n.3, 3 n.2, 5 n.9, 25 n.3, 38 n.27, 42, 48 n.46, 56 n.2, 57, 66 n.27, 123, 125, 130 n.5, 134, 135, 232–4, 254, 269
  Growing-Block view 255–6
mind-body problem xii
mind-dependence 9, 218
mind-independence 150–1, 241

Misak, C. 21 n.27, 85 n.8, 195 n.11, 215 n.44
Mitova, V. 149
Mlodinow, L. 103–4
model of reality (MOR) 75
Mohammadian, M. 34 n.17
Molière 153
money pumps 191
Moore, A.W. xi, 144 n.26, 211 n.38
Moore, G.E. 47 n.40
MOR, *see* model of reality (MOR)
Morgenbesser, S. 34 n.18
Morley, E.W. 105
Moss, J. 215 n.44
Mourelatos, A.P.D. 56 n.11
Moving Spotlight, the 232–5
Mumma, J. 108 n.9

Nagel, T. xii, 69, 98, 204
naïve starting points 24–6
naturalism 84–6
Nefarious Presentism 231
Nelkin, D. 166
Nelson, M. 175–6 n.36
Nerlich, G. 13
Neurath, O. 91 n.20
Newtonian physics 82–3
Newtonian picture of space and time 104–8, *106*
Nicholson, D. 80, 87, 89, 90
'no ownership' view of the self 68 n.30
Nolan, D. 34 n.16, 146
Norton, J. 25 n.3, 130 n.5, 134, 234

Oakeley, H. 226 n.5
Oaklander, L.N. 3 n.3, 20, 224
objective inertial reference frame 113
Odysseus 204
Olson, E.T. 10–12, 261 n.13
O'Neill, J. 198
onomatopoeia 47
ontological asymmetry 242–3
ontological restraint 243–4
open future
  branching time 169–70
  causal fizz 170–1
  characterising 166–71
  semantically 167–9
O'Rourke, M. 101 n.32
Orwell, G. 223
  Nineteen Eighty-Four 228
Orwellian Presentism 228, 230
Owen, G.E.L. 122 n.24

Papineau, D. 86 n.11
Parfit, D. 64 n.21, 65, 188–9, 193, 196
Parmenidean Presentism 51–2
past truths true, making 237–41
past/future asymmetry 133, 134, 138, 140, 169, 170, 172, 173, 175, 177, 183, 187, 188, 211, 216, 219, 220, 223, 233
Paul, L.A. 25, 27 n.5, 37, 53, 71, 203
Pearson, O. 219
Peirce, C.S. 21, 34 n.17, 42, 85 n.8, 213, 214 n.43, 215, 249
perceptual experience 23, 26, 30–2, 37, 41–3, 48–50
perdurantism 54, 55, **57**
Pereboom, D. 163 n.7
Perović, K. 177, 226, 254, 257, 258, 269, 271
persistence, rival accounts of 54–8
Pettit, P. 69
Pettigrew, R. 204 n.27
Phillips, I. 11
physics, processes in 91–3
Pica, P. 108 n.9
Pickering, A. 93 n.24
Pilgrim, B. 73
Plantinga, A. 240
Plato 249
  Phaedrus 101 n.32
Polanyi, K. 240 n.20
potentiality 84 n.7, 172–86, 188, 195, 198–202, **208**, 213, 216–20, 223, 225 n.2, 233–5, 242, 243, 251, 253, 254, 255 n.2, 256–61, 263–6, 268, 273
  asymmetry of reference 175–6
  causal fizz and 177–81
  cross-temporal 173
  distinguished from actuality 173–5, 183, 185
  evolving universe 182
  Growing-Block view 226
  temporal value asymmetry 199–202
  for tenses **208**
Power, S.E. 48 n.46, 49 n.47, 50

pragmatic arguments 245–7
pragmatic transcendental argument 214
pragmatism 249–52
presence 46–8
Presentism 50, 226–31, 234, 240, 259, 264, 265
  Ersatzer 228–9
  Haecceitist 229
  Lucretian 230
  Nefarious 231
  Orwellian 230
  Parmenidean 51–2
  Surrogate 228
  train-carriage 242, 244–6
  Upstanding 231
Price, H. 3 n.4, 5 n.8, 11, 45 n.38, 101 n.33, 108 n.10, 130 n.5, 133 n.10, 155 n.40, 155 n.41, 250
  on actuality/potentiality distinction 183
  on causation 134, 136, 140, 143–4, 146
Prior, A.N. 11, 80 n.1, 206 n.31
process ontology 64, 92
processes 80–4
  in biology 87–91
  in physics 91–3
productive causation 179
  asymmetries of records and agency 154–7
progressive/perfective distinction 84 n.7
pro-rate group 11–12
Prosser, S. 23, 39
  on experience 30–2
  indiscriminability argument 27–9
  unintelligibility argument 29–30, 32
pure science 97–8
purely tensed views 272
Putnam, H. 33, 99–101, 103, 113, 249
Pyrrhonian scepticism 155, 156

qualitative change and movement 43–4
quantum gravity 86, 103, 104, 108 n.10, 119–25
Quantum Mechanics 235
Quine, W.V. 21, 35, 54, 56, 79, 80 n.1, 85, 86
Quinean orthodoxy 245
Quine-Putnam indispensability argument 84–5 n.8

Ramsey, F. 8, 195 n.11
Rashbrook, O. 48 n.46
Raven, M.J. 11, 12
Raz, J. 215
real-from-a-scientific-point-of-view 251
real-from-our-point-of-view 251
realism
  about causation 141–54
  metaphysical 247–9
  quasi-realism 151
records, asymmetry of 138–41, 147, 148, 151–7, 235
referentially fit 241
referentially successful 241
Reichenbach, H. 46, 95, 111–12, 155
Reichenbachian simultaneity 110–11, *111*
Reid, T. 77
relativism 18
relativity
  frame 104, 105, 109
  General Theory of Relativity 103, 114–20, *116*, *117*, 123, 124, 141
  and McTchange 103–25
  Newtonian picture of space and time 104–8
  Special Theory of Relativity 76, 101, 103, 108–16, *111*, *112*, 119
Rennick, S. 184 n.45
Rescher, N. 90–3
Resnik, M. 84
response-dependent properties 9–10
retentionalism 49–51
Richard, M. 271
Rideout, D.P. 120
robust change 6
Röck, T. 25
Rorty, R. 248
Rosenkranz, S. 226
  Growing-Block view 270–1
Routley, R. 245 n.29
Russell, B. 6 n.10, 18, 150, 240 n.23
  on problem of induction 152
Russo, F. 150
Ryanasiewicz, R. 105 n.6

Sainsbury, R.M. 122 n.24
Salmon, W. 92, 179
Sanford, D.H. 7

Sanson, D. 254, 259
Saunders, S. 169, 235
Savitt, S.F. 114
Schaffer, J. 87 n.13
Schectman, M. 197
Schellenberg, S. 40
Schuster, M.M. 24
science 79–102
  applied 97–8
  biology, processes in 87–91
  commitment 84–6
  discovery 94–7
  embedded perspective of 80, 94, 98–102
  justification 94–7
  physics, processes in 91–3
  presuppositions of 93–102
  processes 80–4
  pure 97–8
  view from nowhere 98–102
second law of thermodynamics 134, 136, 141, 143, 152, 153, 211
second-dimension 'solution' 13–14
Sedley, D.N. 162 n.5, 199 n.21, 219 n.50, 227 n.10, 230 n.16
self-plagiarism 186
Sellars, W. xi n.2, xii
semantically open futures 167–9
sense experience 21
  common sense and 33–5
Setiya, K. 198, 209, 215, 218–19
Seymour, B. 192 n.4
Shakespeare, W
  Henry V 161
  Julius Caesar 240
  Midsummer Night's Dream, A 240
shape of a life 195–9
Shardlow, J. 25 n.3
Shebliske, W.L. 48 n.45
Shrinking Block, the 234–5, 246, 247
Shrinking Tree view 169, 233–4
Sider, T. 6 n.10, 54, 85, 101 n.32, 230, 249
Siegel, S. 41
Simons, P. 55–6
simpliciter 6, 49, 60, 72, 119, 174, 206, 253, 264, 270
Sklar, L. 105 n.5, 110, 155 n.40
Skow, B. 6, 13, 18, 27 n.5
Slater, M.H. 101 n.32
Slovic, P. 24–5 n.1
Smart, J.J.C. 3 n.3, 7 n.12, 11
Smith, D.C. 150
Smith, J.E. 94 n.28
Smith, N.J.J. 184 n.45
Socrates 147
SOFism 72
Sorabji, R. 5, 164 n.12
Sorkin, R.D. 98, 119, 120
spatial analogy 217
Spaulding, S. 41 n.33
Speaks, J. 30 n.13, 31
special concern for the future, practical necessity of 68–9
Special Theory of Relativity 76, 101, 103, 108–16, *111*, *112*, 119
Spelke, E. 108 n.9
static shrub view 169
stationarity 192–3, 195, 197, 198
Stebbing, L.S. 103, 141
Stein, H. 114
Stevens, C. (now Yusuf Islam) 203
Steward, H. 56 n.9, 56 n.11, 81–2, 164 n.14
Stoicism 164 n.12
Stoics 164
Story, G.W. 192 n.4
Stout, R. 56 n.9, 56 n.11, 81–3
stream of consciousness 48
Strong Forecasting 198–9
Strulovici, B. 202
substance ontology 63, 64, 88, 269
succession 45–6
Sullivan, M. 189, 193–5, 198–9, 210
superassertibility 146, 147, 149 n.34
Surrogate Presentism 228
survey of temporal ontologies 225–*36*
Swinburne, R. 105 n.4, 118 n.19
Sydney Time Studies, The xi n.3
Sylvan, R. 244, 245 n.29
synchronic unity of agency 70

table thumping 92 n.22
Tallant, J. xi n.3, 3 n.2, 5 n.9, 20, 38 n.27, 123, 125, 226 n.8, 230, 231
Tan, P. 260–1
Taylor, K. 241–3
temporal experience

duration over time 44–5
extension through time 44–5
flow 46–8
jerk 46–8
order and succession 45–6
presence 46–8
qualitative change and movement 43–4
whoosh 46–8
temporal value asymmetry 188–205
actuality 199–202
changing preferences, choosing for 202–5
decision theory 189–92, **190**, **191**
diachronic rationality 192–5
potentiality 199–202
shape of a life 195–9
temporal variation
McTchange *versus* 3–22, 25, 29, 32, 33, 38, 44, 73, 80, 91–3, 97, 99, 101, 113, 262, 1229 n.1
response-dependent properties of 9–10
temporaryism 271
tensed sense of 'is' 224–5
tenseless sense of 'is' 224–5
'thank goodness that's over!' argument 187–220
action and 216–20
actuality and 216–20
agency and 210–11
belief and 210–11
Billy Pilgrim 212–13 (*see also* Vonnegut, K.)
control and 210–11
fatalism as equanimity 213–15
Mellor and MacBeath's responses 206–10, **208**
passion and 216–20
potentiality and 216–20
temporal value asymmetry (*see* temporal value asymmetry)
Tralfamadorians 212–13 (*see also* Vonnegut, K.)
Thébault, K.P.Y. 124, 125
Theravada Buddhism 5 n.6
Thin Red line, the 169
thing ontology 63
things-in-process 88
Thoma, J. 204
Thomas, E. 88 n.15, 105 n.5, 226 n.5
Thomasson, A. 239 n.19
time 129–57
anti-rate group 11–13
dynamic and static views, difference between 15–16
no rate 15
passage 10–11
perception 40
pro-rate group 11–12
second-dimension 'solution' 13–14
and space, asymmetry between 5
travel to the past 184–6
timeless sense of 'is' 224–5
Todd, P. 167
Tooley, M. 226, 261 n.15
Growing-Block view 270
topological transformation 106–7
Torre, S. 168
Torrengo, G. 18, 19, 43 n.36, 47, 48, 51 n.55, 131, 234–5
train-carriage presentism 242, 244–6
Tralfamadorians, *see* Vonnegut, K.
transeunt causation 180 n.40
transformative experiences 203
transphenomenality 240
truthmaking 237, 243, 244, 260, 261
Tversky, A. 24–5 n.1

Uhlenbeck, G.E. 134
Ullmann-Margalit, E. 34 n.18
unintelligibility argument 29–30, 32
unity of agency 69–71, 100
diachronic 68–71
synchronic 70
Upstanding Presentism 231
Urminsky, O. 192 n.5
utility function 190

Valberg, J.J. 46–7, 175–6
van der Ham, I.J.M. 108 n.9
van Inwagen, P. 12 n.19, 163 n.6, 163 n.8, 267 n.20
Velleman, J.D. 45, 51, 59–61, 64–7, 197, 198, 212 n.40, 215
Vendler, Z. 56 n.10
Vetter, B. 172 n.27, 173, 174
Vihvelin, K. 184 n.45

Vineberg, S. 192 n.6
Vlaev, I. 192 n.4
Vlastos, G. 122 n.24
Von Wright, G.H. 165 n.16
Vonnegut, K. 51 n.54, 73
  Slaughterhouse 5 188

Wallace, D. 169, 235
Weak Forecasting 199
Weaver, C.G. 141, 143
Weyl, H. 49
Wheeler-DeWitt equation 124
Whitehead, A.N. 6 n.11
Whiteheadian views 269
whooshiness 46–8
width of cosmological role 149–50, 154, 238, 241
Wiggins, D. 88 n.16
Williams, D.C. 20, 46, 121 n.23, 262, 274 n.25
Williamson, J. 150
Wilson, D.S. 75, 235
Witherall, A. 244 n.26
Wittgenstein, L. 113
Wittmann, M. 26, 48 n.46
Wolf, S. 67, 70, 198
Woodward, J. 10, 37, 96, 138, 181 n.41
world-as-constraint-on-theory 249
Wright, C. 85 n.8, 145 n.28, 228
  on realism about causation 146–51
Wüthrich, C. 121
Wylie, A. 94–5

Yaghmaie, A. 97, 98
Yagisawa, T. 244

Zagzebski, L.T. 159–60, 177, 183
  inconsistent triad 171–2
Zahavi, D. 49 n.49
Zauberman, G. 192 n.5
Zeno's arrow paradox 122
Zhuangzi, Z. 249
Zimmerman, D.W. 4 n.4, 24, 59, 226 n.8
Zupko, J. 34 n.18

www.ingramcontent.com/pod-product-compliance
Lightning Source LLC
LaVergne TN
LVHW020631240726
844103LV00003B/271